This monograph presents a comprehensive description of the theoretical foundations and experimental applications of spectroscopic methods in plasma physics research.

The first three chapters introduce the classical and quantum theories of radiation, with detailed descriptions of line strengths and high density effects. The next chapter describes theoretical and experimental aspects of spectral line broadening. The following five chapters are concerned with continuous spectra, level kinetics and cross sections, thermodynamic equilibrium relations, radiative energy transfer, and radiative energy losses. The book concludes with three chapters covering the basics of various applications of plasma spectroscopy to density and temperature measurements and to the determination of some other plasma properties. Over one thousand references guide the reader not only to original research covered in the chapters, but also to experimental details and instrumentation.

This will be an important text and reference for all those working on plasmas in physics, optics, nuclear engineering, and chemistry, as well as astronomy, astrophysics and space physics.

**CAMBRIDGE MONOGRAPHS ON PLASMA PHYSICS**

General Editors: M.G. Haines, K.I. Hopcraft, I.H. Hutchinson, C.M. Surko and K. Schindler

PRINCIPLES OF PLASMA SPECTROSCOPY

# Principles of
# Plasma Spectroscopy

### Hans R. Griem
University of Maryland at College Park

CAMBRIDGE
UNIVERSITY PRESS

PUBLISHED BY THE PRESS SYNDICATE OF THE UNIVERSITY OF CAMBRIDGE
The Pitt Building, Trumpington Street, Cambridge, United Kingdom

CAMBRIDGE UNIVERSITY PRESS
The Edinburgh Building, Cambridge CB2 2RU, UK
40 West 20th Street, New York NY 10011–4211, USA
477 Williamstown Road, Port Melbourne, VIC 3207, Australia
Ruiz de Alarcón 13, 28014 Madrid, Spain
Dock House, The Waterfront, Cape Town 8001, South Africa

http://www.cambridge.org

First published 1997
First paperback edition 2005

Typeset in 10/13pt Times

*A catalogue record for this book is available from the British Library*

*Library of Congress Cataloguing in Publication data*

Griem, Hans R.
Principles of plasma spectroscopy / Hans R. Griem.
p.   cm. – (Cambridge monographs on plasma physics ; 2)
Includes bibliographical references and index.
ISBN 0 521 45504 9 (hardback)
1. Plasma spectroscopy.   I. Title.   II. Series.
QC718.8.S6G75   1997
530.4′46–dc21   96–37158   CIP

ISBN 0 521 45504 9 hardback
ISBN 0 521 61941 6 paperback

# Contents

# Preface

This book was written for the benefit of young researchers in diverse disciplines ranging from experimental plasma physics to astrophysics, and graduate students wanting to enter the interdisciplinary area of research now generally called plasma spectroscopy. The author has attempted to develop the theoretical foundations of the numerous applications of plasma spectroscopy from first principles. However, some familiarity with atomic structure and collision calculations, with quantum-mechanical perturbation theory and with statistical mechanics of plasmas is assumed. The emphasis is on the quantitative emission spectroscopy of atoms and ions immersed in high-temperature plasmas and in weak radiation fields, where multi-photon processes are not important.

As in the author's previous books on plasma spectroscopy and spectral line broadening written, respectively, over three and two decades ago, various applications are discussed in considerable detail, as are the underlying critical experiments. Hopefully, the reader will find the numerous references useful and current up to the latter part of 1995. They provide advice concerning access to basic data, which are needed for the implementation of many of the experimental methods, and to descriptions of instrumentation.

The author has once more benefited from his experience in teaching special lecture courses at the University of Maryland and recently also at the Ruhr University in Bochum and some of its neighboring institutions. His thanks are not only due to the advanced graduate students and fellow scientists in these courses but also to many colleagues at these and other institutions, who so generously provided advice, criticism and new research results. Of all these colleagues, I owe particular thanks to Ray Elton and to John Hey for their sustained and critical help with the manuscript. Any remaining errors or deficiencies are entirely my responsibility, and I would be grateful if they were communicated to me.

I am most grateful to Dorothea F. Brosius, who so patiently and professionally prepared numerous drafts and the final computerized manuscript. Very special thanks go to her and to my wife Irmgard for all the help and encouragement.

1996, College Park, Maryland

# Symbols

| | | |
|---|---|---|
| $\mathscr{L}(\omega)$ | line shape operator | 63 |
| $m$ | (general) quantum number | 15 |
| $m(m_e)$ | electron mass | 4 |
| $M$ | radiator (atom or ion) mass | 54 |
| $M, M'$ | magnetic quantum numbers | 36 |
| $n$ | principal quantum number | 3 |
| $n$ | (general) quantum number | 15 |
| $n_b$ | number of bound electrons | 199, 201 |
| $n_c$ | maximum boundstate principal quantum number | 197, 200, 201 |
| $n_{c\ell}$ | principal quantum number of collision limit | 181 |
| $n_{cr}$ | critical quantum number for PLTE | 215 |
| $n_{max}$ | maximum principal quantum number | 141 |
| $n_\omega$ | number of photons per mode | 13 |
| $n_0$ | maximum principal quantum number in ion-sphere model | 206 |
| $n_\ell^*$ | effective principal quantum number | 43 |
| $n(\omega)$ | refractive index | 33 |
| $N$ | number density | 9 |
| $N_e(N)$ | electron density | 66 |
| $N_m$ | population density in state $m$ | 22 |
| $N_p(N_i)$ | density of perturbing ions | 69 |
| N | newton (unit of force) | 2 |
| $p$ | probability of reemission | 228 |
| $p(\mathbf{r}_1, \mathbf{r}_2, \cdots)$ | configuration space distribution | 68 |
| $\mathbf{p}$ | mechanical momentum | 12 |
| $P$ | projection operator | 61 |
| $P$ | radiated power per unit volume | 247 |
| $P_a$ | absorbed power | 5 |
| $P_e$ | emitted power | 4 |
| $P_e$ | photon escape probability | 237 |
| $P_{nm}$ | transition probability ($m \rightarrow n$) | 15 |
| $P(\mathbf{F})$ | probability distribution for field vectors | 68 |
| $P(\omega)$ | spectral power | 57 |
| PLTE | partial local thermodynamic equilibrium | 156 |
| $\mathbf{P}$ | volume polarization | 9 |
| $\mathbf{P}$ | canonical momentum | 11 |
| $\mathbf{P}^{(1)}$ | dipole transition operator | 38 |
| $q_0$ | maximum momentum transfer | 174 |
| $q_\omega, q_\omega^*$ | destruction, creation operators | 12 |
| $Q$ | heat | 190 |
| $Q$ | average ion charge | 199 |

| | | |
|---|---|---|
| $u(s,0)$ | interaction representation time evolution operator | 59 |
| $U$ | interaction Hamiltonian | 60 |
| $U_n$ | time-independent wave function | 15 |
| $U_\Omega$ | spectral energy density of wave field | 123 |
| $\mathbf{v}$ | velocity | 2 |
| $w$ | half (HWHM) line width | 55 |
| $W$ | Racah coefficient | 38 |
| $W$ | number of microscopic distributions | 189 |
| $W_{mn}$ | transition probability for absorption | 22 |
| $W(F)$ | microfield distribution function | 58 |
| $W(\mathbf{F}' \to \mathbf{F})$ | transition rate between microfield states | 85 |
| W | weber (unit of magnetic flux) | 2 |
| $\mathbf{x}$ | displacement vector | 3 |
| $X_{mg}$ | excitation rate coefficient (ground state to state $m$) | 160 |
| $z$ | effective nuclear charge | 3 |
| $z$ | nuclear charge | 40 |
| $z$ | charge state label | 158 |
| $z$ | ionic charge | 159 |
| $z_{eff}$ | effective ionic charge | 271, 320 |
| $z_n$ | number of bound electrons in $n$-th shell | 159 |
| $z_p$ | charge of perturbing ions | 69 |
| $z'$ | effective charge | 246 |
| $Z$ | number of bound electrons (per atom or ion) | 48 |
| $Z_a$ | internal partition function | 190, 193 |
| $Z_n$ | effective nuclear charge for $n$-th shell | 253 |
| $Z_{PL}$ | Planck-Larkin partition function | 207 |
| | | |
| $\alpha$ | fine-structure constant | 3 |
| $\alpha$ | ion broadening parameter | 92 |
| $\alpha$ | phase angle | 123 |
| $\alpha$ | recombination rate coefficient | 158 |
| $\alpha$ | $\alpha$-particle | 301 |
| $\alpha_d$ | effective dielectronic recombination coefficient | 180 |
| $\beta$ | reduced micro-fieldstrength | 68 |
| $\beta$ | ratio of ionization and thermal energies | 168 |
| $\beta_1$ | escape factor | 237 |
| $\gamma$ | decay (damping) rate | 6 |
| $\gamma$ | set of quantum numbers | 37 |
| $\gamma_{nm}$ | (radiative) damping constant | 27 |
| $\gamma(u)$ | effective Gaunt factor | 173 |
| $\Gamma_{ii}$ | ion-ion coupling parameter | 73 |
| $\delta_\ell(p)$ | phase shift of free electron wave function | 208 |
| $\Delta E_z$ | reduction of ionization energy | 140, 195 |

| | | |
|---|---|---|
| $\Delta Z_1$ | correction to free electron partition function | 209 |
| $\Delta\omega$ | frequency detuning | 23 |
| $\epsilon_d$ | dielectric function | 154 |
| $\epsilon_{n\ell}$ | energy eigenvalue (in Rydberg units) | 40 |
| $\epsilon_0$ | dielectric constant of vacuum | 2 |
| $\epsilon'$ | parameter in approximate source function | 234 |
| $\epsilon''$ | quenching probability | 235 |
| $\epsilon(K, \Delta\omega)$ | plasma dielectric | 79 |
| $\epsilon(\omega)$ | (volume) emission coefficient | 144 |
| $\eta$ | scaled electron density ($N_e/z^7$) | 162 |
| $\kappa_R$ | Rayleigh scattering coefficient | 224 |
| $\kappa(\omega)$ | absorption coefficient | 33 |
| $\kappa'(\omega)$ | effective absorption coefficient | 142 |
| $\lambda$ | wavelength | 7 |
| $\Lambda$ | multipole order of electron-radiator interaction | 93 |
| $\mu$ | reduced mass of electron | 203 |
| $\mu_0$ | magnetic permeability of vacuum | 2 |
| $\nu_{coll}$ | effective collision frequency | 7 |
| $\nu(\Delta\omega)$ | effective frequency for microfield changes | 86 |
| $\xi$ | Biberman factor | 138 |
| $\rho$ | charge density | 2 |
| $\rho_i$ | probability (density of state) operator | 57 |
| $\rho_{min}$ | minimum impact parameter | 83 |
| $\rho_{E_n}$ | density of states | 15 |
| $\rho_0$ | Weisskopf radius | 102 |
| $\rho_D'$ | (electron and ion) Debye radius | 50 |
| $\boldsymbol{\rho}$ | (vector) impact parameter | 102 |
| $\sigma_a$ | absorption cross section | 6 |
| $\sigma_{i,f}$ | inelastic cross sections | 67 |
| $\sigma_{nm}$ | photoionization cross section | 134 |
| $\sigma_s$ | scattering cross section | 8 |
| $\sigma_{exc}$ | excitation cross section | 160 |
| $\sigma_{ion}$ | ionization cross section | 158 |
| $\sigma_{res}$ | resonance scattering cross section | 8 |
| $\sigma_{Th}$ | Thomson scattering cross section | 8 |
| $\sigma_{mn}^a$ | absorption cross section ($n \to m$) | 21 |
| $\sigma_{nm}^i$ | induced emission cross section ($m \to n$) | 22 |
| $\tau(\omega)$ | optical depth | 230 |
| $\phi$ | scalar potential | 2 |
| $\phi$ | Bates and Damgaard correction factor | 43 |
| $\bar{\chi}$ | averaged ionization energy | 169 |
| $\omega$ | angular frequency | 4 |

# 1
# Classical theory of radiation

Although most of the electromagnetic radiation from many natural and laboratory plasmas is atomic in origin and therefore subject to quantum effects, it remains useful to introduce some of the basic radiative processes via classical theory. Other important foundations of plasma spectroscopy are atomic physics and plasma physics, especially the statistical mechanics of ionized gases. Generally, the basic theory is well established in these parent disciplines. However, the large variety of processes contributing to emission or absorption spectra often requires more or less drastic simplifications in their theoretical description or computer modeling. Critical experiments are playing an essential role in checking the reliability of various models and in delineating the region of their applicability.

Plasma spectroscopy, although being a highly specialized subfield, is at the same time a very interdisciplinary science. It not only owes its origins largely to astronomy, but also returns to astronomy and astrophysics methods of analysis of spectra and a multitude of basic data, which have both been subjected to experimental scrutiny. The state of stellar plasmas is significantly influenced by radiation, and the latter is more or less controlled by radiative energy transfer. Internally consistent treatments of the states of matter and radiation first developed by astronomers are now also becoming important for the description of plasma experiments.

This give-and-take between the various sciences and scientists, experimentalists, theoreticians, computer modelers, and observers, is responsible for much of the progress made in our understanding and working knowledge of plasma radiation physics in the three decades since the first monograph on plasma spectroscopy (Griem 1964), an early review (Cooper 1966) and an introductory text (Marr 1968) were written. This progress is not only reflected in astrophysical data tables (Allen 1973,

1

Lang 1980), but also in a number of other monographs covering the physics of atoms and ions in plasmas (Sobel'man, Vainshtein and Yukov 1981, Janev, Presnyakov and Shevelko 1985, Shevelko and Vainshtein 1993, Lisitsa 1994, Kobzev, Iakubov and Popovich 1995). References to tables of spectroscopic data can be found in the introduction to chapter 3.

## 1.1  Electromagnetic equations and fields from moving charges

The International System (S.I.) of units will be generally used here, although most results will be expressed in terms of fundamental constants such that evaluation, e.g., in cgs units will be straightforward. To discuss experimental results, wavelengths will usually be in Angström units and densities per $cm^3$.

Maxwell's equations lead to the inhomogeneous wave equations for vector and scalar potentials, namely,

$$\frac{1}{c^2}\ddot{\mathbf{A}} - \nabla^2 \mathbf{A} = \mu_0 \rho \mathbf{v} \tag{1.1}$$

and

$$\frac{1}{c^2}\ddot{\phi} - \nabla^2 \phi = \frac{1}{\epsilon_0}\rho, \tag{1.2}$$

if the Lorentz condition

$$\nabla \cdot \mathbf{A} + \epsilon_0 \mu_0 \dot{\phi} = 0 \tag{1.3}$$

is also imposed. Here the vacuum values of dielectric constant and magnetic permeability are $\epsilon_0 = (4\pi \times 9 \times 10^9)^{-1} = 8.85 \times 10^{-12}\, C^2 N^{-1}\, m^{-2}$ and $\mu_0 = 4\pi \times 10^{-7}\, W A^{-1}\, m^{-1}$, and $\rho$ and $\mathbf{v}$ are charge densities and velocities. Also, the units are coulomb (C) for charge, newton (N) for force and weber (W) for magnetic flux ($1\,W = 1$ tesla $m^2$), while the current is in amperes (A), not to be confused with the vector potential.

A special solution of the inhomogeneous wave equations is given by the Lienard-Wiechert potentials (Jackson 1962, Landau and Lifshitz 1951), which account for retardation effects and assume point charges as sources of the field. The electric and magnetic fields then follow from

$$\mathbf{E} = -\nabla\phi - \dot{\mathbf{A}} \tag{1.4}$$

and

$$\mathbf{B} = \nabla \times \mathbf{A}, \tag{1.5}$$

and in our case we have to add the general solution of the homogeneous wave equation for **A**. As shown, e.g., in Heitler (1954), a moving electron produces fields according to the above equations given by

$$
\mathbf{E} = -\frac{e}{4\pi\epsilon_0}\left\{\left(1-\frac{v^2}{c^2}\right)\left(\mathbf{r}+\mathbf{v}\frac{r}{c}\right)\right.
$$
$$
\left.-\frac{1}{c^2}\mathbf{r}\times\left[\left(\mathbf{r}+\mathbf{v}\frac{r}{c}\right)\times\dot{\mathbf{v}}\right]\right\}\left(r+\frac{1}{c}\mathbf{v}\cdot\mathbf{r}\right)^{-3}, \tag{1.6}
$$

$$
\mathbf{B} = \frac{1}{rc}\mathbf{E}\times\mathbf{r}, \tag{1.7}
$$

where $\mathbf{r}$ is the vector from the field point to the charge. The magnetic field is always transverse. The electric field becomes purely transverse only at large distances, i.e., in the wave zone, where the first term in (1.6) is negligible. In the wave zone, the fields are therefore proportional to the acceleration.

## 1.2  Emission of radiation

Since Maxwell's equations are linear, fields from a number of closely spaced point charges $e_k$ at positions $\mathbf{r}_k = \bar{\mathbf{r}} + \mathbf{x}_k$ can be obtained by superposition of fields according to (1.6) and (1.7). Assuming the average distance $|\bar{\mathbf{r}}|$ from the field point to the charge cluster to be much larger than the $|\mathbf{x}_k|$, and $v/c$ to be small, the far field then becomes

$$
\mathbf{E} = \frac{1}{4\pi\epsilon_0}\bar{\mathbf{r}}\times\bar{\mathbf{r}}\times\sum_k e_k\ddot{\mathbf{x}}_k/c^2|\bar{\mathbf{r}}|^3, \tag{1.8}
$$

$$
\mathbf{B} = \frac{1}{4\pi\epsilon_0}\bar{\mathbf{r}}\times\sum_k e_k\ddot{\mathbf{x}}_k/c^3|\bar{\mathbf{r}}|^2. \tag{1.9}
$$

The dominant contributions to the far field are seen to be proportional to the second derivative of the electric dipole moment of the charge cluster. If retardation effects are accounted for more accurately and all first order terms in $v/c$ are included, additional electric quadrupole terms arise in (1.8) and (1.9) which are usually only important if the dipole moment vanishes. Furthermore, if magnetic dipoles are considered as sources, corresponding to small current loops on the right-hand side of (1.1), expressions analogous to (1.8) and (1.9) are obtained with $\mathbf{E}$ and $\mathbf{B}$ exchanged and the magnetic dipole moment replacing the electric dipole moment (Jackson 1962). For atomic electrons, this magnetic dipole moment is typically smaller than electric dipole moments by a factor $\sim \alpha z/n^2$, where $z$ is the effective nuclear charge and $n$ the principal quantum number for the active electron (Cowan 1981, Sobel'man 1992). The first factor is the fine-structure constant, $\alpha = e^2/4\pi\epsilon_o\hbar c \approx 1/137$.

To calculate the radiated power, we need the Poynting vector,

$$\mathbf{S} = \mathbf{E} \times \mathbf{H} = -\frac{1}{(4\pi)^2 \epsilon_0} |\bar{\mathbf{r}} \times \sum_k e_k \ddot{\mathbf{x}}_k|^2 \bar{\mathbf{r}} \ / \ c^3 |\bar{\mathbf{r}}|^5$$

$$= -\frac{1}{(4\pi)^2 \epsilon_0} |\sum_k e_k \ddot{\mathbf{x}}_k|^2 \bar{\mathbf{r}} \sin^2\theta \ / \ c^3 |\mathbf{r}|^3. \qquad (1.10)$$

The minus sign appears because $\bar{\mathbf{r}}$ is from the field point to the center of the charge distribution, and $\theta$ in the second version is the angle between $\bar{\mathbf{r}}$ and the second derivative of the dipole moment. Frequently one is interested only in the total radiated power, which follows by integrating $\mathbf{S}$, e.g., over a large spherical surface surrounding the charge distribution. This power is

$$P_e = \frac{1}{6\pi\epsilon_0 c^3} |\sum_k e_k \ddot{\mathbf{x}}_k|^2$$

$$= \frac{2}{3}\frac{mr_0}{c}|\sum_k \ddot{\mathbf{x}}_k|^2. \qquad (1.11)$$

In the second version, which only holds for electrons, the classical electron radius $r_0 = e^2/4\pi\epsilon_0 mc^2 \approx 2.818 \times 10^{-15}$m is used to simplify the expression.

Finally, for a harmonic oscillator, i.e., $\sum e_k \mathbf{x}_k = e\mathbf{x}(t) = e\mathbf{x}_0 \cos \omega_0 t$, etc., the time-averaged power is

$$\bar{P}_e = \frac{1}{3c}mr_0\omega_0^4|\mathbf{x}_0|^2. \qquad (1.12)$$

This is the quantity of primary interest in quantitative spectroscopy.

Another case of great interest involves radiation from free electrons in magnetized laboratory and natural plasmas (Bekefi 1966). The accelerated motion in this case causes cyclotron, cyclotron-harmonic, and synchrotron radiation, in the order of increasing electron energy toward relativistic energies. To describe this radiation, one must return to (1.6) and (1.7), inserting the helical motion in a magnetic field. The interested reader is referred to Bekefi's (1966) book.

### 1.3   Absorption by harmonic oscillators

If a harmonic oscillator is initially not excited but, say, at $t = 0$ exposed to incident electromagnetic waves, it will be driven into oscillations. Interaction with waves at $t > 0$ will then usually lead to an increase of the oscillator's energy, i.e., to absorption or loss of wave energy. An appropriate equation of motion for the oscillator is

$$\ddot{\mathbf{x}} + \omega_0^2\mathbf{x} = \frac{e}{m}\sum \mathbf{E}_\omega \cos(\omega t + \delta_\omega). \qquad (1.13)$$

Here the $\mathbf{E}_\omega$ are the wave amplitudes of the initial waves and the $\delta_\omega$ their phases, and the sum is over frequencies, modes, and phases. The special solution required here is

$$\mathbf{x}(t) = \frac{e}{m} \sum \frac{\mathbf{E}_\omega}{\omega_0^2 - \omega^2} \left[\cos(\omega t + \delta_\omega) - \cos(\omega_0 t + \delta_\omega)\right]. \tag{1.14}$$

Writing the wave field as $\sum' \mathbf{E}'_\omega \cos(\omega t + \delta')$, the absorbed power or work done by the field is

$$
\begin{aligned}
dP_a &= e\dot{\mathbf{x}}(t) \cdot \sum{}' \mathbf{E}'_\omega \cos(\omega t + \delta'_\omega) \\
&= \frac{e^2}{m} \sum E_\omega^2 \left[-\frac{\omega}{\omega_0^2 - \omega^2} \sin(\omega t + \delta_\omega) + \frac{\omega_0}{\omega_0^2 - \omega^2} \sin(\omega_0 t + \delta_\omega)\right] \\
&\quad \times \cos(\omega t + \delta_\omega).
\end{aligned} \tag{1.15}
$$

(Note that the free oscillations included in Griem 1964 need not be considered here.) On the average over many oscillations, the first term in the square-bracketed expression does not contribute. The average absorbed power is, therefore, using standard trigonometric formulas for the $\sin(\omega_0 t + \delta_\omega)\cos(\omega t + \delta_\omega)$ product and averaging over phases

$$
\begin{aligned}
d\bar{P}_a &= \frac{e^2}{2m} \sum E_\omega^2 \frac{\omega_0}{\omega_0^2 - \omega^2} \frac{1}{\tau} \int_0^\tau \sin[(\omega_0 - \omega)t]dt \\
&= \frac{e^2}{2m} \sum E_\omega^2 \frac{\omega_0}{\omega_0 + \omega} \frac{1 - \cos[(\omega_0 - \omega)\tau]}{\tau(\omega_0 - \omega)^2}.
\end{aligned} \tag{1.16}
$$

For large $\tau$ the last factor is proportional to the Dirac delta function $\delta(\omega - \omega_0)$. Replacing $\sum$ by $\int d\omega$ and using $\int_{-\infty}^\infty dx(1 - \cos x)/x^2 = \pi$, the total absorbed power thus becomes

$$\bar{P}_a = \frac{\pi e^2}{4m} E_\omega^2. \tag{1.17}$$

The spectral density $E_\omega^2$ of the waves' electric field is related to the spectral energy flux $\overline{S(\omega)}$, i.e., the Poynting vector magnitude per angular frequency interval, through

$$
\begin{aligned}
\overline{S(\omega)} &= |\mathbf{E}_\omega \times \mathbf{H}_\omega| \\
&= \frac{1}{\mu_0 c} E_\omega^2 \overline{\cos^2(\omega t + \delta_\omega)} \\
&= c\epsilon_0 E_\omega^2/2,
\end{aligned} \tag{1.18}
$$

using (1.10) and $c^2 = 1/\epsilon_0\mu_0$. In terms of $\overline{S(\omega)}$, the absorbed power is finally

$$\bar{P}_a = \frac{\pi e^2}{2\epsilon_0 mc} \overline{S(\omega)}, \tag{1.19}$$

and we infer that the absorption cross section $\sigma(\omega)$ is related to the first factor through

$$\int \sigma_a(\omega)d\omega = \frac{\pi e^2}{2\epsilon_0 mc} = 2\pi^2 r_0 c. \tag{1.20}$$

This classical formula agrees with the quantum mechanical result if the line absorption oscillator strength introduced in the next chapter is $f = 1$.

It is also noteworthy that the results obtained here depend on the random phases of the wave modes. One could, with different assumptions for the phases and initial conditions, even obtain a negative power, a classical analog of the induced emission process first postulated by Einstein (1917) for his derivation of Planck's law.

## 1.4　Radiation damping

In our discussion of linear oscillators and their effects on electromagnetic waves, we so far implicitly assumed undamped oscillations of the oscillator. This is clearly inconsistent with the emission of radiation at a rate according to (1.12). To conserve energy, the energy of the oscillator, $\frac{1}{2}m\omega_0^2|x_0|^2$, must decrease to make up for the radiated power. This suggests

$$\frac{1}{2}m\omega_0^2\frac{d}{dt}|x_0|^2 = -P_e = -\frac{1}{3c}mr_0\omega_0^2|x_0|^2 \tag{1.21}$$

if the rate of change in the amplitude remains small enough not to invalidate (1.12).

The characteristic decay rate of the square of the amplitude is then estimated by

$$\gamma = -\frac{1}{|x_0|^2}\frac{d}{dt}|x_0|^2 = \frac{2}{3c}r_0\omega_0^2 \tag{1.22}$$

provided $\gamma \ll \omega_0$ so that the time variation of $x_0$ is very slow.

Because the radiation fields now have an $\exp[(-i\omega_0 - \frac{1}{2}\gamma)t]$ dependence on time rather than the $\exp(-i\omega_0 t)$ dependence for constant amplitudes, their Fourier components $\int_0^\infty dt \exp[i(\omega - \omega_0)t - \frac{1}{2}\gamma t]$ are given by $[\frac{1}{2}\gamma - i(\omega - \omega_0)]^{-1}$. The emitted power per frequency interval is proportional to the absolute value squared of these Fourier components, i.e.,

$$\frac{dP_e}{d\omega} = \frac{1}{3c}mr_0\omega_0^4\overline{|x_0|^2}\frac{\gamma/2\pi}{(\gamma/2)^2 + (\omega - \omega_0)^2} \tag{1.23}$$

if one uses a time average of $|x_0|^2$. Moreover, $\gamma/2\pi$ is introduced as normalization factor so that the spectrally integrated power remains the same.

The last factor in (1.23) is the normalized line shape function accounting for the damping of the oscillator by radiation. Collisions can cause a similar effect, if their duration is short. The line shape then remains the same (Lorentzian), but an effective collision frequency $v_{coll}$ must be added to $\gamma$. (Details concerning this collision or impact broadening and other spectral line broadening mechanisms are discussed in chapter 4.)

A more rigorous calculation of radiation damping involves the solution of the equation of motion of the harmonic oscillator in its own radiation field. The reaction of this field then results in damping of the oscillations and a very small frequency shift (Jackson 1962). These results agree with the above estimate only in the limit $r_0/\lambda \approx r_0\omega_0/c \ll 1$, for a given wavelength $\lambda$. According to (1.22), this is equivalent to $\gamma \ll \omega_0$.

In the same regime, which covers all atomic radiation quite comfortably, one can describe the effects of radiation damping also by introducing an effective friction force into the equation of motion of the oscillator. The work done by this force $\mathbf{F}_r$ must correspond to the radiated energy, i.e.,

$$-\int \mathbf{F}_r \cdot \dot{\mathbf{x}}dt = \frac{2}{3c}mr_0 \int |\ddot{\mathbf{x}}|^2 dt, \qquad (1.24)$$

using (1.11). Integrating $\ddot{\mathbf{x}} \cdot \ddot{\mathbf{x}}$ by parts and averaging over an interval containing many periods gives

$$\mathbf{F}_r = \frac{2}{3c}mr_0 \dddot{\mathbf{x}} \approx -\frac{2}{3c}mr_0\omega_0^2\dot{\mathbf{x}} = -\gamma m\dot{\mathbf{x}}. \qquad (1.25)$$

This is again as expected and shows that $\gamma m$ is analogous to a coefficient of friction.

## 1.5 Scattering of radiation

The forced oscillations of a harmonic oscillator induced by the electric field of an electromagnetic field also cause emission of radiation. This is given by (1.12), if only the total scattered power is desired, or by the product of this expression and $3\sin^2\theta/8\pi$ if the differential scattered power per unit solid angle is required. The angle $\theta$ is here between the direction of observation and the electric field vector of the primary wave, and $3/8\pi$ is needed to retain the total scattered power. Finally, the $\sin^2\theta$ dependence is obtained from (1.10).

To calculate the quantity equivalent to $\mathbf{x}_0$ in (1.12) one considers the equation of motion

$$\ddot{\mathbf{x}} + \gamma\dot{\mathbf{x}} + \omega_0^2\mathbf{x} = \frac{e}{m}\mathbf{E}_0\exp(i\omega t) \qquad (1.26)$$

for the forced oscillator with radiation damping according to (1.25). The stationary solution for $\mathbf{x}(t)$ is

$$\mathbf{x} = \frac{(e/m)\mathbf{E}_0 \exp(i\omega t)}{\omega_0^2 - \omega^2 + i\gamma\omega}. \tag{1.27}$$

Substitution into (1.12), and again replacing the amplitude of the incident wave by the incident spectral energy flux according to (1.18), results in a scattered power

$$P_s = \frac{8\pi}{3}r_0^2\frac{\omega^4}{(\omega_0^2 - \omega^2)^2 + \gamma^2\omega^2}S(\omega). \tag{1.28}$$

Here we omit the average sign over $S(\omega)$.

As in the case of absorption, the factor of $S(\omega)$ is a cross section, namely for scattering by a single oscillator

$$\sigma_s = \frac{8\pi}{3}r_0^2\frac{\omega^4}{(\omega_0^2 - \omega^2)^2 + \gamma^2\omega^2}. \tag{1.29}$$

If an array of scatterers is involved, and if their positions are correlated, the different scattered waves will interfere. The result of such coherent scattering, therefore, depends critically on the relative phases of the various scattered waves.

For free electrons, that is $\omega_0 = 0$, we obtain the total Thomson scattering cross section

$$\sigma_{Th} = \frac{8\pi}{3}r_0^2. \tag{1.30}$$

For $\omega \approx \omega_0$ the approximation $\omega_0^2 - \omega^2 \approx 2\omega(\omega_0 - \omega)$ leads to the resonance scattering cross section

$$\sigma_{res} = \frac{2\pi}{3}r_0^2\frac{\omega_0^2}{(\omega_0 - \omega)^2 + (\gamma/2)^2}. \tag{1.31}$$

Its frequency integral is $\int \sigma_{res}d\omega = 4\pi^2r_0^2\omega_0^2/3\gamma = 2\pi^2r_0c$, using (1.22). This integral is, therefore, the same as that of the absorption cross section given by (1.20). This suggests that in the present picture it makes no difference whether we speak of absorption followed by emission, or of resonant scattering or resonance fluorescence, at least if we average over directions and neglect collisions, etc.

Finally, if the angular dependence of the scattering is required, we must re-instate the factor $\sin^2\theta$ discussed at the beginning of this section and then obtain with the additional normalization factor $8\pi/3$ for the differential cross section

$$\frac{d\sigma_s}{d\Omega} = \frac{3}{8\pi}\sigma_s\sin^2\theta = r_0^2\frac{\omega^4\sin^2\theta}{(\omega_0^2 - \omega^2)^2 + \gamma^2\omega^2}. \tag{1.32}$$

From this result those for $\omega_0 = 0$ and $\omega \approx \omega_0$ again follow easily.

## 1.6 Optical refractivity

In most natural and laboratory plasmas, densities are low enough that electromagnetic waves have phase velocities so close to $c$ that refraction of light rays and deviations from vacuum wavelengths can be ignored. (Actually, tabulated wavelengths for lines in and near the visible range are usually measured in air.) However, in very dense plasmas and also for wavelengths near strong lines, deviations of the refractive index from $n = 1$ can be important. This is frequently due to the polarization of free electrons but, especially in partially ionized gases, in some cases also due to bound electrons.

Excluding spectral regions very near resonance with lines, i.e., neglecting radiative and other damping, the dipole moments of the harmonic oscillators representing atoms, ions and electrons in a plasma are from (1.27)

$$e\mathbf{x} = \frac{(e^2/m)\mathbf{E}_0 \exp(i\omega t)}{\omega_0^2 - \omega^2} \tag{1.33}$$

in a wave field of amplitude $\mathbf{E}_0$ and frequency $\omega$. For $N$ such dipoles per unit volume this corresponds to a volume polarization

$$\mathbf{P} = Ne\mathbf{x} = N\frac{(e^2/m)\mathbf{E}_0 \exp(i\omega t)}{\omega_0^2 - \omega^2}, \tag{1.34}$$

neglecting the Lorentz-Lorenz correction for fields from the oscillators (Jackson 1962) in comparison to the original wave field. With these approximations the displacement vector is

$$\mathbf{D} = \epsilon_0 \mathbf{E} + \mathbf{P} \equiv \epsilon_0 n^2 \mathbf{E} = \epsilon_0 \left[ 1 + \frac{(e^2/\epsilon_0 m)N}{\omega_0^2 - \omega^2} \right] \mathbf{E}. \tag{1.35}$$

It is consistent with the above approximations to write $n^2 - 1 = (n+1)(n-1) \approx 2(n-1)$ so that we finally obtain

$$n = 1 + \frac{(e^2/2\epsilon_0 m)N}{\omega_0^2 - \omega^2} = 1 + \frac{2\pi r_0 c^2 N}{\omega_0^2 - \omega^2}. \tag{1.36}$$

An important special case is that of free electrons, i.e., $\omega_0 = 0$, giving

$$n = 1 - \frac{2\pi r_0 c^2 N_e}{\omega^2} = 1 - \frac{1}{2}\left(\frac{\omega_{pe}}{\omega}\right)^2 = 1 - \frac{r_0}{2\pi}\lambda^2 N_e. \tag{1.37}$$

Here $\omega_{pe}$ is the electron plasma frequency. In plasmas, the phase velocity is therefore normally larger than $c$, unless contributions from bound electrons according to (1.36) dominate and we have $\omega < \omega_0$. (Remember, however, that this expression is not valid for $\omega \approx \omega_0$, see section 2.10.) Finally, in magnetized plasmas, the optical properties become anisotropic and depend on frequency in a much more complicated way (Bekefi 1966).

# 2

# Quantum theory of radiation

Atoms and ions containing residual bound electrons do not quite resemble the simple harmonic oscillator model used so successfully in the classical theory of radiation. However, replacing the atoms or ions with sets of harmonic oscillators of a great number of discrete resonance frequencies and having various amplitudes, together with the results of classical radiation theory, go a long way toward a quantitative description of emission or absorption spectra. The set of resonance frequencies is obtained from measured or calculated energy levels using Ritz's combination principle. The amplitudes are associated with matrix elements of appropriate quantum mechanical operators between wave functions of the two energy eigenstates involved at a given frequency. In other words, quantities of the emitters, absorbers, or scatterers are described quantum-mechanically, whereas the electromagnetic field is treated classically.

Such semi-classical description of matter-electromagnetic field interactions became unnecessary very early in the development of quantum theory. It will therefore not be discussed in any detail. Instead, we will begin immediately with the combined theory of matter and radiation (Heitler 1954, Dirac 1958, Loudon 1983).

## 2.1   Quantum theory of particles and fields

There are various ways to also quantize the electromagnetic fields (Cohen-Tannoudji, DuPont-Roc and Grynberg 1989), of which that performed on the combined Hamiltonian equations of motion for the field-matter system is followed here. The nonrelativistic Hamiltonian for an electron (with negative charge of magnitude $e$ and mass $m$) in such field is

$$H = \frac{1}{2m}(\mathbf{P} + e\mathbf{A})^2 - e\phi + E_r, \tag{2.1}$$

11

where **P** is the canonical momentum, namely in terms of the mechanical momentum and the vector potential

$$\mathbf{P} = \mathbf{p} - e\mathbf{A}. \tag{2.2}$$

The electrostatic potential can be assumed to be entirely due to the nucleus and to any other electrons, while the field energy is

$$E_{r'} = \sum n_\omega \hbar\omega \tag{2.3}$$

in terms of the numbers of photons per mode and the photon energies. Also, for simplicity, we will here consider one-electron atoms or ions and assume that relativistic effects are not important.

To see whether the above Hamiltonian is consistent with (classical) charged particle dynamics, we may use the Hamiltonian equations of motion, i.e., for the $i$-th component

$$\dot{P}_i = \frac{d}{dt}(p_i - eA_i) = -\frac{\partial H}{\partial q_i} = -\frac{e}{m}\sum_j (p_j + eA_j)\frac{\partial A_j}{\partial q_i} + e\frac{\partial \phi}{\partial q_i}. \tag{2.4}$$

Written in terms of the fields from (1.4) and (1.5), this gives for the rate of change of the mechanical momentum

$$\begin{aligned}
\dot{p}_i = \dot{P}_i + e\frac{d}{dt}A_i &= -\frac{\partial H}{\partial q_i} + e\sum_k \frac{\partial A_i}{\partial q_k}\dot{q}_k + e\frac{\partial A_i}{\partial t} \\
&= -e\sum_{j,k}\left(\dot{q}_j\frac{\partial A_j}{\partial q_i} - \dot{q}_k\frac{\partial A_i}{\partial q_k}\right) + e\frac{\partial \phi}{\partial q_i} + e\frac{\partial A_i}{\partial t} \\
&= -e(\mathbf{v}\times\mathbf{B} + \mathbf{E}) \tag{2.5}
\end{aligned}$$

if we also use $(p_j + eA_j)/m = \partial H/\partial P_j = \dot{q}_j$ in our nonrelativistic treatment. This is the familiar equation of motion of an electron subject to the Lorentz force.

In the quantum theory of radiation, the vector potential is expanded into Fourier components

$$\mathbf{A}(\mathbf{x}, t) = \sum [q_\omega(t)\mathbf{A}_\omega(\mathbf{x}) + q_\omega^*(t)\mathbf{A}_\omega^*(\mathbf{x})], \tag{2.6}$$

the coefficients $q_\omega$ and $q_\omega^*$ being operators. For **A** to represent transverse waves in vacuum, it must fulfill the wave equation, i.e., the homogeneous version of (1.1) and also the Lorentz condition (1.3) with $\dot{\phi} = 0$. With (2.6) one obtains

$$\nabla^2\mathbf{A}_\omega + \frac{\omega^2}{c^2}\mathbf{A}_\omega = 0, \tag{2.7}$$

$$\nabla \cdot \mathbf{A}_\omega = 0, \tag{2.8}$$

$$\ddot{q}_\omega + \omega^2 q_\omega = 0 \tag{2.9}$$

and the same relations for the complex conjugate quantities.

Defining field "coordinates" and "momenta" as

$$Q_\omega = q_\omega + q_\omega^* \tag{2.10}$$

and

$$P_\omega = -i\omega(q_\omega - q_\omega^*), \tag{2.11}$$

one can show that the harmonic oscillator Hamiltonian

$$H_\omega = \frac{1}{2}(P_\omega^2 + \omega^2 Q_\omega^2) \tag{2.12}$$

indeed gives the proper field equations. This is seen from the canonical equations of motion

$$\dot{P}_\omega = -\partial H_\omega / \partial Q_\omega = -\omega^2 Q_\omega \tag{2.13}$$

and

$$\dot{Q}_\omega = \partial H_\omega / \partial P_\omega = P_\omega, \tag{2.14}$$

which give, e.g.,

$$\ddot{Q}_\omega + \omega^2 Q_\omega = 0, \tag{2.15}$$

consistent with (2.9). The Hamiltonian formalism therefore also applies to the field, if the Fourier components $A_\omega$ are solutions of (2.7) and (2.8) for appropriate boundary conditions.

From the solutions of the Schrödinger equation for the linear oscillator, one obtains for the field energy (per mode)

$$E_\omega = (n_\omega + \frac{1}{2})\hbar\omega \tag{2.16}$$

with the "1/2" suppressed in (2.3). According to (2.15), the matrix elements of the Hermitian operator $Q$ are analogous to those of the coordinate of the linear oscillator. Because of this fact and the requirement that $\langle n_\omega | H | n_\omega \rangle = E_\omega$ the only nonvanishing $Q$ matrix elements are therefore

$$\begin{aligned}
\langle n_\omega | Q_\omega | n_\omega + 1 \rangle &= \langle n_\omega + 1 | Q_\omega^* | n_\omega \rangle \\
&= \left[ \frac{\hbar(n_\omega + 1)}{2\omega} \right]^{1/2} \exp(-i\omega t).
\end{aligned} \tag{2.17}$$

This matrix and the corresponding $P_\omega$ matrix from (2.14) also obey the commutation rule

$$\langle n_\omega | Q_\omega P_\omega - P_\omega Q_\omega | n_{\omega'} \rangle = i\hbar \delta_{n_\omega n_{\omega'}}. \tag{2.18}$$

For applications in quantitative spectroscopy, it is more convenient to use matrix elements of $q_\omega$ and $q_\omega^*$, which follow from (2.10), (2.11) and (2.17) as

$$\langle n_\omega | q_\omega | n_\omega + 1 \rangle = \left[ \frac{\hbar(n_\omega + 1)}{2\omega} \right]^{1/2} \exp(-i\omega t) \qquad (2.19)$$

and

$$\langle n_\omega + 1 | q_\omega^* | n_\omega \rangle = \left[ \frac{\hbar(n_\omega + 1)}{2\omega} \right]^{1/2} \exp(i\omega t). \qquad (2.20)$$

All other matrix elements of $q_\omega$ and $q_\omega^*$ vanish, hence the names destruction and creation operator, respectively, for these operators.

## 2.2   Radiative transition probabilities

The states of the matter-field system can generally be approximated by product wave functions for the separate systems of the atom or ion and the field. These factored (unperturbed) wave functions are eigenfunctions of the unperturbed Hamiltonian, obtained from (2.1) by setting $\mathbf{A} = 0$. The interaction Hamiltonian is therefore

$$
\begin{aligned}
H_i &= H_{i1} + H_{i2} \\
&= -i\frac{\hbar e}{m} \nabla \cdot \sum_\omega (q_\omega \mathbf{A}_\omega + q_\omega^* \mathbf{A}_\omega^*) \\
&\quad + \frac{e^2}{2m} \sum_{\omega,\omega'} (q_\omega q_{\omega'} \mathbf{A}_\omega \cdot \mathbf{A}_{\omega'} + q_\omega q_{\omega'}^* \mathbf{A}_\omega \cdot \mathbf{A}_{\omega'}^* \\
&\quad + q_\omega^* q_{\omega'} \mathbf{A}_\omega^* \mathbf{A}_{\omega'} + q_\omega^* q_{\omega'}^* \mathbf{A}_\omega^* \mathbf{A}_{\omega'}^*).
\end{aligned}
\qquad (2.21)
$$

To obtain this explicit expression, the momentum was replaced by the operator $-i\hbar\nabla$, $\nabla$ being the gradient operator for the electron coordinates, and the vector potential was expanded using (2.6). Because of the interaction between the atomic system and the fields, any initial state of the total system well represented by one of the unperturbed product eigenfunctions will generally evolve into some superposition of such unperturbed functions of the same total energy as the initial state.

Normally $H_i$ can be considered a small quantity so that the time-dependent Schrödinger equation corresponding to the total Hamiltonian can be solved using Dirac's time-dependent perturbation theory. The solutions are written as sums of wave functions of the unperturbed Hamiltonian with time-dependent coefficients $c_{mn}(t)$

$$\psi_m(t) = \sum_n c_{mn}(t) U_n \exp(-iE_n t/\hbar), \qquad (2.22)$$

where the $U_n$ are time-independent (spatial) wave functions of the complete system in the absence of the perturbation and $E_n$ the corresponding energies. Substitution into the time-dependent Schrödinger equation and orthogonality of the spatial wave functions, i.e., $\int U_m^* U_n d\tau = \delta_{mn}$, gives the set of coupled ordinary and linear differential equations

$$\dot{c}_{mn}(t) = -\frac{i}{\hbar} \sum_{n'} c_{mn'} \langle n|H_i|n'\rangle \exp[-i(E_{n'} - E_n)t/\hbar], \qquad (2.23)$$

which is usually solved iteratively.

Remembering that $|c_{mn}(t)|^2$ is the probability of finding the system at time $t > 0$ in state $n$ if it was initially in state $m$ and assuming that no other states are populated at $t = 0$, one sets $c_{mn}(t) = \delta_{mn}$ and uses this as zero-order solution in the sum in (2.23). Integration then gives a first-order solution for $n \neq m$

$$\begin{aligned} c_{mn}(t) &= -\frac{i}{\hbar} \langle n|H_i|m\rangle \int_0^t \exp[-i(E_m - E_n)t/\hbar]dt \\ &= \frac{\langle n|H_i|m\rangle}{E_m - E_n} \{\exp[-i(E_m - E_n)t/\hbar] - 1\} \end{aligned} \qquad (2.24)$$

and therefore a probability for the system to be in state $n$ of

$$|c_{mn}(t)|^2 \approx 2\frac{|\langle n|H_i|m\rangle|^2}{(E_m - E_n)^2} \left[1 - \cos(E_m - E_n)t/\hbar\right]. \qquad (2.25)$$

Since the spectrum of the total system is always continuous, it is reasonable to integrate this expression over some interval of $E_n$ values, or to sum over the corresponding set of $n$ values,

$$\sum_n |c_{mn}(t)|^2 \approx \int |c_{mn}(t)|^2 \rho_{E_n} dE_n, \qquad (2.26)$$

where $\rho_{E_n}$ is the number of final states per energy interval. To the extent that $\rho_{E_n}$ and $\langle n|H_i|m\rangle$ can be considered constant, the integrand is seen to be essentially a $\delta$-function upon substitution of (2.25). We therefore obtain, as expected from Fermi's golden rule,

$$\int |c_{mn}(t)|^2 \rho_{E_n} dE \approx \frac{2\pi}{\hbar} |\langle n|H_i|m\rangle|^2 \rho_{E_n} t. \qquad (2.27)$$

The transition probability per unit time between the degenerate states $m$ and $n$ is therefore

$$P_{nm} = \frac{2\pi}{\hbar} |\langle n|H_i|m\rangle|^2 \rho_{E_n}, \qquad (2.28)$$

provided the various approximations made are indeed valid.

Although conservation of total energy was implicit, the range $\Delta E_n$ permitted in the summation or integration consistent with factoring out

$|\langle n|H_i|m\rangle|^2 \rho_{E_n}$ must be large enough to give $\Delta E_n t/\hbar > 1$ for $t$ values small enough not to invalidate perturbation theory. This restriction implies $\sum_n P_{nm} t < 1$ and therefore $\Delta E_n \gg \sum_n P_{nm}\hbar$ which, in turn, can be related to the radiative width of the atomic level involved (section 2.7). Normally this width is entirely negligible in plasma spectroscopy, as are any changes in the interaction Hamiltonian matrix elements and the density of states over relevant energy intervals.

If matrix elements $\langle n|H_i|m\rangle$ between some pairs of degenerate states of the total system do not exist, e.g., because of symmetry considerations, the iterative solution of (2.23) must be carried out to higher order by reinserting the first-order solution for $c_{mn'}(t)$, i.e., essentially (2.24) with $n$ replaced by $n'$, into (2.23). As discussed by Heitler (1954), the integration constant "1" has to be omitted because the $c_{mn'}(t)$ are too small and varying too rapidly for us to insist on precise initial conditions. The new equation becomes in this way

$$\dot{c}_{mn}(t) = -\frac{i}{\hbar} \sum_{n'} \frac{\langle n|H_i|n'\rangle\langle n'|H_i|m\rangle}{E_m - E_{n'}} \exp\left[-i(E_m - E_n)t/\hbar\right] \qquad (2.29)$$

and is seen to be entirely analogous to (2.23) with the zero-order approximation $c_{mn'} = \delta_{mn}$ and with $\langle n|H_i|m\rangle$ replaced by an effective interaction matrix element

$$\langle n|H_i'|m\rangle = \sum_{n'} \frac{\langle n|H_i|n'\rangle\langle n'|H_i|m\rangle}{E_m - E_{n'}}. \qquad (2.30)$$

The corresponding transition probability per unit time is accordingly

$$P_{nm}' = \frac{2\pi}{\hbar} \left| \sum_{n'} \frac{\langle n|H_i|n'\rangle\langle n'|H_i|m\rangle}{E_m - E_{n'}} \right|^2 \rho_{E_n}. \qquad (2.31)$$

Transition rates of even higher order are not important in plasma spectroscopy but could, of course, be obtained by continuing the iteration procedure. Also, the case $E_{n'} = E_m$ must be treated separately (section 2.9). It is entirely possible that contributions from $H_{i2}$ according to (2.21) may give first-order transition probabilities similar in magnitude to any second-order transition probability from $H_{i1}$ in cases where the $H_{i1}$ matrix element between initial and final states is zero (section 2.8). We finally note that the rate equation approach used here is not appropriate for strong and coherent electromagnetic fields, whose effects are the subject of a monograph by Shore (1990). The study of these fields, usually produced by lasers, has given rise to a large and important new research area, nonlinear optics (Bloembergen 1965, Letokhov and Chebotayev 1977).

## 2.3 Density of final states of the radiation field and normalization of the vector potential

Assuming a cubic normalization volume with sides of length $L$ and periodic boundary conditions, the number of modes in wave number interval $k, k + dk$ and solid angle range $d\Omega$ is given by the standard expression

$$dN_{k\Omega} = \left(\frac{L}{2\pi}\right)^3 k^2 dk d\Omega \qquad (2.32)$$

for each of the two directions of polarization of the electromagnetic waves. With the photon energy contribution to $E_n$ written as $\hbar c k = \hbar \omega$ and assuming a multiplicity (statistical weight) $g_n$ of the final state of the particle system, the density of states for a given polarization and for $\mathbf{k}$ vector directions within solid angle interval $d\Omega$ follows as

$$d\rho_{E_n} = g_n \frac{dN_{k\Omega}}{dE_n} = \frac{g_n}{\hbar} \left(\frac{L}{2\pi c}\right)^3 \omega^2 d\Omega. \qquad (2.33)$$

Note that $k = \omega/c$ was used here, implying a refractive index $n_r$ of 1. In general, a factor $n_r^2 c/v_g$ also containing the group velocity $v_g = dk/d\omega$ should be inserted which in plasmas is usually well approximated by $n_r = [1 - (\omega_{pe}/\omega)^2]^{1/2}$ for $\omega$ larger than the electron plasma frequency (Bekefi 1966). See also sections 1.6, 2.10, and 5.5 and note that the correction factor varies smoothly near $\omega_{pe}$ for highly collisional plasmas, but essentially vanishes at smaller frequencies for which there are no electromagnetic eigenmodes.

Before turning to the calculation of transition probabilities for specific processes, one needs to evaluate the vector potentials occurring in the interaction matrix elements defined by (2.21). These potentials must be solutions of (2.7) and (2.8). As already implied by the determination of the density of states, they therefore obey

$$\mathbf{A}_\omega = A_\omega \mathbf{a}_\omega \exp(-i\mathbf{k} \cdot \mathbf{r}). \qquad (2.34)$$

The $A_\omega$ here are the amplitudes of the vector potentials associated with a given mode of the field, $\mathbf{r}$ is the coordinate vector, and the $\mathbf{a}_\omega$ are unit vectors perpendicular to $\mathbf{k}$ which indicate the electric field polarization. Consider now the energy density operator of the electromagnetic field

$$
\begin{aligned}
E_r/L^3 &= \epsilon_0 \mathbf{E} \cdot \mathbf{E} \\
&= \epsilon_0 \dot{\mathbf{A}} \cdot \dot{\mathbf{A}} \\
&= \epsilon_0 \sum \omega^2 A_\omega^2 \left(q_\omega e^{-i\mathbf{k}\cdot\mathbf{r}} + q_\omega^* e^{i\mathbf{k}\cdot\mathbf{r}}\right)^2 \\
&\to \epsilon_0 \sum \omega^2 A_\omega^2 \left(q_\omega q_\omega^* + q_\omega^* q_\omega\right),
\end{aligned}
\qquad (2.35)
$$

where the first step is based on the equality of electrical and magnetic energies and the next steps involve (1.4) and (2.6), the solution of (2.9), and (2.34). The final step uses (2.19) and (2.20), i.e., selection rules for $q$ and $q^*$, and all cross terms between different modes were omitted. This is consistent with using number states for the radiation field, because in this presentation phases are left undetermined (Loudon 1983, Cohen-Tannoudji et al. 1989).

Using the $q$ and $q^*$ matrix elements from (2.19) and (2.20) then gives

$$
\begin{aligned}
\langle n_\omega | E_r | n_\omega \rangle &= \epsilon_0 L^3 \sum \omega^2 A_\omega^2 \left( \langle n_\omega | q_\omega | n_\omega + 1 \rangle \langle n_\omega + 1 | q_\omega^* | n_\omega \rangle \right. \\
&\quad \left. + \langle n_\omega | q_\omega^* | n_\omega - 1 \rangle \langle n_\omega - 1 | q_\omega | n_\omega \rangle \right) \\
&= \epsilon_0 L^3 \sum \frac{1}{2} \hbar \omega (n_\omega + 1 + n_\omega) A_\omega^2 \\
&= \epsilon_0 L^3 \sum \hbar \omega (n_\omega + \tfrac{1}{2}) A_\omega^2.
\end{aligned}
\tag{2.36}
$$

The normalized amplitude of the vector potential must therefore be taken as

$$
A_\omega = (\epsilon_0 L^3)^{-1/2}
\tag{2.37}
$$

for field and photon energies to be consistent with each other. This result and (2.21) and (2.34) finally allow us to write the two contributions to the interaction Hamiltonian as

$$
\begin{aligned}
H_{i1} &= -i \frac{\hbar e}{\sqrt{\epsilon_0} m_e L^{3/2}} \nabla \cdot \sum_\omega [q_\omega \exp(-i\mathbf{k} \cdot \mathbf{r}) \\
&\quad + q_\omega^* \exp(i\mathbf{k} \cdot \mathbf{r})] \mathbf{a}_\omega
\end{aligned}
\tag{2.38}
$$

and

$$
\begin{aligned}
H_{i2} &= \frac{e^2}{2\epsilon_0 m_e L^3} \sum_{\omega,\omega'} \{ q_\omega q_{\omega'} \exp[-i(\mathbf{k}+\mathbf{k}') \cdot \mathbf{r}] \\
&\quad + q_\omega q_{\omega'}^* \exp[-i(\mathbf{k}-\mathbf{k}') \cdot \mathbf{r}] + q_\omega^* q_{\omega'} \exp[i(\mathbf{k}-\mathbf{k}') \cdot \mathbf{r}] \\
&\quad + q_\omega^* q_{\omega'}^* \exp[i(\mathbf{k}+\mathbf{k}') \cdot \mathbf{r}] \} \mathbf{a}_\omega \cdot \mathbf{a}_{\omega'}
\end{aligned}
\tag{2.39}
$$

with $\mathbf{k} = \mathbf{k}_\omega$ and $\mathbf{k}' = \mathbf{k}_{\omega'}$. For the relatively weak radiation fields encountered in most applications of plasma spectroscopy, matrix elements of $H_{i1}$ are generally much larger than those of $H_{i2'}$.

## 2.4    Spontaneous emission

The most important process in plasma spectroscopy is the emission from excited atoms or ions of photons into field modes that are empty, corresponding to $n_\omega = 0$ and resulting in $n_\omega = 1$. The relevant $q^*$ matrix element is therefore given by (2.20) for $n_\omega = 0$. Using it and (2.28), (2.33) and (2.38) we obtain for the transition probability per unit time for a photon to be emitted into the element of solid angle $d\Omega$

$$dA_{nm} = \frac{e^2 \omega}{8\pi^2 \epsilon_0 \hbar c^3} g_n \left| \langle n | i \frac{\hbar}{m_e} \nabla \cdot \mathbf{a} \exp(i\mathbf{k} \cdot \mathbf{r}) | m \rangle \right|^2 d\Omega. \tag{2.40}$$

Note here that the $n$ and $m$ now stand only for the atomic quantum numbers of final and initial states, respectively. Also, note that the matrix element involves integrations over the electron's coordinates and, for two- and more-electron atoms or ions, also a sum over these electrons.

The exponential factor can normally be replaced by the first two terms in its power series expansion, $1 + i\mathbf{k} \cdot \mathbf{r}$. The first term gives the transition probability (omitting "per unit time" from now on) for electric dipole transitions,

$$
\begin{aligned}
dA_{nm} &= \frac{e^2 \omega}{8\pi^2 \epsilon_0 \hbar c^3} g_n \left| \langle n | i \frac{\hbar}{m_e} \nabla \cdot \mathbf{a} | m \rangle \right|^2 d\Omega \\
&= \frac{e^2 \omega^3}{8\pi^2 \epsilon_0 \hbar c^3} g_n \left| \langle n | \mathbf{r} \cdot \mathbf{a} | m \rangle \right|^2 d\Omega,
\end{aligned}
\tag{2.41}
$$

if we replace the operator $-i\hbar \nabla$ by $m\dot{\mathbf{r}} \stackrel{\triangle}{=} i\omega m\mathbf{r}$ using the time dependence of the unperturbed states of our nonrelativistic Hamiltonian.

Normally one is interested only in the total transition probability, regardless of the polarization. Since the $\mathbf{a}$ and $\mathbf{k}$ vectors are perpendicular, we have for any polarization in the plane defined by $\mathbf{r}$ and $\mathbf{k}$

$$\mathbf{r} \cdot \mathbf{a} = |\mathbf{r}| \sin\theta, \tag{2.42}$$

if $\theta$ is the angle between $\mathbf{r}$ and $\mathbf{k}$. As for the classical oscillator, the emission per unit solid angle is proportional to $\sin^2\theta$.

Modes polarized perpendicularly to this plane do not contribute, see (1.10), and one can substitute $\bar{\mathbf{r}} \sim \mathbf{k}$, $\ddot{\mathbf{x}}_k \sim \mathbf{r}$, $\mathbf{E} \sim \mathbf{a}$. With $\int \sin^2\theta d\Omega = 2\pi \int \sin^3\theta d\theta = 8\pi/3$ and writing $\mathbf{r}$ in terms of its components $x_i$, the total transition probability for spontaneous electric dipole emission finally is therefore

$$A_{nm} = \frac{4}{3\hbar c} m_e r_0 \omega^3 g_n \sum_i |\langle n | x_i | m \rangle|^2, \tag{2.43}$$

if we again introduce the classical electron radius $r_0 = e^2/4\pi\epsilon_0 m_e c^2$. After multiplication with the photon energy, the emitted power is seen to closely resemble the semi-classical result in (1.12). The factor 4 can be interpreted by saying that there are two matrix elements (Unsöld 1955), $\langle n | x_i | m \rangle$ and $\langle m | x_i | n \rangle$, for each pair of states, while $g_n$ represents the number of oscillators available for a given initial state $m$ because of the degeneracy of the final states.

Occasionally, the expression for the spontaneous transition probability
may have to be modified for dispersion effects, especially at frequencies
not much larger than the electron plasma frequency. Neglecting dispersion
due to bound electrons, section 2.10, the correction factor for (2.43) is in
the simplest approximation $n_r = [1 - (\omega_{pe}/\omega)^2]^{1/2}$, as discussed following
(2.33). More generally, this factor is $n_r^2 c/v_g$, and of course again essentially
zero for $\omega < \omega_{pe}$. We also imply here that the electron cyclotron frequency
obeys $\omega_{ce} \ll \omega_{pe}$ and that the plasma acts as a transparent dielectric. If
absorption is important, a generalization of the correction factor due to
Barnett, Huttner and Loudon (1992) remains valid also near the plasma
frequency.

For pairs of states having no dipole matrix elements, matrix elements
of the operator arising from the second term in the expansion of $\exp(i\mathbf{k} \cdot \mathbf{r})$
may also have to be considered. It leads to electric quadrupole transitions,
see, e.g., Cowan (1981) for the corresponding transition probability, whose
rate is smaller than dipole transition rates by a factor $\sim (\mathbf{k} \cdot \mathbf{r})^2$. With
$k = \omega/c$ and assuming the effective range of the coordinate integral to be
determined by the final state wavefunction, i.e., $r \approx n^2 a_0/z$, with $n$ here
being a principal quantum number and $z$ an effective charge, $(\mathbf{k} \cdot \mathbf{r})^2$ can
be estimated to be of the order $(n^2 a_0 \omega/zc)^2$. But one typically also has
$\omega \approx z^2 e^2/8\pi\epsilon_0 a_0 \hbar n^2$ and therefore $|(\mathbf{k} \cdot \mathbf{r})^2| \approx (\alpha z/2)^2$, which is very small
except for high $z$. (Recall $\alpha/2 \approx 1/274$.) As already discussed in section
1.2, the situation is similar for magnetic dipole transitions (Cowan 1981).
For transition probabilities of higher multipole transitions the reader is
referred to Shore and Menzel (1968), Heckmann and Träbert (1989), and
Sobel'man (1992).

## 2.5  Absorption

In analogy to (2.41), the transition probability $dW_{mn}$ for absorption of
photons from a mode corresponding to directions in solid angle element
$d\Omega$ can be obtained from (2.28), (2.33) and (2.38) and the relevant $q$-matrix
element, in this case $\langle n_\omega - 1|q_\omega|n_\omega\rangle$. Using (2.19) instead of (2.20) one has
in this way for electric dipole transitions

$$dW_{mn} = \frac{e^2 \omega}{8\pi^2 \epsilon_0 \hbar c^3} g_m \left| \langle m|i\frac{\hbar}{m_e}\nabla \cdot \mathbf{a}|n\rangle \right|^2 n_\omega d\Omega \qquad (2.44)$$

for the induced transition rate due to one mode from lower level $n$ to
upper level $m$. This rate is, as to be expected, proportional both to the
statistical weight of the upper level and to the number of photons initially
present. The latter quantity can be related to the directional intensity of

the radiation through

$$I(\omega,\Omega)d\omega d\Omega = \sum \hbar\omega c n_\omega L^{-3} dN_{\omega\Omega}$$

$$= \sum \frac{\hbar\omega c n_\omega k^2}{(2\pi)^3} dk d\Omega$$

$$= \sum \frac{\hbar\omega^3 n_\omega}{(2\pi)^3 c^2} d\omega d\Omega, \qquad (2.45)$$

where the sum is over the modes, of both polarizations, contributing within $d\omega$ and $d\Omega$.

Performing this summation over modes and integrating over $\Omega$ in (2.44), replacing $-i\hbar\nabla/m_e$ by $\omega\mathbf{r}$, and averaging $(\mathbf{r}\cdot\mathbf{a})^2$ over directions, i.e., using $\frac{1}{2}\int\cos^2\theta\sin\theta\,d\theta = 1/3$, the transition probability for absorption becomes

$$W_{mn} = \frac{16\pi^3\alpha}{3\hbar} g_m \sum_i |\langle m|x_i|n\rangle|^2_{av} I(\omega). \qquad (2.46)$$

The average is over magnetic quantum numbers, and the fine-structure constant again combines the atomic constants. Moreover, $I(\omega,\Omega) = I(\omega)$ was assumed to be isotropic. Multiplying (2.46) with the photon energy, one obtains the absorbed power and the factor of $I(\omega)$ in such an expression, but with an additional factor $d\Omega/4\pi$ to represent a narrow beam of radiation, can be interpreted as the frequency-integrated absorption cross section, i.e.,

$$\int \sigma^a_{mn}(\omega)d\omega = \frac{4\pi^2\alpha\omega}{3} g_m \sum_i |\langle m|x_i|n\rangle|^2_{av}$$

$$= 2\pi^2\alpha\frac{\hbar}{m} f_{mn}$$

$$= 2\pi^2 r_0 c f_{mn}. \qquad (2.47)$$

Here $f_{mn}$ is the absorption oscillator strength

$$f_{mn} = \frac{2m_e\omega}{3\hbar} g_m \sum_i |\langle m|x_i|n\rangle|^2_{av} \qquad (2.48)$$

and for $f_{mn} = 1$, (2.47) agrees with the classical result in (1.20).

Absorption due to other than electric dipole transitions is generally negligible in plasma spectroscopy, as are effects due to anisotropies in the orientations of the atomic dipole moments. Strongly magnetized plasmas could provide an exception to the latter statement, or plasmas in which level populations are controlled by laser pumping or by electron beams.

As another comment, note that in (2.45) any deviations of the refractive index from 1 were neglected. In cases where such deviations are important, the actual group velocity $v_g$ should be used in the first step and the inverse square of the phase velocity multiplied with $c/v_g$ in the final step, so that the absorption cross section must be divided by the square of the refractive index. However, this factor is cancelled by the factor $n_r^2$ discussed following (2.33), leaving $c/v_g \approx [1 - (\omega_p/\omega)^2]^{-1/2}$ as the final correction factor for plasma dispersion due to free electrons (Bekefi 1966).

## 2.6  Induced emission

In deriving the transition probability for spontaneous transitions, namely (2.43), $q^*$ matrix elements corresponding to $n_\omega = 0$ in (2.20) were used. For $n_\omega \geq 1$, additional terms arise from $|\langle n_\omega + 1|q^*|n_\omega\rangle|^2 \sim (n_\omega + 1)$ which are proportional to $n_\omega$, the "1" corresponding to spontaneous transitions. The $n_\omega$ term causes induced (or stimulated) emission with a transition probability entirely analogous to that for absorption, namely for isotropic radiation

$$W_{nm} = \frac{16\pi^3\alpha}{3\hbar} g_n \sum_i |\langle n|x_i|m\rangle|^2_{av} I(\omega). \tag{2.49}$$

The only change with respect to (2.46) is that the transition rate is now proportional to $g_n$, the statistical weight of the lower state. Therefore the cross section for induced emission obeys a formula similar to (2.47), except that in

$$\int \sigma^i_{nm}(\omega) d\omega = 2\pi^2 r_0 c f_{nm} \tag{2.50}$$

the emission oscillator strength must be used. It is related to the absorption oscillator strength in (2.48) through

$$f_{nm} = \frac{g_n}{g_m} f_{mn}. \tag{2.51}$$

Before turning to higher order interactions or more accurate solutions of the time-dependent Schrödinger equation, it is appropriate to verify that the transition probabilities obtained so far are consistent with Planck's law for thermal radiation by reversing Einstein's (1917) original argument for the existence of the stimulated emission process. In equilibrium, the net rate of radiative transitions between a pair of states must balance, i.e.,

$$N_n W_{mn} = N_m(A_{nm} + W_{nm}) \tag{2.52}$$

for population densities of atoms or ions $N_n$ and $N_m$ in lower and upper levels. In thermal equilibrium, relative populations are governed by statistical weights (degeneracies) and Boltzmann factors, i.e.,

$$\frac{N_m}{N_n} = \frac{g_m}{g_n} \exp\left(-\frac{E_m - E_n}{kT}\right) = \frac{g_m}{g_n} \exp\left(-\frac{\hbar\omega}{kT}\right). \tag{2.53}$$

(The Boltzmann constant $k$ here must not be confused with the wave number $k$ used earlier in this chapter.) Together with (2.43), (2.46) and (2.49) these two equilibrium conditions completely determine the isotropic intensity of a thermal (blackbody) radiator as

$$I_T(\omega) = \frac{\hbar\omega^3}{4\pi^3c^2}\left[\exp\left(\frac{\hbar\omega}{kT}\right) - 1\right]^{-1}, \tag{2.54}$$

which indeed agrees with Planck's law of thermal or blackbody radiation.

From the discussions of plasma dispersion effects on the various transition probabilities near the ends of sections 2.4 and 2.5, and from an extension of the latter to induced emission, it follows that Planck's radiation law must be multiplied by $n_r^2 \approx 1 - (\omega_{pe}/\omega)^2$ if the electron plasma frequency is approached from above. Below the plasma cutoff, the intensity would be exactly zero in this approximation. Bound electron contributions (section 2.10) to $n_r^2$ are usually neglected.

## 2.7 Natural line broadening

As mentioned in section 2.2, the time-dependent perturbation theory used so far to obtain the various transition probabilities is not valid for long times, i.e., times for which probabilities for completed radiative transitions are no longer small. In view of the general relationship between frequency uncertainties $\Delta\omega$ and observation time $\Delta t$, namely $\Delta\omega\Delta t \approx 1$, a more accurate solution of the time-dependent Schrödinger equation (2.23) is needed to obtain results that retain their validity for $\Delta\omega$ of the order of the radiative, also called natural, line width.

To demonstrate the general method (Heitler 1954) it is sufficient to consider a three-level atom and to assume that spontaneous transitions can occur only between levels 3 and 2 and between 2 and 1, the ground state. Moreover, absorption and induced emission will be neglected. Initially, at $t = 0$, the atom is in state 3, corresponding in (2.22) to $c_{33}(0) = c_3(0) = 1$, $c_{32}(0) = c_2(0) = 0$, and $c_{31}(0) = c_1(0) = 0$, while all field modes are empty. At large times, we expect $|c_1(\infty)| = 1$ and $c_3(\infty) = c_2(\infty) = 0$, and two photons, $\hbar\omega_{32}$ and $\hbar\omega_{21}$, to be present. The individual photon energies are approximate, and their distributions can be investigated by solving the rate equations corresponding to (2.23) for the present model problem, namely

$$\dot{c}_3 = -\frac{i}{\hbar}\sum H_{32}\exp(i\Delta\omega_{32}t)c_2, \tag{2.55}$$

$$\dot{c}_2 = -\frac{i}{\hbar}H_{23}\exp(-i\Delta\omega_{32}t)c_3 - \frac{i}{\hbar}\sum H_{21}\exp(i\Delta\omega_{21}t)c_1, \tag{2.56}$$

and

$$\dot{c}_1 = -\frac{i}{\hbar} H_{12} \exp(-i\Delta\omega_{21}t)c_2. \tag{2.57}$$

The first equation accounts for the decay of state 3 by emission of any of the possible photons near $\hbar\omega_{32}$ with frequency detunings $\Delta\omega_{32}$. The second equation, for the coefficient of a system wavefunction with the atom in state 2 and one photon present, begins with a population term related to the emission of this photon $\hbar(\omega_{32} + \Delta\omega_{32})$. The second term describes the decay of state 2 by emission of any photon near $\hbar\omega_{21}$. The last equation describes the population of state 1 after emission of a second specific photon $\hbar(\omega_{21} + \Delta\omega_{21})$. The $H_{32}$, etc., are abbreviations for the interaction matrix elements $\langle n|H_i|n'\rangle$, and the summations are over frequency or photon energy ranges larger than the resulting line widths but much smaller than, e.g., $\hbar\omega_{32}$.

To find a solution of the coupled system consistent with the initial and time-asymptotic values of the $c_i(t)$, which for $i = 1$ and 2 are also functions of the frequency detunings, one begins by assuming

$$c_3 = \exp(-\alpha t) \tag{2.58}$$

and substituting this trial solution into (2.56) and (2.57). These equations are then analogous to those for two-coupled linear oscillators, $c_3$ playing the role of a forcing term. Trial solutions for the other coefficients consistent with the initial conditions are written as

$$c_2 = a\{\exp[-(i\Delta\omega_{32} + \alpha)t] - \exp(-\beta t)\} \tag{2.59}$$

and

$$c_1 = b\{\exp[-(i\Delta\omega_{31} + \alpha)t] - 1\} + c\{\exp[-(i\Delta\omega_{21} + \beta)t] - 1\}, \tag{2.60}$$

and the question is whether the constants $a$, $b$, and $c$, in regard to time, can be chosen such that the system of differential equations for the coefficients $c_i(t)$ is fulfilled.

To verify the trial solutions, (2.59) and (2.60) are substituted in (2.57). This yields with $\Delta\omega_{31} = \Delta\omega_{32} + \Delta\omega_{21}$

$$b = \frac{H_{12}}{\hbar} \frac{a}{\Delta\omega_{31} - i\alpha} \tag{2.61}$$

and

$$c = -\frac{H_{12}}{\hbar} \frac{a}{\Delta\omega_{21} - i\beta}, \tag{2.62}$$

so that the sum occurring in (2.56) can be written with (2.60) as

$$
\frac{i}{\hbar} \sum H_{21} \exp(i\Delta\omega_{21}t) c_1
$$

$$
= \frac{i}{\hbar} \sum H_{21} H_{12} \left\{ \frac{\exp[-(i\Delta\omega_{32} + \alpha)t] - \exp(i\Delta\omega_{21}t)}{\Delta\omega_{21} + \Delta\omega_{32} - i\alpha} \right.
$$

$$
\left. - \frac{\exp(-\beta t) - \exp(i\Delta\omega_{21}t)}{\Delta\omega_{21} - i\beta} \right\} a. \tag{2.63}
$$

The sum over field modes near $\omega_{21}$ can be replaced by an integral over frequency detunings $\Delta\omega_{21}$ after introducing $\hbar\rho_E(\hbar\omega_{21})$ as the number of modes per unit frequency interval and factoring out slowly varying factors. This gives

$$
\frac{i}{\hbar} \sum H_{21} \exp(i\Delta\omega_{21}t) c_1
$$

$$
= \frac{i}{\hbar} \sum{}' H_{21} H_{12} \rho_E(\hbar\omega_{21})
$$

$$
\times \left\{ \exp\left[-(i\omega_{32} + \alpha)t\right] \right.
$$

$$
\times \int \frac{1 - \exp[(i\Delta\omega_{21} + i\Delta\omega_{32} + \alpha)t]}{\Delta\omega_{21} + \Delta\omega_{32} - i\alpha} d\Delta\omega_{21}
$$

$$
\left. - \exp(-\beta t) \int \frac{1 - \exp[(i\Delta\omega_{21} + \beta)t]}{\Delta\omega_{21} - i\beta} d\Delta\omega_{21} \right\} a, \tag{2.64}
$$

the remaining sum being over polarizations and solid angles.

Both integrals can be expressed for $t > 0$ by

$$
E(x, y, t) = \int_{-\infty}^{\infty} \frac{1 - \exp[(iz + ix + y)t]}{z + y - iy} dz
$$

$$
= \int_{-\infty}^{+\infty} \frac{dz'}{z' - iy} - \int_{-\infty}^{\infty} \frac{\exp(iz'' + y'')}{z'' - iy''} dz'' \tag{2.65}
$$

if the frequency range involved is assumed large in comparison to the decay constants $\alpha$ and $\beta$. The first integral gives $\pi i$ and the second integral follows from contour integration to $2\pi i$, resulting in

$$
E(x, y, t) = -\pi i. \tag{2.66}
$$

Use of this result in (2.64) and comparison with (2.28) shows that the sum in (2.56) can be expressed through

$$
\frac{i}{\hbar} \sum H_{21} \exp(i\Delta\omega_{21}t) c_1 = \frac{1}{2} A_{12}\, a\{\exp[-(i\Delta\omega_{32} - \alpha)t] - \exp(-\beta t)\}, \tag{2.67}
$$

where $A_{12}$ is the spontaneous transition probability for 2 to 1 transitions. Using this expression in (2.56) and also using (2.58), (2.59) and (2.60), one obtains

$$a = \frac{H_{23}}{\hbar} \frac{1}{\Delta\omega_{32} - i\alpha + iA_{12}/2} \tag{2.68}$$

and

$$\beta = \frac{1}{2}A_{12}. \tag{2.69}$$

The sum in (2.55) can according to these results and (2.59) be written as

$$\frac{i}{\hbar} \sum H_{32} \exp(i\Delta\omega_{32}t)c_2 = \frac{i}{\hbar^2} \sum H_{32}H_{23} \exp(-\alpha t)$$
$$\times \frac{1 - \exp[(i\Delta\omega_{32} + \alpha - A_{12}/2)t]}{\Delta\omega_{32} - i\alpha + iA_{12}/2}$$
$$= \frac{1}{2}A_{23} \exp(-\alpha t), \tag{2.70}$$

with the final result obtained using the same steps as taken in (2.63) to (2.67). Here $A_{23}$ is the spontaneous transition probability for 3 to 2 transitions, and the only remaining constant, $\alpha$, is found after substitution of (2.58) and (2.70) into (2.55) to be

$$\alpha = \frac{1}{2}A_{23}. \tag{2.71}$$

The relative probability for the two photons to be emitted at detunings $\Delta\omega_{32}$ and $\Delta\omega_{21}$ and for the atom to be in its ground state "1", long after excitation into state "3", is given by

$$|c_1(\infty)|^2 = |b + c|^2 = (H_{21}H_{12}H_{32}H_{23}/\hbar^4)$$
$$\times [(\Delta\omega_{32} + \Delta\omega_{21})^2 + (A_{23}/2)^2]^{-1}[(\Delta\omega_{21}^2 + (A_{12}/2)^2]^{-1}, \tag{2.72}$$

according to (2.60), (2.61), (2.62), (2.68), (2.69) and (2.71) and $\omega_{31} = \omega_{32} + \omega_{21}$. The frequency distribution of the resonance line near $\omega_{21}$ follows by averaging over $\Delta\omega_{32}$. It is a Lorentz profile corresponding to a damping constant equal to $A_{12}$, as one would expect from classical theory (section 1.4). However, since $\Delta\omega_{21}$ appears in both frequency-dependent factors, averaging over it is more involved, i.e., omitting the prefactor

$$\int_{-\infty}^{+\infty} \left[(\Delta\omega_{32} + \Delta\omega_{21})^2 + (A_{23}/2)^2\right]^{-1} \left[(\Delta\omega_{21})^2 + (A_{12}/2)^2\right]^{-1} d\Delta\omega_{21}$$
$$= \int_{-\infty}^{+\infty} \left[(\Delta\omega_{32} + \Delta\omega_{21}) + \frac{i}{2}A_{23}\right]^{-1} \left[(\Delta\omega_{32} + \Delta\omega_{21}) - \frac{i}{2}A_{23}\right]^{-1}$$
$$\times \left(\Delta\omega_{21} + \frac{i}{2}A_{12}\right)^{-1} \left(\Delta\omega_{21} - \frac{i}{2}A_{12}\right)^{-1} d\Delta\omega_{21}$$

$$= 2\pi i \left[ (iA_{23})^{-1} \left( -\Delta\omega_{32} + \frac{i}{2}A_{23} + \frac{i}{2}A_{12} \right)^{-1} \right.$$

$$\times \left( -\Delta\omega_{32} - \frac{i}{2}A_{23} - \frac{i}{2}A_{12} \right)^{-1} + (iA_{12})^{-1}$$

$$\left. \times \left( \Delta\omega_{32} + \frac{i}{2}A_{23} + \frac{i}{2}A_{12} \right)^{-1} \left( \Delta\omega_{32} - \frac{i}{2}A_{23} - \frac{i}{2}A_{12} \right)^{-1} \right]$$

$$= 2\pi \frac{A_{12} + A_{23}}{A_{12}A_{23}} \frac{1}{(\Delta\omega_{32})^2 + \frac{1}{4}(A_{12} + A_{23})^2}. \tag{2.73}$$

This Lorentz profile therefore has a width corresponding to the sum of the radiative decay rates. This important result can be generalized (Heitler 1954) through an effective damping constant

$$\gamma_{nm} = \sum_{m'} A_{m'n} + \sum_{n'} A_{n'm} \tag{2.74}$$

in situations where spontaneous decays to various levels $m'$ and $n'$ are possible from upper and lower levels of the line. If absorption and induced emission rates are comparable to the sum of spontaneous rates, these rates must be also added. This is straightforward if the radiation field has a flat spectrum in the vicinity of the line. Otherwise spectral line shapes of emission coefficients, etc., become functions of the spectrum of the radiation field (Griem 1989) and are not Lorentzian. For strong and resonant radiation fields one can no longer use perturbation theory to describe the ensuing power broadening (Karplus and Schwinger 1948, Townes and Schawlow 1955, Lisitsa and Yakovlenko 1975a and b).

Natural line broadening is almost always completely negligible in applications of plasma spectroscopy. Possible exceptions are inner shell x-ray transitions and lines involving autoionizing levels, or resonance lines of highly charged ions. Their nonradiative decay rates (autoionization or Auger rates) should also be added in (2.74), but the resulting profiles are generally not Lorentzian because of interference between the various decays (Fano 1961). However, by far the most important spectral line broadening processes in plasmas are Doppler effects and radiator-perturbing particle interactions; the latter are discussed in chapter 4.

## 2.8  Scattering of radiation

Of various higher-order processes involving photons and atoms or ions with bound electrons, scattering of photons on such bound electron systems is most important. Since two photons participate in this process, the second term $H_{i2}$ in (2.21) for the interaction Hamiltonian must now

be included. The first term contributes to the same order via the effective second-order interaction Hamiltonian in (2.30), so that we need to
consider the matrix elements

$$\langle n'|H_s|n\rangle = \sum_{n''} \frac{\langle n'|H_{i1}|n''\rangle\langle n''|H_{i1}|n\rangle}{E_n - E_{n''}} + \langle n'|H_{i2}|n\rangle. \tag{2.75}$$

The states involved are those of the complete system, $n$ and $n'$ for initial
and final states, each having one photon, and $n''$ for intermediate states
having either no or two photons.

Using (2.38) and (2.39) and the $q$ and $q^*$ matrix elements according to
(2.19) and (2.20) for $n_\omega = 0$, and the dipole approximation, the $H_s$ matrix
elements are explicitly

$$\begin{aligned}
\langle n'|H_s|n\rangle =\ & \frac{2\pi r_0 \hbar c^2}{L^3 m_e (\omega\omega')^{1/2}} \exp[i(\omega - \omega')t] \Big[ m_e\langle n'|\mathbf{a}_\omega \cdot \mathbf{a}_{\omega'}|n\rangle \\
& + \sum_{n''} \frac{\langle n'|\mathbf{p}\cdot\mathbf{a}_\omega|n''\rangle\langle n''|\mathbf{p}\cdot\mathbf{a}_{\omega'}|n\rangle}{E_n - E_{n''} + \hbar\omega} \\
& + \sum_{n''} \frac{\langle n'|\mathbf{p}\cdot\mathbf{a}_{\omega'}|n''\rangle\langle n''|\mathbf{p}\cdot\mathbf{a}_\omega|n\rangle}{E_n - E_{n''} - \hbar\omega'} \Big],
\end{aligned} \tag{2.76}$$

if we use $-i\hbar\nabla = \mathbf{p}$ and write the $H_{i2}$ term first. It is entirely due to the
$q_\omega^* q_\omega$ term in (2.39), while the second-order $H_{i1}$ interactions contribute two
terms, one due to absorption of photon $\hbar\omega$ followed by emission of $\hbar\omega'$,
the other involving first the emission of $\hbar\omega'$ and then the absorption of
$\hbar\omega$. This difference is reflected in the two energy denominators, which are
expressed in terms of atomic level energies and photon energies. Similarly,
the matrix elements on the right-hand side of (2.76) are only between
atomic states.

We can calculate the rate of scattering from (2.28) and (2.33) or, rather,
the corresponding differential scattering cross section by division with
$c/L^3$, the flux corresponding to a single initial photon,

$$\begin{aligned}
\frac{d\sigma_{n'n}}{d\Omega} =\ & r_0^2 \frac{\omega'}{\omega} \Big| \langle n'|\mathbf{a}_\omega \cdot \mathbf{a}_{\omega'}|n\rangle + \frac{m_e}{\hbar} \sum_{n''} \omega_{n''n'}\omega_{n''n} \\
& \times \Big[ \frac{\langle n'|\mathbf{x}\cdot\mathbf{a}_\omega|n''\rangle\langle n''|\mathbf{x}\cdot\mathbf{a}_{\omega'}|n\rangle}{\omega - \omega_{n''n}} \\
& - \frac{\langle n'|\mathbf{x}\cdot\mathbf{a}_{\omega'}|n''\rangle\langle n''|\mathbf{x}\cdot\mathbf{a}_\omega|n\rangle}{\omega' + \omega_{n''n}} \Big] \Big|^2.
\end{aligned} \tag{2.77}$$

In this expression the $p$-matrix elements were replaced by coordinate
matrix elements, i.e., by their components along the two polarization

vectors, and the atomic energy intervals by the corresponding frequencies. The statistical weight factor $g_{n'}$ does not appear, because the matrix elements depend on magnetic quantum numbers which must therefore be explicitly included in actual calculations. Except for the notation, this formula for the differential scattering cross section is the same as Dirac's (1958) equation (82), but appears to disagree with his equation (85). This equation resembles very closely the Kramers-Heisenberg (1925) dispersion formula and involves the use of commutators, which are singular for $\omega_{n''n} = 0$. It is also consistent with Heitler's (1954) result, but is not valid near resonances (section 2.9).

If the final state has a different energy than the initial state (i.e., in Raman scattering), no completely general simplifications of (2.77) seem possible. Normally one neglects the first term and, at least for groundstate scattering, also the second term in the sum which then is not at all resonant. Assuming further that only one intermediate energy level is important, an approximate cross section is

$$\frac{d\sigma_{n'n}}{d\Omega} \approx \frac{g_{n'} r_0^2 \omega'}{4 g_{n''} \omega} \cos^2\phi \frac{|\omega_{n''n'} \omega_{n''n}|}{(\omega - \omega_{n''n})^2} f_{n''n'} f_{n''n}. \tag{2.78}$$

Here the absorption oscillator strengths for the transitions from initial and final states, according to (2.48) and (2.51), were introduced by averaging over magnetic sublevels and assuming, e.g., $|\langle n''|z|n \rangle|^2 = |\langle n''|x|n \rangle|^2$.

The $\cos^2\phi$ factor stems from the scalar products with the two polarization vectors. It corresponds to the $\sin^2\theta$ factor in the classical formula for dipole scattering on a harmonic oscillator, (1.32), discussed at the end of section 1.5. The angle $\theta$ was defined there as the angle between the direction of polarization of the incoming photon, i.e., $\mathbf{a}_\omega$, and the wave vector $\mathbf{k}'$ of the outgoing photon. For $\mathbf{a}_{\omega'}$ polarization vectors in the $\mathbf{a}_\omega$, $\mathbf{k}'$ plane, we therefore have $\phi + \theta = \pi/2$ so that the two angular factors are equal. This leaves the $\mathbf{a}_{\omega'}$ polarization at right angles to the $\mathbf{a}_\omega$, $\mathbf{k}'$ plane. Returning to (2.76) for the scattering matrix element, one immediately sees that the first term vanishes for this case and that the remaining matrix elements are of the type $\langle n'|z|n'' \rangle \langle n''|x|n \rangle$, etc. They therefore also vanish, provided that the magnetic quantum numbers of initial and final states are the same. This vanishing was implicitly assumed by the use of (2.33), which gives the density of modes associated with a single polarization. For final states with different magnetic quantum numbers, one obtains scattering also into the mode polarized perpendicularly to the $\mathbf{a}_\omega$, $\mathbf{k}'$ plane (Penney 1969) with the angular distribution given by $\sin^2\phi$. For further discussions and a different definition of the angular factors see Sakurai (1967), who defines the polarization vectors relative to the plane determined by the $\mathbf{k}$-vectors of initial and final photons.

For further comparison with the classical theory, consider now elastic scattering, i.e., $n' = n$, recalling that conservation of energy requires

$$\omega' = \omega - (E_{n'} - E_n)/\hbar \qquad (2.79)$$

for the scattering process to occur, i.e., $\omega' = \omega$ in this case. Going back to (2.77) and expressing the matrix elements in terms of absorption oscillator strengths, one obtains per bound electron

$$\frac{d\sigma}{d\Omega} = r_0^2 \cos^2\phi \left[ 1 + \frac{1}{2} \sum_{n''} f_{n''n} \left( \frac{\omega_{n''n}}{\omega - \omega_{n''n}} - \frac{\omega_{n''n}}{\omega + \omega_{n''n}} \right) \right]^2$$

$$= r_0^2 \cos^2\phi \left[ 1 + \sum_{n''} f_{n''n} \frac{\omega_{n''n}^2}{\omega^2 - \omega_{n''n}^2} \right]^2 . \qquad (2.80)$$

For a single intermediate state and $f_{n''n} = 1$, this reduces near the corresponding line to the classical result for a harmonic oscillator of resonance frequency $\omega_0 = \omega_{n''n}$, i.e., to the differential cross section corresponding to (1.29) in the limit of negligible damping. For frequencies much larger than any contributing $\omega_{n''n}$, the first term in the square bracket dominates. It represents the Thomson scattering on free electrons with a total cross section given by (1.30). For $\omega^2 \ll \omega_{n''}^2$, one obtains the Rayleigh scattering cross section corresponding to (8.6).

The sum over immediate states also extends over the continuum, and any observed scattering depends on the relative phases of waves scattered on different atoms, ions, or free electrons (Bekefi 1966, Sheffield 1975).

### 2.9  Resonance fluorescence

In the preceding section, resonances between frequencies of the electromagnetic field and atomic energy level spacings were assumed not to be too close. In the classical theory, finite resonance scattering cross sections were obtained through the inclusion of radiative damping (1.29). To derive the corresponding quantum-mechanical result (Weisskopf 1933), e.g., for scattering on an atom in ground state 1, one needs to solve a system of equations for the $c$-coefficients in the time-dependent Schrödinger equation similar to the system considered in section 2.7 for the natural line broadening problem. It is also clear from section 2.8 that the $H_{i2}$ interaction can be neglected in the present context, as can the second-order term corresponding to emission of the scattered photon before absorption of the incoming photon. This would not necessarily be true for scattering on an excited atom or ion.

The $c$-coefficients or probability amplitudes required in this case are $c_1$ for the initial state of the system, atomic ground state 1 plus photons $\hbar\omega$,

$c_2$ for the excited state 2 produced by the absorption of one $\hbar\omega$, and $c_1'$ for the atomic ground state after emission of another photon $\hbar\omega'$. The differential equations for these coefficients are in analogy to (2.55), (2.56) and (2.57)

$$\dot{c}_1 = -\frac{i}{\hbar} \sum H_{12} \exp(i\Delta\omega_{12}t) c_2, \tag{2.81}$$

$$\dot{c}_2 = -\frac{i}{\hbar} H_{21} \exp(-i\Delta\omega_{12}) c_1 - \frac{i}{\hbar} \sum H_{21}' \exp(-i\Delta\omega_{12}'t) c_1', \tag{2.82}$$

$$\dot{c}_1' = -\frac{i}{\hbar} H_{12}' \exp(i\Delta\omega_{12}'t) c_2, \tag{2.83}$$

with $\Delta\omega_{12}$ and $\Delta\omega_{12}'$ representing the detunings of photons 1 and 2 from the resonance value. The first equation accounts for the depopulation of the ground state by the absorption of any of the incoming photons. The second equation balances the photo-excitation by a specific photon with the de-excitation accompanied by the emission of any of the $\hbar\omega'$ photons. The final equation describes the return to the ground state after emission of a specific photon $\hbar\omega'$. The sums are over $\Delta\omega_{12}$ and $\Delta\omega_{12}'$ values large enough to cover any of the line widths but small enough for the $H_{12}$, etc., matrix elements not to change significantly.

Appropriate initial conditions are $c_1(0) = 1$, $c_2(0) = c_1'(0) = 0$. They suggest to begin with

$$c_1 = \exp(-\alpha t) \tag{2.84}$$

and

$$c_2 = b\{\exp[(-i\Delta\omega_{12} - \alpha)t] - \exp(-\beta t)\} \tag{2.85}$$

as trial solutions. Substituting the last expression into (2.83) and integrating results in

$$c_1' = -b\frac{H_{12}'}{\hbar} \left\{ \frac{\exp[(i\Delta\omega_{12}' - i\Delta\omega_{12} - \alpha)t] - 1}{\Delta\omega_{12}' - \Delta\omega_{12} + i\alpha} \right.$$
$$\left. - \frac{\exp[(i\Delta\omega_{12}' - \beta)t] - 1}{\Delta\omega_{12}' + i\beta} \right\}. \tag{2.86}$$

Inserting these trial solutions into (2.82) leads to the consistency conditions

$$(i\Delta\omega_{12} + \alpha)b = \frac{i}{\hbar}H_{21} + \sum \frac{iH_{12}'H_{21}'}{\hbar^2} \frac{1 - \exp[(i\Delta\omega_{12} - i\Delta\omega_{12}' + \alpha)t]}{\Delta\omega_{12} - \Delta\omega_{12}' - i\alpha} b \tag{2.87}$$

and

$$\beta = -\sum \frac{iH'_{12}H'_{21}}{\hbar^2} \frac{1 - \exp[(-i\Delta\omega'_{12} + \beta)t]}{\Delta\omega'_{12} + i\beta}. \tag{2.88}$$

The sums or integrals over $\Delta\omega'_{12}$ are entirely analogous to those evaluated in section 2.7. Therefore (2.87) and (2.88) can be written as

$$(i\Delta\omega_{12} + \alpha)b = \frac{i}{\hbar}H_{21} + \frac{1}{2}A_{12}b \tag{2.89}$$

and

$$\beta = \frac{1}{2}A_{12} \tag{2.90}$$

in terms of the spontaneous decay rate of the excited state. The constant $b$ (with respect to time) is therefore

$$b = \frac{iH_{21}}{\hbar} \frac{1}{i\Delta\omega_{12} + \alpha - A_{12}}, \tag{2.91}$$

and with (2.81), (2.84), (2.85), (2.90) and (2.91) we find

$$\alpha = \sum \frac{iH_{12}H_{21}}{\hbar^2} \frac{1 - \exp[(i\Delta\omega_{12} + \alpha - A_{12}/2)t]}{\Delta\omega_{12} - i\alpha + iA_{12}/2} \tag{2.92}$$

as an implicit condition for the last remaining constant. Since in plasmas spectral lines are almost always much wider than natural line widths, it is usually appropriate to evaluate the sum over $\Delta\omega_{12}$ by the same method as used for the sums over $\Delta\omega'_{12}$. This gives

$$\alpha = \frac{1}{2}W_{21} \tag{2.93}$$

in terms of the transition probability for absorption in (2.46). (Recall that this relation was for isotropic radiation fields. Otherwise a factor $\Delta\Omega/4\pi$ should be applied to allow for the actual solid angle interval $\Delta\Omega$ containing incoming photons.)

The probability for reaching the final state of the system is from (2.86), (2.90), (2.91) and (2.93) for this broadband illumination

$$|c'_1(\infty)|^2 = \frac{|H'_{12}|^2|H_{21}|^2\hbar^{-4}}{[(\Delta\omega'_{12} - \Delta\omega_{12})^2 + (W_{21}/2)^2][(\Delta\omega'_{12})^2 + (A_{12}/2)^2]}. \tag{2.94}$$

Since $\Delta\omega_{12}$ varies by assumption over a large interval compared with $A_{12}$, scattered light has a Lorentzian frequency spectrum with the same natural width as the corresponding emission line. Summing or integrating over $\Delta\omega_{12}$, the total probability for the emission of a scattered photon becomes with $\sum |H_{21}|^2\rho_e = (\hbar/2\pi)W_{21}$ (golden rule)

$$\sum |c'_1(\infty)|^2 = \frac{|H'_{12}|^2\hbar^{-2}}{(\Delta\omega'_{12})^2 + (A_{12}/2)^2}. \tag{2.95}$$

Comparison with, e.g., (2.78) then suggests that $(A_{12}/2)^2$, or its generalization accounting also for other radiative or collisional transitions, should be inserted in the denominator there to obtain results that remain valid near and at resonance. An entirely analogous situation exists for resonant Raman scattering (Penney 1969).

So far it therefore appears that resonance fluorescence could just as well be treated by considering absorption and re-emission as independent processes. However, this is only correct if the incoming spectrum is indeed broad. Otherwise, scattering must be treated as a two-photon process (Heitler 1954) to maintain, e.g., energy conservation. For a very narrow $\omega$ spectrum also the $\omega'$ spectrum will be narrow, even compared with the natural line width. Very interesting situations can occur if rare collisions lead to changes in the energies of the scattered photons (chapter 8). Furthermore, if the incoming photons correspond to strong and resonant fields, the perturbation theory must be replaced by solutions used in nonlinear laser spectroscopy (Bloembergen 1965, Letokhov and Chebotayev 1977, Shore 1990, Mukamel 1995).

## 2.10  Optical refractivity

From the occurrence of the refractive index $n(\omega)$ and the absorption coefficient $\kappa(\omega)$ in the expression

$$
\begin{aligned}
E(x,t) &= A \exp\left[-i\left(\omega t - \frac{2\pi}{\lambda}x\right) - \frac{1}{2}\kappa x\right] \\
&= A \exp\left[-i\frac{\omega}{c}(ct - x)\right] \exp\left[i\frac{\omega}{c}\left(n - 1 + i\frac{\kappa c}{2\omega}\right)x\right] \quad (2.96)
\end{aligned}
$$

for a plane wave field propagating in an isotropic medium follows that the refractivity $n(\omega) - 1$ and absorption coefficient $\kappa(\omega)$, i.e., product of absorption cross section and density of absorbing atoms, can be written as

$$
n(\omega) - 1 = \mathrm{Re} f(\omega) \quad (2.97)
$$

and

$$
\kappa(\omega) = \frac{2\omega}{c}\mathrm{Im} f(\omega). \quad (2.98)
$$

Here $f(\omega)$ is a complex function with real and imaginary parts related by the integral relation

$$
\mathrm{Re} f(\omega) = \frac{1}{2\pi}\int_{-\infty}^{+\infty}\frac{\mathrm{Im} f(\omega')}{\omega' - \omega}d\omega'. \quad (2.99)
$$

The corresponding Kramers-Kronig relations to be used here are

$$n(\omega) - 1 = \frac{c}{4\pi} \int_{-\infty}^{+\infty} \frac{\kappa(\omega')d\omega'}{\omega'(\omega' - \omega)}. \tag{2.100}$$

The absorption coefficient due to transitions from state 1 to 2 is according to (2.47) and assuming a Lorentzian line shape of width $\gamma$ between half intensity points

$$\kappa(\omega) = \frac{\pi r_0 c \gamma f_{21} N_1}{(\omega - \omega_{12})^2 + (\gamma/2)^2}, \tag{2.101}$$

if $N_1$ is the density in the lower state and $f_{21}$ the absorption oscillator strength. After multiplication with $\omega/\omega_{12}$, which leads only to insignificant changes in the region of interest, and insertion into (2.100) one can do the integral analytically,

$$n(\omega) - 1 = -\pi \frac{r_0 c^2 f_{21} N_1}{\omega_{12}} \frac{\omega - \omega_{12}}{(\omega - \omega_{12})^2 + (\gamma/2)^2}. \tag{2.102}$$

This formula is naturally only valid for $|\omega - \omega_{12}| \ll \omega_{12}$ and may have to be corrected for induced emission, other lines, etc. However, it is not essential that only natural line broadening contributes, as long as the absorption and emission line profiles are Lorentzian. In other cases of homogeneous line broadening (chapter 4), (2.100) must be evaluated numerically. Comparison of (2.102) with the classical formula (1.36), in which $\gamma$ was neglected, shows agreement for $|\omega - \omega_{12}| \ll \omega_{12} = \omega_0$ and $\gamma = 0$. It also suggests that (2.102) can be generalized to cover a broader frequency range by replacing $\omega_{12}$ in the first factor by $\frac{1}{2}(\omega + \omega_{12})$. This extension of (2.102) is

$$\begin{aligned}
n(\omega) - 1 &= -2\pi \frac{r_0 c^2 f_{21} N_1}{(\omega + \omega_{12})^2} \frac{\omega^2 - \omega_{12}^2}{(\omega - \omega_{12})^2 + (\gamma/2)^2} \\
&\approx -2\pi \frac{r_0 c^2 f_{21} N_1}{\omega^2 - \omega_{12}^2},
\end{aligned} \tag{2.103}$$

the second version holding only for frequencies outside the line in question. These results, which should of course be summed over initial and final states, are strikingly similar to those for light scattering (section 2.8), which were therefore also called dispersion formulas and can be traced back to a correspondence principle derivation by Kramers and Heisenberg (1925).

# 3

# Oscillator and line strengths

The quantitative description of atomic radiation reviewed in the previous chapter requires knowledge of energy levels and of level populations, which will be the principal subjects of chapters 6 and 7, and of oscillator strengths or of the closely related line strengths. Their calculation and measurement are the main topics of the present chapter. As to energy levels and wavelengths, a large body of high quality empirical data is available (Moore 1949-1958, Martin, Zalubas and Hagen 1978, Cowan 1981, Kelly 1987a and b, Bashkin and Stoner 1975, 1978, 1981), and atomic structure calculations (Cowan 1981, Sobel'man 1992) are usually of an accuracy that is sufficient for most plasma spectroscopy applications. Our knowledge of oscillator strengths, and therefore also of transition probabilities for spontaneous transitions and line strengths, has also greatly improved since the predecessor of this monograph (Griem 1964) was written. These advances are the result of improved experiments and computations and of critical evaluations of data (Wiese, Smith and Glennon 1966, Wiese, Smith and Miles 1969, Fuhr and Wiese 1995, Martin, Fuhr and Wiese 1988, Fuhr, Martin and Wiese 1988). Especially complete and accurate data are now available for atoms and ions of carbon, nitrogen and oxygen (Wiese, Fuhr and Deters 1995). Astrophysical data tables (Allen 1973, Lang 1980) should also be consulted. Moreover, the National Institute of Standards and Technology provides access to atomic and molecular wavelengths and intensity information, to energy level and transition probability data, and to various bibliographies via Internet (NIST Physics Laboratory Home Page at http://physics.nist.gov).

Before introducing the general theoretical and experimental methods used to acquire these data, it should be pointed out that several computer codes are available for the generation of oscillator strengths and related quantities (Cowan 1968, 1981, Berrington et al. 1987, Cunto et al. 1993). For good theoretical reasons it is not possible to provide error estimates of general validity. However, it is widely accepted that large oscillator strengths are more reliable than small oscillator strengths, with errors ranging from a few percent, or less, to uncertainties by an order of magnitude or more. Given this situation, the accuracy of semiclassical calculations (More and Warren 1991, D'yachkov and Pankratov 1994) is most remarkable. Such calculations are very efficient.

The relationship between transition probabilities and oscillator strengths is obtained by comparing (2.43) and (2.48), namely

$$A_{nm} = \frac{2r_0\omega^2}{c}\frac{g_n}{g_m}f_{mn} = \frac{2r_0\omega^2}{c}f_{nm}. \tag{3.1}$$

If emission oscillator strengths are used instead of the absorption oscillator strengths, the simpler second version is obtained from (2.51). However, it is traditional to employ absorption oscillator strengths. Another possibility is to use line strengths, i.e.,

$$
\begin{aligned}
S_{nm} &= g_m f_{nm}\left(\frac{3\hbar}{2m_e\omega}\right) \\
&= g_n f_{mn}\left(\frac{3\hbar}{2m_e\omega}\right) \\
&= g_m g_n \sum_i |\langle m|x_i|n\rangle|_{av}^2 \\
&= \sum_M \sum_{M'} \sum_i |\langle m|x_i|n\rangle|^2,
\end{aligned}
\tag{3.2}
$$

again using (2.48) and writing the average over magnetic quantum numbers $M$ and $M'$ of levels $m$ and $n$ explicitly in order to obtain the final version. Extending the sum over coordinates $i$ to include the coordinates of all electrons generalizes these formulas to two- and more-electron atoms or ions.

Using line strengths commends itself not only because of their symmetry with respect to initial and final states, but also because it allows us to write

$$\sum_M A_{nm} = g_m A_{nm} = \frac{4m_e r_0 \omega^3}{3\hbar c}S_{nm}. \tag{3.3}$$

For statistically populated nearby upper levels and nearly equal frequencies, this quantity is therefore a direct measure of the relative line intensities or strengths. Finally, if $J$ is the total angular momentum quantum number, the statistical weight of the corresponding level is

$$g_J = 2J + 1. \tag{3.4}$$

Because of corrections in relations involving matrix elements and frequencies due to nuclear motion (Bethe and Salpeter 1957), there is, e.g., for one-electron atoms and ions a correction factor $(1 + m/M)^{-1}$ in (3.1), which allows for the reduced electron-nuclear mass factor in the various Rydberg constants (Goldwire 1968). Such corrections are of no practical importance in plasma spectroscopy.

### 3.1 Relative line strengths

To make full use of Racah algebra (Racah 1942, 1943), i.e., of the spherical symmetry of isolated atoms or ions, it is advantageous to consider the square root of the line strength and to replace the Cartesian components of the electron coordinate vectors by spherical components, i.e.,

$$r^{(1)}_{\pm 1} = \mp (x \pm iy) \tag{3.5}$$

and

$$r^{(1)}_0 = z \tag{3.6}$$

and to consider the $r^{(1)}_q$ as the components of a spherical (Racah) tensor of rank 1, $\mathbf{r}^{(1)}$. In terms of the sum of these tensors over the electrons of a given atom or ion, the line strength is therefore (Cowan 1981)

$$S_{nm} = \sum_M \sum_{M'} \sum_q |\langle \gamma JM | \sum_i r^{(1)}_q(i) | \gamma' J'M' \rangle|^2. \tag{3.7}$$

The quantum numbers $J$ and $M$ are as just explained, and $\gamma$ stands for all other quantum numbers that may be required to specify the eigenstates of the system.

Since $\mathbf{r}^{(1)}$ is a vector operator, electric dipole line strengths can have nonzero values only between states of opposite parity (Laporte's rule). Further selection rules are imposed by the Wigner-Eckart theorem, which gives

$$\langle \gamma JM | \sum_i r^{(1)}_q(i) | \gamma' J'M' \rangle = (-1)^{J-M} \begin{pmatrix} J & 1 & J' \\ -M & q & M' \end{pmatrix}$$

$$\times \langle \gamma J \| \sum_i \mathbf{r}^{(1)}(i) \| \gamma' J' \rangle \tag{3.8}$$

in terms of Wigner's 3-$j$ symbols. These symbols are proportional to the Clebsch-Gordan coefficients used in the older literature and to the reduced matrix element of the tensor operator. This reduced matrix element is independent of all magnetic quantum numbers, including $q$. From the properties of the 3-$j$ symbols follow the general selection rules for electric dipole radiation

$$\Delta J \equiv J - J' = 0, \pm 1, \tag{3.9}$$

but $J' = J = 0$ is not allowed, and

$$\Delta M \equiv M - M' = -q = 0, \mp 1. \tag{3.10}$$

The $\Delta M$ selection rules are related to linear polarization for $M' = M$ and circular polarizations for $M' = M \pm 1$ (Cowan 1981, Sobel'man 1992).

To extract further information from atomic structure theory, it is necessary to expand the wave functions $|\gamma J\rangle$ in terms of some basis set $|\beta J\rangle$, namely

$$|\gamma J\rangle = \sum_{\beta} Y_{\beta J}^{\gamma} |\beta J\rangle \tag{3.11}$$

in Cowan's notation, always realizing that, e.g., the set of state vector components defined by this relation is independent of the magnetic quantum number $M$. The reduced matrix element in (3.8), usually expressed in units of $a_0$, may then be written as

$$\langle \gamma J \| \sum_i \mathbf{r}^{(1)}(i) \| \gamma' J' \rangle = \langle \gamma J \| \mathbf{P}^{(1)} \| \gamma' J' \rangle$$

$$= \sum_{\beta, \beta'} Y_{\beta J}^{\gamma} \langle \beta J \| \mathbf{P}^{(1)} \| \beta' J' \rangle Y_{\beta' J'}^{\gamma'}, \tag{3.12}$$

with the $Y$-coefficients to be determined by finding the energy eigenvectors of the atom or ion in question which correspond to $\gamma, J$ or $\gamma', J'$. The inner matrix in the double matrix product would normally correspond to the dipole transition matrix element in some pure-coupling representation, e.g., $LS$-coupling, depending on which basis functions are used for the expansion in (3.11).

Because the dipole transition operator does not act on spin coordinates, the inner matrix element for the case of $LS$ coupled $|\beta J\rangle$, etc., wave functions can be greatly simplified using the method of uncoupling. The general result is (Cowan 1981)

$$\langle \gamma L S J \| \mathbf{P}^{(1)} \| \gamma' L' S' J' \rangle = \delta_{ss'} (-1)^{L+S+J'+1}$$

$$\times [(2J+1)(2J'+1)]^{1/2} \begin{Bmatrix} L & S & J \\ J' & 1 & L \end{Bmatrix} \langle \gamma L S \| \mathbf{P}^{(1)} \| \gamma' L' S' \rangle, \tag{3.13}$$

where $\{ \ \}$ is a 6-$j$ symbol. It is proportional to the Racah $W$ coefficients used and partially tabulated in Griem (1964). More extensive tables of 3-$j$ and 6-$j$ symbols can be found in Rotenberg et al. (1959) and in Cowan (1981). Besides the obvious $\Delta S = 0$ selection rule, the 6-$j$ symbols give the additional rule

$$\Delta L = L - L' = 0, \pm 1, \tag{3.14}$$

but again exclude $L = L' = 0$. These selection rules do not generally apply to the actual line or oscillator strengths, because the eigenvectors in (3.11), etc., tend to involve components corresponding to $|\beta J\rangle$ basis states that have various $LS$ values. Such a situation is called intermediate coupling.

## 3.2 Absolute line strengths for one-electron atoms

For one-electron atoms or ions and for atoms or ions with only one active electron outside completely filled inner shells (section 3.3), (3.7), (3.8), (3.12) and (3.13) result in line strengths whose square root is

$$
\begin{aligned}
S^{1/2} &= \langle n\ell sj\|\mathbf{r}^{(1)}\|n'\ell'sj'\rangle \\
&= (-1)^{\ell+j'+3/2}[(2j+1)(2j'+1)]^{1/2} \\
&\quad \times \begin{Bmatrix} \ell & s & j \\ j' & 1 & \ell' \end{Bmatrix} \langle n\ell\|\mathbf{r}^{(1)}|n'\ell'\rangle
\end{aligned} \tag{3.15}
$$

in terms of the usual single electron quantum numbers $n$, $\ell$, $j$ and $s$. The 3-$j$ symbols from (3.8) together with the sum over the tensor components $q$ in (3.7) disappear because of their orthogonality and sum rules (Cowan 1981), and the transformation in (3.11) is the identity transformation in case of isolated one-electron systems. The first-order tensor operator $\mathbf{r}^{(1)}$ can be factored into radial and angular parts,

$$
\mathbf{r}^{(1)} = r\mathbf{C}^{(1)}, \tag{3.16}
$$

where $r$ is the usual radial coordinate and $\mathbf{C}^{(1)}$ the Racah tensor of first rank. Its reduced matrix elements are (Cowan 1981)

$$
\langle \ell\|\mathbf{C}^{(1)}\|\ell'\rangle = (-1)^{\ell}[(2\ell+1)(2\ell'+1)]^{1/2} \begin{pmatrix} \ell & 1 & \ell' \\ 0 & 0 & 0 \end{pmatrix}, \tag{3.17}
$$

and the reduced matrix element in (3.15) can, with the help of a special relation for the 3-$j$ symbol in (3.17), be expressed as

$$
\langle n\ell\|\mathbf{r}^{(1)}\|n'\ell'\rangle = (-1)^{\ell+\ell_>}(\ell_>)^{1/2} \int_0^\infty rR_{n\ell}R_{n'\ell'}dr, \tag{3.18}
$$

with $\ell_>$ being the larger of $\ell' = \ell \pm 1$ and $\ell$. For $\ell' = \ell$ the one-electron matrix element is zero, and the $R_{n\ell}(r)$ are the radial wave functions of the bound electron. (Note that a factor $r$ is included in the $R_{n\ell}$ which are normalized to $\int R_{n\ell}^2(r)dr = 1$.)

The radial wave functions are the solutions of the radial Schrödinger equation

$$
-\frac{d^2}{dr^2}R_{n\ell} - [\frac{2z}{r} - \frac{\ell(\ell+1)}{r^2} + \epsilon_{n\ell}]R_{n\ell} = 0 \tag{3.19}
$$

in Rydberg energy units and writing $r$ in units of the Bohr radius $a_0$. For actual one-electron systems with nuclear charge $Z = z$, the potential is, of course, indeed given by $-z/r$ and the energy by $-z^2/n^2$. For atoms or ions with other electrons in closed inner shells, $-z/r$ must be replaced by an effective potential which also accounts for the electron-electron interactions. The energy parameters $\epsilon_{n\ell}$ must be determined self-consistently to ensure that the wave functions fulfill appropriate conditions at $r = 0$ and $r \to \infty$.

For hydrogen and hydrogen-like ions, wave functions and radial matrix elements can be calculated analytically (Bethe and Salpeter 1957). Note that the $R_{n\ell}$ functions there correspond to our $R_{n\ell}/r$ and the $R_{n\ell}^{n'\ell'}$ to the radial integral in (3.18). Because of their possible use in estimates for two- and more-electron ions, corresponding values for the squares of the reduced matrix elements of $\mathbf{r}^{(1)}$ are listed in table 3.1 for transitions between states with $n \leq 4$ and $n' \leq 8$. Corresponding oscillator strengths for hydrogen, averaged or summed over $j$-values associated with a given $\ell$ value, are listed in table 3.2, together with values also averaged over $\ell$ and summed over $\ell'$. These $\ell$-averaged oscillator strengths are to be used in the various relations with statistical weights $g_n = 2n^2$ rather than the $(2j + 1)$ factors for resolved lines. For hydrogen, one needs to use the latter factors and corresponding oscillator strengths only if fine structure is resolved, or if states of the same principal quantum number are not statistically populated (chapter 6). More extensive tabulations of hydrogen oscillator strengths can be found in the astrophysical literature (Green, Rush and Chandler 1957, Goldwire 1968).

For one-(valence)electron atoms and ions containing other bound electrons in closed-shell configurations, the line strengths can also be calculated from (3.15); but $-z/r$ in (3.19) must now be replaced by an effective potential, as mentioned below this equation. Of the various theoretical methods, the Hartree-Fock self-consistent field method is most generally used (Cowan 1981). It also allows calculations for more general situations (section 3.3) if configuration interactions and intermediate coupling are allowed for. However, because of remaining approximations, e.g., in the relativistic corrections to the Hamiltonian and in electron-electron correlation energies, results of such calculations are not necessarily close to exact values. This can be seen by comparing results obtained using the three theoretically equivalent forms of the transition operator (Cowan 1981).

For quick estimates, the Coulomb approximation method (Bates and Damgaard 1949, Oertel and Shomo 1968, Klarsfeld 1989) remains useful. It provides matrix elements as functions of effective principal quantum numbers, or of the differences of their so-called quantum defects. The basic idea is to take the potential in (3.19) to be Coulomb-like, but to

Table 3.1. Squares of reduced matrix elements of the coordinate operator for hydrogen in atomic units.

| Initial State | 1s | 2s | 2p | | 3s | 3p | | 3d | |
|---|---|---|---|---|---|---|---|---|---|
| Final State | np | np | ns | nd | np | ns | nd | np | nf |
| n = 1 | — | — | 1.67 | — | — | 0.3 | — | — | — |
| n = 2 | 1.666 | 27.00 | 27.00 | — | 0.9 | 9.2 | — | 22.5 | — |
| n = 3 | 0.267 | 9.18 | 0.88 | 45.04 | 162.0 | 162.0 | 202.4 | 101.2 | — |
| n = 4 | 0.093 | 1.64 | 0.15 | 5.84 | 29.9 | 6.0 | 114.4 | 1.7 | 313.8 |
| n = 5 | 0.044 | 0.60 | 0.052 | 1.90 | 5.1 | 0.9 | 17.6 | 0.23 | 33.0 |
| n = 6 | 0.024 | 0.29 | 0.025 | 0.82 | 1.9 | 0.33 | 6.0 | 0.08 | 9.6 |
| n = 7 | 0.015 | 0.17 | 0.014 | 0.48 | 0.9 | 0.16 | 2.8 | 0.03 | 4.2 |
| n = 8 | 0.010 | 0.10 | 0.009 | 0.30 | 0.5 | 0.09 | 1.6 | 0.02 | 2.4 |
| Asymptotic | $4.7/n^3$ | $44/n^3$ | $3.7/n^3$ | $117.2/n^3$ | $169/n^3$ | $28/n^3$ | $496/n^3$ | $5/n^3$ | $594/n^3$ |

| Initial State | 4s | 4p | | 4d | | 4f | |
|---|---|---|---|---|---|---|---|
| Final State | np | ns | nd | np | nf | nd | ng |
| n = 1 | — | 0.09 | — | — | — | — | — |
| n = 2 | 0.15 | 1.66 | — | 2.9 | — | — | — |
| n = 3 | 6.0 | 29.8 | 3.4 | 57.0 | — | 209.4 | — |
| n = 4 | 540.0 | 540.0 | 864.0 | 432.0 | 756.0 | 504.0 | — |
| n = 5 | 72.6 | 21.2 | 243.8 | 9.3 | 593.4 | 5.50 | 1256.0 |
| n = 6 | 11.9 | 2.9 | 38.6 | 1.3 | 80.7 | 0.64 | 110.4 |
| n = 7 | 5.7 | 1.4 | 15.4 | 0.5 | 25.8 | 0.16 | 29.2 |
| n = 8 | 2.1 | 0.6 | 6.4 | 0.2 | 11.7 | 0.08 | 12.0 |
| Asymptotic | $445/n^3$ | $102/n^3$ | $1310/n^3$ | $33/n^3$ | $1488/n^3$ | $12/n^3$ | $1572/n^3$ |

Table 3.2.  Absorption (or emission) oscillator strengths for hydrogen and hydrogenic ions.†

| Initial State | *1s* | *2s* | *2p* | | *2* | *3s* | *3p* | | *3d* | | *3* |
|---|---|---|---|---|---|---|---|---|---|---|---|
| Final State | *np* | *np* | *ns* | *nd* | *n* | *np* | *ns* | *nd* | *np* | *nf* | *n* |
| *n* = 1 | — | — | — | — | −0.1040 | — | −0.026 | — | — | — | −0.0088 |
| *n* = 2 | 0.4162 | — | — | — | — | −0.041 | −0.145 | — | −0.417 | — | −0.2848 |
| *n* = 3 | 0.0791 | 0.435 | 0.014 | 0.696 | 0.6408 | — | — | — | — | — | — |
| *n* = 4 | 0.0290 | 0.103 | 0.0031 | 0.122 | 0.1193 | 0.484 | 0.032 | 0.619 | 0.011 | 1.016 | 0.8420 |
| *n* = 5 | 0.0139 | 0.042 | 0.0012 | 0.044 | 0.0447 | 0.121 | 0.007 | 0.139 | 0.0022 | 0.156 | 0.1506 |
| *n* = 6 | 0.0078 | 0.022 | 0.0006 | 0.022 | 0.0221 | 0.052 | 0.003 | 0.056 | 0.0009 | 0.053 | 0.0559 |
| *n* = 7 | 0.0048 | 0.013 | 0.0003 | 0.012 | 0.0127 | 0.027 | 0.002 | 0.028 | 0.0004 | 0.025 | 0.0277 |
| *n* = 8 | 0.0032 | 0.008 | 0.0002 | 0.008 | 0.0080 | 0.016 | 0.001 | 0.017 | 0.0002 | 0.015 | 0.0160 |

†From Bethe and Salpeter (1957) and Unsöld (1955). The negative entries are emission oscillator strengths.

use Coulomb wave functions corresponding to the actual, i.e., measured energies of the $n\ell$ levels. One thus accepts singular behavior at $r = 0$; but errors in $r$-matrix elements from wave function errors in the region near the origin are usually small, unless there are substantial cancellations in the radial integral (small $\ell$-values). The effective principal quantum numbers are defined by

$$n_\ell^* = z(\epsilon_{n\ell})^{-1/2} \tag{3.20}$$

if $z$ is now the nuclear charge screened by the inner electrons, i.e., $z = 1$ for neutral atoms, $z = 2$ for singly ionized atoms, etc. The reduced matrix element used here is expressed in terms of a function $\phi(n_{\ell-1}^*, n_\ell^*, \ell)$ as follows

$$
\begin{aligned}
\langle n'\ell' \| \mathbf{r}^{(1)} \| n\ell \rangle &= \frac{3n_>}{2z} [\ell_> |n_>^2 - \ell_>^2|]^{1/2} \phi(n_<, n_>, \ell_>) \\
&= -\langle n\ell \| \mathbf{r}^{(1)} \| n'\ell' \rangle.
\end{aligned} \tag{3.21}
$$

Note that the reduced matrix elements form an antisymmetric matrix, that $\ell_>$ is the larger of the two angular momenta, and that $n_>$ is the corresponding effective quantum number $n_\ell^*$, $n_<$ being equal to $n_{\ell-1}^*$. The $\phi$ functions for $\ell_> \le 6$, inclusive of the phase factors which may be important for applications to complex atoms or ions, were tabulated by Oertel and Shomo (1968), who also calculated analogous functions for quadrupole matrix elements. For $n_< - n_> = 0$, the $\phi$-functions are all equal to 1, so that we find for so-called $\Delta n = 0$ transitions

$$|\langle n\ell' \| \mathbf{r}^{(1)} \| n\ell \rangle|^2 = \frac{9n_>^{*2}}{4z^2} \ell_> |n_>^2 - \ell_>^2|, \tag{3.22}$$

which is consistent with (63.5) of Bethe and Salpeter (1957) for the radial integral on the right-hand side of (3.18), but generalized to $z \ne 1$ and effective principal quantum numbers, rather than integers. The additional factor is associated with the particular definition of the reduced matrix element.

## 3.3  Line strengths for two- and more-electron atoms

Since the wave functions obtained in atomic structure calculations are superpositions of antisymmetrized, angular-momentum coupled product wave functions, considerable analysis is required to reduce the corresponding transition matrix elements to single-electron matrix elements involving initial and final spatial wave functions that are the factors in the multi-electron product wave functions. No general results can be obtained if configuration interaction (Cowan 1981) must be included, or if intermediate coupling is important, e.g., in (3.11). Remember only that in either case one essentially superimposes transition amplitudes, which opens up the possibility for cancellation effects and apparent violations of some selection rules.

Neglecting these effects, Racah algebra and fractional parentage expansions allow the development of expressions for line strengths, etc., for many transition arrays (Rohrlich 1959, Cowan 1981, Sobel'man 1992). The original work involved Racah coefficients, in contrast to the more symmetrical 6-$j$ symbols in the recent literature.

The next simplest situation to having only one electron outside closed shells are transitions between states of the configurations $\ldots \ell_1^w \ell_2$ and $\ldots \ell_1^w \ell_2'$, where $w$ is the number of $\ell_1$ electrons. Using uncoupling formulas to uncouple $S$ from $L$, and $\ell_2$ and $\ell_2'$ from $L_1$, the parent's orbital angular momentum, one can write the reduced matrix elements on the right-hand side of (3.13) as

$$\langle (\ldots \alpha_1 L_1, \ell_2) L \| \mathbf{r}_N^{(1)} \| (\ldots \alpha_1' L_1', \ell_2' L' \rangle =$$
$$\delta_{\alpha_1 L_1 S_1, \alpha_1' L_1' S_1'} (-1)^{-L + L_1 + \ell_2' - 1}$$
$$\times [(2L + 1)(2L' + 1)]^{1/2}$$
$$\times \left\{ \begin{matrix} L_1 & \ell_2 & L \\ 1 & L' & \ell_2' \end{matrix} \right\} \times \langle n\ell_2 \| \mathbf{r}_N^{(1)} \| n' \ell_2' \rangle. \tag{3.23}$$

The subscript $N$ indicates the coordinates of the active electron, i.e., $\mathbf{r}_N^{(1)}$ corresponds to $\mathbf{P}^{(1)}$ in (3.12) and the wave functions are not symmetrized, with electrons $1, 2, \ldots, N - 1$ in some standard order. For $w = 0$, i.e., the case with only one electron outside closed shells, the 6-$j$ symbol multiplied with the square root reduces to $(-1)^{\ell + \ell' + 1}$, and (3.13) and (3.23) give the same result as (3.15), except that the radial wave functions in (3.18) will be different. Analogous formulas can be derived for other coupling schemes (Cowan 1981), all giving different selection rules. Another application is the prediction of relative intensities, if level populations are according to statistical weights and frequencies or wavelengths are fairly close to each other.

For transition arrays involving more complex (but single) configurations, in particular also active electrons in subshells containing two or more equivalent electrons, the reduction of (3.13) to single electron matrix elements leads to several factors besides the "line" prefactor of the $LS$ matrix elements in (3.13). For single configurations only one electron can make a transition. The most general transition array is then

$$\ell_1^{w_1} \ldots \ell_i^n \ldots \ell_j^{k-1} \ldots \ell_q^{w_q} \rightarrow \ell_1^{w_1} \ldots \ell_i^{n-1} \ldots \ell_j^k \ldots \ell_q^{w_q}, \tag{3.24}$$

and the square root of the line strengths can be written (Cowan 1981) as

$$S_{LS}^{1/2} = D_1 D_2 D_3 \ldots D_q \langle \ell_i || \mathbf{P} || \ell_j \rangle. \tag{3.25}$$

The first factor takes care of the coordinate permutations required to construct antisymmetric wave functions from the "ordered" and angular momentum-coupled wave functions, to which Racah's methods can be applied. Its square is, as to be expected, given by the product of the number of electrons in the active subshells

$$D_1^2 = nk, \tag{3.26}$$

and the corresponding phase factor is important only if configuration interactions or intermediate coupling must be taken into consideration. The second factor, for $n, k > 2$, contains the coefficients of fractional parentage for $\ell_i^n \alpha_i L_i S_i$ in terms of "parent" terms $\ell^{n-1} \alpha_i' L_i' S_i'$ and for $\ell_j^k$, etc., which allow the selection of the active electron from its "siblings." The third factor uncouples the spin $S$ from $L$ and $L'$. It corresponds to the line factor in (3.13). The other factors, expressed in terms of 6-$j$ and 9-$j$ symbols, account for various couplings and uncouplings and also shift the electrons between subshells. The resulting expressions are complicated but suitable for use in computer programs (Cowan 1981). Often some of the factors can be omitted and the general expression simplified considerably (Shore and Menzel 1968). Particularly transparent are results for transition arrays $\ell_1^n \ell_2^{k-1} - \ell_1^{n-1} \ell_2^k$, and any closed inner subshells, and even more so, results for $\ell_1^n - \ell_1^{n-1} \ell_2$. See equations (14.88) and (14.90) of Cowan (1981).

## 3.4 Sum rules

Appropriate sums of oscillator or line strengths obey rules which serve various purposes, including consistency checks of measurements (section 3.6) or calculations. One of the rules concerns the sum of all line strengths associated with all transitions $\gamma J - \gamma' J'$ for given $J, J'$ but summed over both $\gamma$ and $\gamma'$. This sum is equal to the corresponding $\beta, \beta'$ sum for all transitions between the pure coupling states in (3.11). It also holds for multiconfiguration wave functions, for which line strengths are shifted between pairs of configurations, such that the sum of all included arrays remains constant.

Most commonly used is the line oscillator strength

$$
\begin{aligned}
f_\ell(\gamma' J' - \gamma J') &= \frac{2 m_e \omega a_0^2}{3 \hbar (2J + 1)} S'(\gamma J - \gamma' J') \\
&= \frac{\hbar \omega}{3 E_H (2J + 1)} S'(\gamma J - \gamma' J') \tag{3.27}
\end{aligned}
$$

from (3.2), but using the notation of section 3.1. A factor $a_0^2$ was inserted so that $S'$ and $r_q^{(1)}(i)$ in (3.7) are now expressed in atomic units, and we made use of the symmetry of line strengths between initial states $\gamma, J$ and final states, $\gamma', J'$. Remember also that the line oscillator strength is summed over final $M'$ values and does not depend on the initial magnetic quantum number $M$, which can therefore be omitted in the argument. The final version is obtained by reducing the constants to $E_H = 13.6\,\text{eV}$, the ionization energy of hydrogen. To the extent that $LS$ coupling provides an accurate presentation, the transition frequencies $\omega$ are nearly independent of $J$ and $J'$. It then makes sense to define multiplet oscillator strengths

$$
\begin{aligned}
f_m(\gamma' L'S - \gamma LS) \\
\equiv \; & \sum_{J'} f_\ell(\gamma' L'S'J' - \gamma LSJ) \\
= \; & \delta_{ss'} \frac{\hbar \omega_{av}}{3 E_H (2L+1)} \; |\langle \gamma LS || \mathbf{P}^{(1)} || \gamma' L'S' \rangle|^2
\end{aligned}
\tag{3.28}
$$

using (3.7), (3.8), (3.12), (3.13) and (3.27), and, most importantly, the normalization of the sum of the squares of 6-$j$ symbols (Cowan 1981). This result does not depend on the $J, M$ of the lower term and can be used to test the appropriateness of $LS$ coupling.

For many cases, wavelengths or frequencies for transitions accompanied by changes in principal quantum numbers are clustered together, i.e., mostly depend on the configurations rather than on the various couplings. This is especially true for highly ionized atoms. It is then very useful for radiation loss calculations, etc., to sum the line strengths over all initial and final states, with the result (Cowan 1968, 1981)

$$
\begin{aligned}
\sum \sum S \; = \; & 2nk \frac{g(\ell_1^{w_1}) g(\ell_2^{w_2}) \dots g(\ell_i^n) \dots g(\ell_j^k)}{(4\ell_i + 2)(4\ell_j + 2)} \\
& \times \ell_> \left( \int r R_{n_i \ell_i} R_{n_j \ell_j} dr \right)^2 .
\end{aligned}
\tag{3.29}
$$

The statistical weights, if all terms for $\ell^w$ are included, are the binomial coefficients

$$
g(\ell^w) = \frac{(4\ell + 2)!}{w!(4\ell + 2 - w)!},
\tag{3.30}
$$

corresponding to the number of ways in which $w$ electrons can be distributed over the $4\ell + 2$ single-electron quantum states without violating the Pauli principle. (Note that $4\ell + 2 - w$ is the number of vacancies, or holes.) By summing the line oscillator strengths, i.e., (3.27) over all final levels $\gamma'J'$ and averaging over all states $\gamma J$ of the initial configuration and using (3.29), an average array oscillator strength

$$\bar{f}_a = \frac{w_i(4\ell_j + 3 - w'_j)}{(4\ell_j + 2)} \bar{f}(n_j\ell_j - n_i\ell_i) \tag{3.31}$$

is obtained in terms of the averaged one-electron oscillator strengths corresponding to the reduced matrix element in (3.18). (If hydrogen wavefunctions are sufficiently accurate, the $f$-values in table 3.2 can be used.) The physical interpretation of (3.31) is that the multi-electron oscillator strength is the product of the one-electron result with the number of electrons $w_i$ in the initial configuration and the probability of holes in the final configuration $(4\ell_j + 3 - w'_j)/(4\ell_j + 2)$. The averaged array oscillator strength is especially appropriate for use in average atom calculations (Green 1964, Zimmerman and More 1980) and for unresolved transition arrays (UTA-s, Bauche-Arnault, Bauche and Klapisch 1979, 1982, 1985).

In one-electron systems or highly excited levels of one-electron ions, i.e., $w_i = w'_j = 1$, it makes sense to average the array oscillator strength over $\ell_i$ and to sum over $\ell_j$. This gives average oscillator strengths $\bar{f}(n_j - n_i)$ between principal quantum states, which for hydrogen are also listed in table 3.2. For almost all transitions they are well represented by a simple analytic formula (Unsöld 1955, Bethe and Salpeter 1957, Zel'dovich and Raizer 1966) which can be derived using correspondence principle arguments.

The most general (Thomas-Reiche-Kuhn) sum rule is even more closely related to the foundations of quantum mechanics (Heisenberg 1925). Returning to the original definitions in (2.48) and (2.51) and notation, and also reversing the replacement of the momentum operator $-i\hbar(\partial/\partial x_k)$ by $im_e\omega x_k$, absorption and emission oscillator strengths can be written as

$$\begin{aligned} f_{mn} = -f_{nm} &= \frac{2}{3}\sum_k \langle n\left|\frac{\partial}{\partial x_k}\right|m\rangle\langle m|x_k|n\rangle \\ &= -\frac{2}{3}\sum_k \langle n|x_k|m\rangle\langle m\left|\frac{\partial}{\partial x_k}\right|n\rangle, \end{aligned} \tag{3.32}$$

with the additional convention that emission oscillator strengths are negative as in table 3.2. Summing over a complete set of states $m$, including an integral over the continuum, yields

$$\sum_m f_{mn} = \frac{2}{3}\sum_k \langle n\left|\frac{\partial}{\partial x_k}x_k\right|n\rangle \tag{3.33}$$

and

$$\sum_m f_{nm} = \frac{2}{3} \sum_k \langle n \left| x_k \frac{\partial}{\partial x_k} \right| n \rangle$$
$$= -\sum_m f_{mn}. \tag{3.34}$$

Subtracting (3.34) from (3.33) gives

$$\sum_m f_{mn} = \frac{1}{3} \sum_k \langle n \left| \frac{\partial}{\partial x_k} x_k - x_k \frac{\partial}{\partial x_k} \right| n \rangle = Z \tag{3.35}$$

because of the commutation rule between the $3Z$ momentum and co-ordinate components, if $Z$ is now the number of bound electrons and relativistic corrections are negligible. Separating absorption and emission contributions and also allowing for degeneracy, the sum rule can be written as

$$\sum_{m>n} f_{mn} - \sum_{m<n} \frac{g_m}{g_n} f_{nm} = Z. \tag{3.36}$$

Compared to the usage in chapter 2, the role of the indices $m$ and $n$ is interchanged in the second (emission) term.

More useful would be a sum rule restricted only to the active electrons, e.g., the use of $Z = 1$ for effective one-electron systems, etc. This procedure should be approximately valid if one includes Pauli-forbidden transitions into occupied states. Such contributions would, of course, be cancelled by the inverse transitions, were one to implement the sum rule for the $Z$-electron system.

While the Thomas-Reiche-Kuhn sum rule is independent of the nature of the effective potential used and exact as long as relativistic effects are negligible, there are also special sum rules for the case where the effective potential is a central potential. For one jumping electron and summing or averaging over magnetic quantum numbers, the Wigner (1931) and Kirkwood (1932) sum rule is

$$\sum_{n'} f_{n'n} = \begin{cases} -\dfrac{1}{3} \dfrac{\ell(2\ell-1)}{2\ell+1}, & n\ell \to n', \ell-1 \\[2ex] \dfrac{1}{3} \dfrac{(\ell+1)(2\ell+3)}{2\ell+1}, & n\ell \to n', \ell+1 \end{cases} \tag{3.37}$$

(Bethe and Salpeter 1957). It shows the general preponderance of transitions in which principal and orbital quantum numbers change in the same sense; but also its validity is limited to atoms for which a nonrelativistic description suffices. For further sum rules, see Dalgarno and Lynn (1957) and Dalgarno (1963, 1966).

### 3.5 Plasma effects on oscillator and line strengths

So far the atoms or ions were assumed to interact separately from each other, and of any other particles, with the radiation field. Because of the long-range Coulomb interactions between plasma particles, one may question this basic assumption already at moderate densities. First of all, there is the influence of the plasma dispersion on the density of state factors in the radiative transition probabilities in sections 2.4 and 2.5. It gives a factor $n_r^2$, where $n_r$ is the real part of the refractive index, and accounts for the various plasma wave cutoffs (Stix 1992, Swanson 1989), of which that at the electron plasma frequency is of most practical importance.

Another concern might be collisions during times of interest for the solution of the time-dependent rate equations, i.e., (2.23), etc., describing the atomic amplitudes in the generally weak radiation field. The effective collision frequencies (chapters 4 and 6) are usually much larger than radiative transition rates. However, since the energy integral in (2.26) and (2.27) should include the entire line profile, whose width is at least as large as the effective collision frequency, the time interval for which accurate solutions of the time-dependent Schrödinger equation are needed is shorter than the time between collisions (Griem et al. 1991a). Collisions therefore normally only affect the line profile (chapter 4) but not its total or integrated intensity.

Because of the frequency dependence of the various radiative processes, e.g., of spontaneous transition probabilities ($\sim \omega^3$) and therefore radiated powers ($\sim \omega^4$), this last conclusion is not entirely correct. However, fractional frequency shifts even in very dense plasmas are usually so small that even the $4\Delta\omega/\omega$ correction corresponding to the $\omega^4$ prefactor of the spontaneous emission power is normally negligible. Depending on definitions of line profiles with extended wings, i.e., $|\hbar\Delta\omega| \gtrsim kT$, a case can be made for significant changes in integrated intensities, if one deals with nearly thermal radiation fields (Zemtsov and Starostin 1993, Cao and Cao 1993, 1995). Nevertheless, it seems preferable to account for such effects in asymptotic expressions for line wings as discussed at the end of section 4.7.

This leaves possible effects on the wave functions needed for the calculation of transition matrix elements. Strictly speaking, the presence of perturbing ions and electrons invalidates the assumption of spherical symmetry which is needed for the application of Racah algebra. Moreover, the effective Hamiltonian, which includes these perturbations, is no longer even against sign changes of the bound electron coordinates. This is due to the presence of electric microfields which cause (dynamic) Stark effects and therefore invalidate Laporte's rule. The ensuing "forbidden" Stark components are then obtained together with the corresponding "allowed" components, such that the spectrally integrated total line intensity is conserved.

While these effects are mostly associated with changes in the angular dependence of boundstate wave functions, high perturber densities naturally also lead to corrections in radial wave functions and matrix elements. One of the models used to assess these effects consists of multiplying the Coulomb potential in (3.19) for the radial wave functions by a Debye shielding factor $\exp(-r/\rho'_D)$, where

$$\rho'_D = [\epsilon_0 k T/e^2(N_e + \sum_z z^2 N_z)]^{1/2}, \tag{3.38}$$

as discussed in more detail in chapter 7. Calculations (Weisheit and Shore 1974) using this shielded Coulomb potential show that radial matrix elements are reduced, which can also be seen by applying perturbation theory, i.e., by writing

$$-\frac{z}{r}\exp(-r/\rho'_D) \approx -\frac{z}{r} + \frac{z}{\rho'_D} - \frac{zr}{2\rho'^2_D} + \cdots \tag{3.39}$$

for the shielded potential. The first correction term upshifts all bound levels equally but does not change the wave function.

The effective perturbation potential is therefore to lowest order in $r/\rho_D$ given by

$$\Delta V = -\frac{zr}{2\rho'^2_D}. \tag{3.40}$$

The first-order corrected radial wave functions then become

$$R^D_{n\ell} = R_{n\ell} + \frac{za_0^2}{\rho'^2_D}\sum_{n'\neq n} R_{n'\ell}\frac{\int R_{n'\ell}rR_{n\ell}dr}{E_{n'\ell} - E_{n\ell}}, \tag{3.41}$$

using standard perturbation theory and converting to Rydberg energy units, etc. Also, since $\Delta V$ is a scalar operator, the only matrix elements to consider are those for $\Delta\ell = 0$. The most important contribution is from $n' = n + 1$, with matrix elements

$$\int R_{n+1,\ell}rR_{n\ell}dr \approx -0.4n^2a_0/z \tag{3.42}$$

for unperturbed hydrogen eigenfunctions. The energy denominator is $E_{n+1} - E_n = -z^2/(n+1)^2 + z^2/n^2 \approx 2z^2/n^3$, so that (3.41) becomes

$$R^D_{n\ell} \approx R_{n\ell} - 0.4\left(\frac{n\,a_0}{z\,\rho'_D}\right)^2\left[\frac{1}{n^2} - \frac{1}{(n+1)^2}\right]^{-1} R_{n+1,\ell}. \tag{3.43}$$

For $n'\ell'$ to $n\ell$ transition matrix elements this gives

$$\int R_{n'\ell'} r R_{n\ell}^D dr \approx \int R_{n'\ell'} r R_{n\ell} dr$$

$$- 0.4 \left[ 1 - \left( \frac{n}{n+1} \right)^2 \right]^{-1} \left( \frac{n^2 a_0}{z \rho'_D} \right)^2$$

$$\times \int R_{n'\ell'} r R_{n+1,\ell} dr, \qquad (3.44)$$

if we neglect the lower state correction. For line strengths, etc., we need the square of this quantity. Its relative correction is

$$\frac{\Delta S^D}{S} \approx -0.8 \left[ 1 - \left( \frac{n}{n+1} \right)^2 \right]^{-1} \left( \frac{n^2 a_0}{z \rho_D} \right)^2 \frac{\int R_{n'\ell'} r R_{n+1,\ell} dr}{\int R_{n'\ell'} r R_{n\ell} dr}$$

$$\approx -0.8 \left( \frac{n^2 a_0}{z \rho'_D} \right)^2 \left[ 1 - \left( \frac{n}{n+1} \right)^2 \right]^{-1} \left( \frac{n}{n+1} \right)^{3/2} \left[ \frac{1 - (n'/n)^2}{1 - (n'/n+1)^2} \right]^2,$$

$$(3.45)$$

with the second version utilizing an approximate formula for the hydrogen matrix elements (Unsöld 1955, Bethe and Salpeter 1957, Zel'dovich and Raiser 1966). As long as the corrections are small, they are typically within a factor $\sim 1.5$ of those obtained by Weisheit and Shore (1974).

Because of strong cancellations in the radial integrals in the perturbation matrix elements, (3.45) should be considered only as a crude error estimate for multi-electron systems, and in any case one should include more perturbing levels for $n \gtrsim 5$. Other models in use for dense plasmas, the ion-sphere model (More 1982, Salzmann and Szichman 1987, Gutierrez, Jouin and Cormier 1994) and the relativistic inferno model (Liberman 1979) have been used for more general situations, but they do not lend themselves to analytical estimates. Some illustrative calculations for high density gold plasmas were made by Liberman and Albritton (1995). However, it seems reasonable to use (3.45) for error estimates, with $\rho'_D$ replaced by the mean ion-ion separation in cases where this is the larger quantity.

The reader should be warned that frequency shifts must be considered as well, which tend to decrease the spacing between high $n$-lines (chapter 4 and section 5.5). As pointed out by More (1994), the changes in radial matrix elements are mostly due to the altered normalization of the upper state wave function, which is closely related to the level spacing. For estimates of line radiation losses, this tends to compensate the reduction in radial matrix elements. A related effect is the lowering of the bound-free continuum limit at high densities (chapter 5), which arises from the second term in the expansion (3.39).

### 3.6   Measurements of radiative transition probabilities

Quantitative plasma spectroscopy (Wiese 1968) has contributed substantially toward the goal of having a reasonably complete set of spontaneous transition probabilities ($A$-values), etc., for atomic and ionic line spectra of elements of interest in astronomy and laboratory applications. Although sometimes line absorption is used to infer the cross section whose frequency integral yields the absorption oscillator strength according to (2.47), most plasma measurements are made in emission. In that case, plasma parameters and size should be selected to avoid significant absorption, so that the emerging intensity is proportional to the product of the desired $A$-value and the integral of the upper level population density along the line of sight (chapter 8). Moreover, spectrally integrated line intensities must be measured with great care in background and line wing corrections (Wiese 1968, Griem et al. 1991a, see also section 10.3).

For the measurement of relative $A$-values within a given ionic species, the most important requirement is that relative upper level populations can be reliably calculated, e.g., from appropriate Boltzmann factors and statistical weights (chapter 7). The critical plasma parameter is therefore the (electron) temperature, which is usually determined spectroscopically (chapter 11) or by Thomson scattering (Sheffield 1975, Hutchinson 1987). However, in order to estimate any deviations from assumed equilibrium population ratios, it is also necessary to know, e.g., the electron density for an evaluation of collisional rates (chapter 6). While some minimum value of the electron density is required, depending on the states involved, to reduce uncertainties in populations to some acceptable level, say, 10%, it is also important to avoid plasma effects as discussed in section 3.5.

To extract absolute $A$-values from plasma measurements, not only an absolute intensity calibration is required, rather than only the frequency dependence of the instrumental response, but the populations and the effective length of the emitting plasma must be determined absolutely. Generally, equilibrium relations are invoked for the populations, e.g., a Saha equation relating upper level densities with the product of electron and (next) ion densities (chapter 7). For single-species plasmas and invoking quasineutrality, it is again sufficient to have measured values of electron temperature and density (Hutchinson 1987, see also chapters 10 and 11). In this case, however, very accurate electron densities are required, because this quantity enters quadratically. For multi-species plasmas, the difficulties of obtaining accurate level populations are more severe. Moreover, for multiply ionized atoms it is usually impractical to have experiments at sufficiently high densities for the thermal equilibrium relations to apply. More involved models (chapter 6) must then be used, but any inferred *A*-values would have additional uncertainties from cross sections and any simplifications made in the kinetic model.

To investigate, e.g., the validity of the *LS*-coupling approximation in oscillator strength calculations, it is sufficient to measure relative line intensities and to have sufficiently high plasma densities for sublevel populations to be statistical. An example for such experiments is a recent measurement of the lines in the 3s-3p and 3p-3d multiplets of the boron-like ions C II - F V and C II - Ne VI, respectively (Glenzer et al. 1994b). Finally, since absolute *A*-value determinations from plasma measurements are so difficult, a combination with lifetime measurements (Heckmann and Träbert 1989) is often preferable. Examples for these combined methods are *A*-value determinations for ultraviolet lines of Ar I (Federman et al. 1992) and N I lines belonging to 3s-3p and 3p-3d multiplets (Musielok, Wiese and Veres 1995). The 3s-3p multiplets adhered to LS coupling predictions for relative *A*-values within $\pm$ 20%, whereas intermediate coupling was found to be important for the 3p-3d multiplets.

Another interesting area of study are possible changes of transition probabilities due to dense plasma effects as discussed near the end of section 3.5. However, claimed observations (Chung et al. 1988, 1989, Suckewer 1991) of such effects are not only inconsistent with these theoretically expected changes, but may also have been subject to large systematic errors associated with emission on broad wings of lines from inhomogeneous plasmas (Huang et al. 1990b, Griem et al. 1991a). Reported small changes of transition probabilities in argon ion laser discharges (Aumayr et al. 1989, 1991, Cao et al. 1993) may require another interpretation.

# 4

# Spectral line broadening

So far we have implicitly assumed that the photon energies related to transitions between bound states of atoms or ions were sharp within the limits of the uncertainty principle, i.e., within the natural line width discussed in section 2.7. Normally this contribution to the width of a spectral line is entirely negligible in applications of plasma spectroscopy, x-ray lines of medium and heavy ions being important exceptions. This is so because Doppler shifts associated with the motions of emitting or absorbing particles and, depending mainly on the density, because level perturbations, etc., caused by plasma particles and fields tend to exceed radiative rates by large factors. Effects of the second kind were called pressure broadening in the earlier literature and were, specifically for plasmas, the subject of monographs (Griem 1974, Sobel'man et al. 1981, 1995) and reviews (Lisitsa 1977, Peach 1981, 1996).

Usually, but not always, these Doppler and pressure broadening effects can be treated independently (see section 4.6). In many cases, and assuming the radiator (or absorber) velocities to be nonrelativistic and their distribution to be Maxwellian, i.e., the relevant one-dimensional velocity distribution to be Gaussian, the corresponding normalized line shape function is also Gaussian, namely

$$L_D(\omega) = \exp[-(\Delta\omega/\omega_D)^2]/\sqrt{\pi}\omega_D \qquad (4.1)$$

with the Doppler broadening parameter given by

$$\omega_D = \left(\frac{2kT}{Mc^2}\right)^{1/2}\omega_0 \qquad (4.2)$$

in terms of the radiators' kinetic temperature $T$, radiator mass $M$, and the frequency detuning $\Delta\omega$ from the rest frame transition frequency $\omega_0$. To obtain the full width at half maximum (FWHM), (4.2) must be multiplied with $2\sqrt{\ell n2} = 1.665$. If wavelength units are preferred, analogous relations are obtained for the normalized line shape $L_D(\lambda)$ by the replacements

$\omega \to \lambda$, $\Delta\omega \to - 2\pi c \Delta\lambda/\lambda_0^2$, $\omega_0 \to \lambda_0$, always assuming $|\Delta\lambda/\lambda_0|$ to be a small quantity, see (4.5) and (4.6) below.

Pressure broadening is less conducive to any general statement, except for a classification according to either the underlying physical mechanism or the basic approximation used in the line profile calculations. Of the various mechanisms, Stark effects caused by the electric fields produced by nearby ions and electrons, or by collective fields associated with plasma waves (section 4.11), are the prevalent cause of pressure broadening in plasmas. However, in partially ionized gases, interactions with neutral perturbers can be important as well, either due to long range Van der Waals forces or, if a resonance condition between radiators and perturbers of the same species is fulfilled, also due to longer range dipole-dipole interactions. These two situations will be discussed in section 4.8 after the more common broadening due to dipole-(radiators), monopole-(perturbers) interactions and due to closely related multipole interactions, i.e., of Stark broadening in the generally accepted but imprecise sense.

The second traditional classification according to the mathematical approximations to a more general theory of line broadening are the Holtsmark (1919) or quasistatic approximation (sections 4.1 and 4.3), the Lorentz (1906) or impact approximation (sections 4.1 and 4.2), and a number of intermediate approximations and computer simulations, which will be discussed in sections 4.1 and 4.5. The corresponding line shape functions have normally no simple analytic form, with the exception of the impact approximation for isolated lines, i.e., lines which are not overlapping other transitions in the same spectrum. These line shapes are Lorentz profiles, namely

$$L(\omega) = \frac{w/\pi}{w^2 + (\Delta\omega - d)^2} \qquad (4.3)$$

in terms of (half) half width (HWHM) $w$ and shift $d$, For theoretical purposes, it is more convenient to use $w$, rather than the FWHM width $2w = \gamma$ briefly used in chapter 2. The latter is more appropriate for measurements, being the separation between half of maximum intensity points. (In the case of doubly peaked lines and provided the profile asymmetry is sufficiently small, one uses the mean of the maxima to define the half intensity points on the outside of the two peaks.)

A schematic representation of Doppler and Lorentzian profiles of equal FWHM widths is shown in figure 4.1, together with a combined profile obtained by the convolution

$$L_c(\omega) = \int_{-\infty}^{+\infty} L_D(\Delta\omega')L(\Delta\omega - \Delta\omega')d\Delta\omega' \qquad (4.4)$$

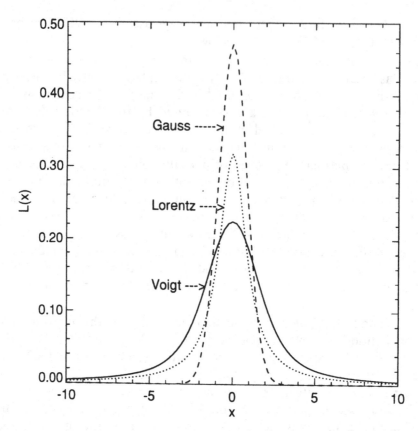

Fig. 4.1. Normalized Gauss (Doppler) and Lorentz (impact) profiles of equal half widths (FWHM width = 2 in *x*-units). Also shown is the Voigt profile resulting from the convolution of these two profiles.

which is, in this case, a Voigt (1912) profile. Such Voigt profiles have been widely tabulated, e.g., by Posener (1959), Finn and Mugglestone (1965), and Hummer (1965). Their widths are well approximated by a simple formula due to Whiting (1968).

The convolution of different line profiles is naturally only a correct procedure if the underlying mechanisms are statistically independent. This is not the case, e.g., if the same collisions are responsible for impact broadening and for the motion of radiating ions, e.g., in certain dense plasmas. In that case the resultant profile can even be narrower than the Doppler profile according to (4.1) and (4.2). Such "collisional narrowing" will be discussed in section 4.6. It was predicted by Dicke (1953).

Another word of caution concerns the conversion of line widths and shifts, e.g., from angular frequency units to wavelengths already mentioned above. As long as they are small compared with the unperturbed frequencies $\omega_0$ or wavelengths $\lambda_0$, differentiation of $\lambda = 2\pi c/\omega$ gives

$$w_\lambda = (\lambda_0^2/2\pi c)w \tag{4.5}$$

$$d_\lambda = -(\lambda_0^2/2\pi c)d \tag{4.6}$$

for the half widths and shifts in wavelength units, e.g., of a profile $L(\lambda)$ analogous to the profile $L(\omega)$ in (4.3). Also, an additional numerical factor arises if different length units are used for $\lambda$ and $c$, as is frequently done.

## 4.1 General theory of pressure broadening

There are various approaches toward a theory of spectral line broadening (see, e.g., Baranger 1962 or Griem 1974) due to interactions between radiators and surrounding particles (perturbers). It is usually assumed that it is sufficient to consider one radiator at a time, i.e., to neglect radiator-radiator interactions. (This is a reasonable assumption except, perhaps, for resonance broadening.) We therefore begin with the formula for the spectral power emitted by a quantum mechanical system due to dipole transitions between initial states $i$ and final states $f$

$$P(\omega) = \frac{4mr_0}{3c}\omega^4 \sum_{if\alpha} \delta(\omega - \omega_{if}^s)|\langle f|x_\alpha|i\rangle|^2\rho_i, \tag{4.7}$$

which follows by multiplication of the spontaneous transition probability in (2.43) with the photon energy $\hbar\omega$ and the probability $\rho_i$ for finding the system (radiator plus perturbers) in the initial state. The $\delta$-function is introduced to ensure energy conservation, i.e.,

$$\hbar\omega = \hbar\omega_{if}^s = E_i^s - E_f^s, \tag{4.8}$$

where $E_i^s$ and $E_f^s$ are energies of the stationary states of the complete system(s), radiator and perturbers. The $\omega^4$ factor is normally treated as constant, except when profile asymmetries are important (section 4.10). It is therefore customary (Baranger 1962) to express the normalized line shape as

$$L(\omega) = \sum_{if\alpha} \delta(\omega - \omega_{if}^s)|\langle f|x_\alpha|i\rangle|^2\rho_i, \tag{4.9}$$

with the normalization condition

$$\sum_{if\alpha} |\langle f|x_\alpha|i\rangle|^2\rho_i = 1. \tag{4.10}$$

In all of these relations the sum is only over states contributing to the line in question, a somewhat ambiguous statement in the case of broad and overlapping lines.

In most situations, (4.9) and (4.10) can also be used for the profile of the cross sections for induced emission, (2.50), and absorption, (2.47), with $i$ and $f$ interchanged in the latter case. However, if Kirchhoff's law is invalid, or if there are phase correlations between the various contributing initial states, absorption and emission profiles can be different. To discuss such situations, a density matrix formulation of line broadening much used in laser physics and reviewed by Cooper (1967) is more appropriate. For very broad lines covering photon energy ranges approaching or exceeding $kT$, it is important to retain Boltzmann factors, etc., in the various line shape formulas in order to retain consistency, e.g., with Planck's law (Zemtsov and Starostin 1993).

If the radiating atom or ion is imbedded in a slowly varying environment, say, of perturbing ions, it may be possible to avoid the complications of a dynamical theory by making the quasistatic approximation, following Holtsmark (1919). For a given perturber configuration, time-independent perturbation theory can then be used to calculate corrections to $\omega_{if}^s$ due to the corresponding ion microfield $F$ and, if necessary, also corrections to the dipole matrix elements. The perturber factor of the density operator $\rho_i$ then yields the probability $W(F)dF$ for the radiator to be in a field in the $F, F + dF$ range so that (4.9) becomes (Smith and Hooper 1967, Dufty 1969)

$$
\begin{aligned}
L_{qs}(\omega) &= \sum_{if\alpha} \int \delta(\omega - \omega_{if}^s(F))|\langle f|x_\alpha(F)|i\rangle|^2 \rho_i^r W(F) dF \\
&= \sum_{if\alpha} |\langle f|x_\alpha(F_\omega)|i\rangle|^2 \rho_i^r W(F_\omega) \left|\frac{d\omega_{if}^s}{dF}\right|^{-1}
\end{aligned}
\tag{4.11}
$$

with $F_\omega$ (not a Fourier component) defined by

$$
\omega_{if}^s(F_\omega) = \omega
\tag{4.12}
$$

and $\rho_i^r$ giving the probabilities of initial radiator states. In the case of the linear Stark effect, i.e., $\omega_{if}^s(F) = \omega_{if} + C_{if}F$, this results in

$$
F_\omega = (\omega - \omega_{if})/C_{if} = \Delta\omega/C_{if},
\tag{4.13}
$$

and the major remaining task is the calculation of $W(F)$ (section 4.3). At very high densities, separation of a perturber factor from the density operator is no longer possible, and a molecular-dynamics simulation of clusters containing the radiator and a sufficiently large number of perturbers may be more appropriate (Kress, Kwon and Collins 1995).

For dynamical theories of line broadening, it is useful to consider the Fourier transform of (4.9), i.e.,

$$
\begin{aligned}
C(s) &= \int_{-\infty}^{+\infty} \exp(-i\omega s)L(\omega)d\omega \\
&= \sum_{if\alpha} \exp(-i\omega_{if}^s s)|\langle f|x_\alpha|i\rangle|^2 \rho_i.
\end{aligned}
\tag{4.14}
$$

Because of $C(-s) = [C(s)]^*$, the inverse transform can be written as

$$
L(\omega) = \frac{1}{\pi}\mathrm{Re}\int_0^\infty \exp(i\omega s)C(s)ds.
\tag{4.15}
$$

Therefore this autocorrelation function $C(s)$ of the light amplitude is needed only for times $s \geq 0$. Following Baranger (1958a), we write $C(s)$ as a trace to facilitate the transition to more convenient representations, namely

$$
\begin{aligned}
C(s) &= \sum_{if\alpha} \langle i|x_\alpha|f\rangle \exp\left[\frac{i}{\hbar}E_f^s s\right]\langle f|x_\alpha|i\rangle \exp\left[-\frac{i}{\hbar}E_i^s s\right]\rho_i \\
&= \sum_\alpha Tr'[x_\alpha t^\dagger(s,0)x_\alpha t(s,0)\rho]_{av},
\end{aligned}
\tag{4.16}
$$

the prime on the trace $(Tr)$ symbol indicating the same restriction as in (4.9) and (4.10). The operator $t(s,0)$ is often called $U(s,0)$ in the original literature. It is shorthand for $\exp(-iH^s s/\hbar)$, with $H^s$ being the Hamiltonian of the radiator-perturber system, while $t^\dagger(s,0)$ is its Hermitian conjugate, i.e., the transpose of $t(-s,0)$. These operators transform the $x_\alpha = x_\alpha(0)$ dipole operators into $x_\alpha(s)$ for $s \neq 0$. Finally, $\rho$ is the statistical or density operator

$$
\rho = \exp(-H^s/kT),
\tag{4.17}
$$

which can usually be assumed to be time independent and to be diagonal in the representation actually used, and the subscript $av$ indicates an average over perturber states. This allows us to write

$$
C(s) = \sum_\alpha Tr'[x_\alpha(0)x_\alpha(s)\rho(0)]_{av},
\tag{4.18}
$$

a compact form for the dipole autocorrelation function first obtained by Baranger (1958a).

Often the lower states, i.e., states $f$ in case of emission, can be taken as unperturbed. This gives instead of (4.16)

$$
C(s) = TrDt_{av}(s,0) \equiv TrD\exp(-\frac{i}{\hbar}Hs)u_{av}(s,0),
\tag{4.19}
$$

with $D$ defined by

$$\langle i|D|i'\rangle = \sum_{\alpha,f}\langle i|x_\alpha|f\rangle\langle f|x_\alpha|i'\rangle \qquad (4.20)$$

and using $\langle f|t^\dagger(s,0)|f'\rangle = \exp(\frac{i}{\hbar}E_f s)\delta_{ff'} = \delta_{ff'}$, if we chose $E_f$ to be at zero energy. The averages are over the initial perturber (and radiator) states, and $u(s,0)$ is the time evolution operator in the interaction representation. Also, $H$ is the unperturbed Hamiltonian. The matrix elements of $u(s,0)$ correspond to the usual $C_{ii'}(t)$ coefficients in Dirac's (1958) time-dependent perturbation theory if we, for the time being, treat the perturbers classically and assume that their paths are not affected by changes in the radiator state. The radiator-perturber Hamiltonian is then a definite function of time, $U(t)$, often called $V(t)$, and the operator $u(s,0)$ obeys the time-dependent Schrödinger equation

$$\begin{aligned}i\hbar\dot{u}(s,0) &= \exp\left(\frac{i}{\hbar}Hs\right)U(s)\exp\left(-\frac{i}{\hbar}Hs\right)u(s,0)\\ &\equiv U'(s)u(s,0),\end{aligned} \qquad (4.21)$$

which corresponds to the system of coupled first-order linear differential equations for Dirac's $C$-coefficients. A completely quantum mechanical formulation of the following general relaxation theory of line broadening is also possible (Smith and Hooper 1967, Smith 1968, Voslamber 1972), as is the inclusion of lower state interactions. See You and Cooper (1994) for a discussion of the correspondences between quantum and classical treatments of "doubled" atoms involved in this case.

We continue here with the discussion of the classical path version (Smith, Cooper and Vidal 1969) of the theory by considering

$$\begin{aligned}u_{av}(s,0) &= \int Q(\mathbf{r})u(\mathbf{r};s,0)d\mathbf{r}\\ &\equiv \int F(\mathbf{r},s)d\mathbf{r}\end{aligned} \qquad (4.22)$$

in terms of the normalized distribution function $Q(\mathbf{r})$ of the 6-$n$ dimensional vector $\mathbf{r}$ consisting, in principle, of the initial coordinates and velocities of all $n$ perturbers. It corresponds to the perturber factor of $\rho(0)$ in (4.18), and its use implies that there are no correlations initially between states of the radiator and the perturber distribution. The operator $u(\mathbf{r};s,0)$ is the radiator's evolution operator, i.e., a solution of (4.21) with initial value $u(\mathbf{r};0,0) = 1$ and $U = U(\mathbf{r},s)$ as interaction Hamiltonian. Since $Q(\mathbf{r})$ is time independent, $F(\mathbf{r},s)$ is clearly also a solution of (4.21). The major accomplishment of the relaxation theory is to obtain the relevant properties of $u_{av}(s,0)$ without explicitly solving the time-dependent Schrödinger equations for the multitude of perturber configurations. This feat was accomplished by the use of a projection operator method (Zwanzig 1961), in which an operator $P$ is defined by

$$Pf(\mathbf{r}) = Q(\mathbf{r}) \int f(\mathbf{r}')d\mathbf{r}'. \qquad (4.23)$$

This allows one to decompose the function $F(\mathbf{r}, s)$ in (4.22) into

$$
\begin{aligned}
F(\mathbf{r}, s) &= Q(\mathbf{r})u(\mathbf{r}; s, 0) \\
&= PF(\mathbf{r}, s) + (1 - P)F(\mathbf{r}, s) \\
&\equiv F_1(\mathbf{r}, s) + F_2(\mathbf{r}, s).
\end{aligned} \qquad (4.24)
$$

A BBGKY method (Voslamber 1972) leads to the same result, and Smith, Cooper and Roszman (1973) provided an even simpler derivation. Because one has

$$
\begin{aligned}
F_1(\mathbf{r}, s) &= PF(\mathbf{r}, s) \\
&= Q(\mathbf{r}) \int F(\mathbf{r}', s)d\mathbf{r}' \\
&= Q(\mathbf{r})u_{av}(s, 0),
\end{aligned} \qquad (4.25)
$$

only the first function is actually needed. By applying $P$ and $(1 - P)$ to (4.21) after multiplication with $Q(\mathbf{r})$, the governing equations for the new functions are seen to be

$$i\hbar(\partial/\partial s)F_1(\mathbf{r}, s) = PU'(\mathbf{r}, s)[F_1(\mathbf{r}, s) + F_2(\mathbf{r}, s)] \qquad (4.26)$$

and

$$i\hbar(\partial/\partial s)F_2(\mathbf{r}, s) = (1 - P)U'(\mathbf{r}, s)[F_1(\mathbf{r}, s) + F_2(\mathbf{r}, s)]. \qquad (4.27)$$

A formal solution for $F_2$ can be obtained in terms of the Green's function

$$G(\mathbf{r}; s, s') = \exp[-\frac{i}{\hbar} \int_{s'}^{s} (1 - P)U'(\mathbf{r}, s)ds], \qquad (4.28)$$

namely with $F_2(\mathbf{r}, 0) = 0$, from $F(\mathbf{r}, 0) = Q(\mathbf{r})$, and time ordering being understood here and in the following,

$$F_2(\mathbf{r}, s) = -\frac{i}{\hbar} \int_0^s G(\mathbf{r}, s, s')(1 - P)U'(\mathbf{r}, s')F_1(\mathbf{r}, s')ds', \qquad (4.29)$$

so that (4.26) becomes an integro-differential equation for the relevant function $F_1$. This equation, written in terms of $F_1(\mathbf{r}, s) = Q(\mathbf{r})u_{av}(s, 0)$ and integrated over $\mathbf{r}$ gives, because of $\int Q(\mathbf{r})d\mathbf{r} = 1$,

$$i\hbar(\partial/\partial s)u_{av}(s, 0) = [U'(\mathbf{r}, s)]_{av}u_{av}(s, 0)$$

$$- \frac{i}{\hbar} \int_0^s [U'(\mathbf{r}, s)G(\mathbf{r}; s, s')(1 - P)U'(\mathbf{r}, s')]_{av}u_{av}(s', 0)ds'. \qquad (4.30)$$

Following Fano (1963) one usually assumes the average interaction $[U'(\mathbf{r}, s)]_{av}$ to vanish. In view of $PU'(\mathbf{r}, s')Q(\mathbf{r}) = Q(\mathbf{r}) \int U'(\mathbf{r}, s') \times Q(\mathbf{r}')d\mathbf{r}' =$

$Q(r)[U'(\mathbf{r}, s')]_{av}$ also the $P$ term under the integral can then be omitted. If we do not make this assumption, $(1 - P)U'(\mathbf{r}, s')$ must be replaced by $U'(\mathbf{r}, s') - [U'(\mathbf{r}, s')]_{av}$. Assuming stationarity, i.e.,

$$[U'(s)\cdots U'(s^{(n)})]_{av} = \exp\left(\frac{i}{\hbar}Hs^{(n)}\right)[U'(s - s^{(n)})\cdots U'(0)]_{av}$$

$$\times \exp\left(-\frac{i}{\hbar}Hs^{(n)}\right) \qquad (4.31)$$

and, therefore, also

$$G(\mathbf{r}; s, s') \to \exp(\frac{i}{\hbar}Hs')G(\mathbf{r}; s - s', 0)\exp(-\frac{i}{\hbar}Hs'), \qquad (4.32)$$

the dynamic relation for $t_{av}(s, 0)$ in (4.16) can provisionally be written as

$$i\hbar(\partial/\partial s)t_{av}(s, 0) = \{H - [U(\mathbf{r}, 0)]_{av}\}t_{av}(s, 0)$$

$$-\frac{i}{\hbar}\int_0^s \exp[-\frac{i}{\hbar}H(s - s')][U'(\mathbf{r}, s - s')G'(\mathbf{r}; s - s', 0)U'(\mathbf{r}, 0)]_{av}t_{av}(s', 0)ds'$$

$$+\frac{i}{\hbar}\int_0^s \exp[-\frac{i}{\hbar}H(s - s')][U'(\mathbf{r}, s - s')G'(\mathbf{r}; s - s', 0)U'_{av}(\mathbf{r}, 0)]_{av}t_{av}(s', 0)ds'.$$

$$(4.33)$$

The prime on the Green's function indicates that $(1 - P)U'(\mathbf{r}; s)$ in (4.28) should be replaced by $U'(\mathbf{r}; s - s') - Q(\mathbf{r})\int U'(\mathbf{r}; 0)\, d\mathbf{r}$. Vanishing of the term arising from the projection operator is generally assumed, as well as that of $[U(\mathbf{r}, s)]_{av}$.

The two conditions for omitting these terms seem to differ, since the average interaction is

$$[U'(\mathbf{r}, 0)]_{av} = \int Q(\mathbf{r})U'(\mathbf{r}, 0)d\mathbf{r}. \qquad (4.34)$$

However, $Q(\mathbf{r})$ is constant relative to the initial spatial coordinates for uniform distributions of initial perturber coordinates, in which case it is sufficient to have

$$\int U'(\mathbf{r}, 0)d\mathbf{r} = 0 \qquad (4.35)$$

for the usual simplifications of (4.33) to be valid. In the vicinity of neutral atoms as radiators, uniform perturber distributions are an excellent approximation, so that the charged perturber contributions by electrons and ions to the average interaction indeed cancel. This is because of quasineutrality of the plasmas of interest here, and because the contributions of various perturbers to $U'(\mathbf{r}, t)$ depend on their signs if perturbing ions act only as point charges.

In the vicinity of radiating ions, on the other hand, $Q(\mathbf{r})$ is not spatially uniform because of the long-range Coulomb interactions which lead to a negative space charge. The corresponding "plasma polarization shift" (Berg et al. 1962) had been the subject of a number of model calculations (Griem 1974). Experimental evidence for or against this effect also had been scant and somewhat controversial, but the corresponding red shifts are now better established both experimentally and theoretically (see sections 4.10 and 10.1). We will nevertheless continue the discussion of the "general" theory omitting the terms related to a nonvanishing average interaction. This omission, together with the neglect of lower state interactions and of initial correlations mentioned below (4.22) and, of course, the use of classical perturber paths, must all be kept in mind in applications of this theory.

The simplified version of (4.33), after taking Laplace transforms of all remaining terms, gives according to (4.15) and (4.19) our desired "general" expression for the line shape

$$
\begin{aligned}
L(\omega) &= \frac{1}{\pi} \mathrm{Re} \int_0^\infty \exp(i\omega s) C(s) ds \\
&= \frac{1}{\pi} \mathrm{Re} Tr D \int_0^\infty \exp(i\omega s) t_{av}(s,0) ds \\
&\equiv \frac{1}{\pi} \mathrm{Re} Tr D\, T(\omega).
\end{aligned}
\tag{4.36}
$$

An algebraic equation for $T(\omega)$ follows from (4.33), still assuming $U'_{av} = 0$, and from $t(0,0) = 1$, $t(\infty,0) = 0$, and using the convolution theorem to factorize the integral term. In this term, we replace $s'$ by the new variable $s'' = s - s'$, except in the $t_{av}(s',0)$ factor, and obtain

$$
\begin{aligned}
&-i\hbar + \hbar\omega T(\omega) \\
&= HT(\omega) - \frac{i}{\hbar} \int_0^\infty \exp[\frac{i}{\hbar}(\hbar\omega - H)s][U'(\mathbf{r},s)G(\mathbf{r}',0)U'(\mathbf{r},0)]_{av} ds \; T(\omega) \\
&\equiv HT(\omega) + \mathscr{L}(\omega) T(\omega).
\end{aligned}
\tag{4.37}
$$

Together with (4.36), this gives the line shape

$$
L(\omega) = -\frac{\hbar}{\pi} \mathrm{Im} Tr D [\hbar\omega - H - \mathscr{L}(\omega)]^{-1},
\tag{4.38}
$$

i.e., a generalized Lorentzian.

Only if $\mathscr{L}(\omega)$ can be replaced by a constant operator, and only if off-diagonal matrix elements of this operator are not important, is the profile actually Lorentzian, with width and shift according to

$$
\hbar w = -\mathrm{Im}\langle i|\mathscr{L}(\omega_0)|i\rangle
\tag{4.39}
$$

and

$$\hbar d = \mathrm{Re}\langle i | \mathscr{L}(\omega_0) | i \rangle. \tag{4.40}$$

Remember again that we neglect final $(f)$ state broadening and use its energy as the origin of the energy scale, so that we have $\omega_0 = E_i/\hbar$. Before discussing the additional conditions required for such simplifications to be reasonably valid, we note that direct calculation of $\mathscr{L}(\omega)$, e.g., for the case of the dipole approximation, i.e.,

$$U(\mathbf{r}, s) \approx -\mathbf{d} \cdot \mathbf{F}(\mathbf{r}, s) \tag{4.41}$$

in terms of the radiator's dipole moment $\mathbf{d}$ and the electric field at the radiator produced by, e.g., the ions characterized by the initial conditions $\mathbf{r}$, would be equivalent to the numerical solution of the time-dependent Schrödinger equation (4.21) for a large number of time sequences of the plasma microfield. Comparison of the subsequent approximations with such computer simulations (see section 4.5) is therefore a useful check of the accuracy of any additional approximations made in semianalytic line shape calculations. The electric fieldstrength $\mathbf{F}(\mathbf{r}, s)$ in (4.41) is not to be confused with the function $F(\mathbf{r}, s)$ in (4.22).

According to (4.33) and (4.37), the line shape operator to be considered is

$$\mathscr{L}(\omega) = -\frac{i}{\hbar} \int_0^\infty \exp[\frac{i}{\hbar}(\hbar\omega - H)s][U'(\mathbf{r}, s)G(\mathbf{r}; s, 0)U'(\mathbf{r}, 0)]_{av} ds, \tag{4.42}$$

and the interaction is the sum of contributions from particles $j$,

$$U'(\mathbf{r}, s) = \sum_j U'_j(s). \tag{4.43}$$

On expansion of the exponential in (4.28) and omitting all projection operators, (4.42) is seen to involve integrals over sums of products of the contributions of various particles. An essential assumption, called the impact approximation by Smith et al. (1969), is now that products involving contributions from different particles average to zero except, perhaps, in the second-order term (section 4.4) for which the so far suppressed time ordering of different collisions can actually be important (Cooper and Smith 1982). The impact approximation line shape operator is

$$\begin{aligned} \mathscr{L}(\omega) &= -\frac{i}{\hbar} n \int_0^\infty \exp[\frac{i}{\hbar}(\hbar\omega - H)s] \\ &\quad \times \left[ U'_1(s) \exp\left(-\frac{i}{\hbar} \int_0^s U'_1(s')ds'\right) U'_1(0) \right]_{av} ds, \end{aligned} \tag{4.44}$$

e.g., for a total number of $n$ perturbing electrons or ions. The average is now only over parameters characterizing one particle, say, $j = 1$. The

above "unified line-shape" approximation is especially appropriate to describe electron effects, for which higher than second-order terms in $U'_1(s)$
may usually be neglected. For the slower moving ions, on the other hand,
overlap between various collisions and higher order terms can be quite
important. Therefore, the impact approximation to the general theory is
often not appropriate for ions as perturbers (section 4.5).

For frequencies near $\omega_0$ and for short durations of effective collisions,
the first exponential factor in (4.44) can be omitted, and we can write

$$
\begin{aligned}
\mathscr{L}(\omega) &\approx n \int_0^\infty \frac{\partial}{\partial s} \left[ \exp\left(-\frac{i}{\hbar} \int_0^s U'_1(s')ds'\right) U'_1(0)\right]_{av} ds' \\
&= n \left[\exp\left(-\frac{i}{\hbar}\int_0^\infty U'_1(s')ds'\right) U'_1(0)\right]_{av} - n[U'_1(0)]_{av} \\
&= n \lim_{T \to \infty} \left[\exp\left(-\frac{i}{\hbar}\int_0^T U'_1(s)ds\right) - 1\right]_{av}.
\end{aligned}
\tag{4.45}
$$

The second term on the second line is zero by assumption, and the third
line can be obtained by expanding the exponentials, the use of stationarity
and term-by-term comparison. Introducing the frequency $f_j$ of collisions
of type $j$ to express the average more explicitly, we have finally

$$
\begin{aligned}
\mathscr{L}(\omega) &\approx i\hbar \sum_j f_j \left[\exp\left(-\frac{i}{\hbar}\int_{-\infty}^{+\infty} U'_j ds\right) - 1\right] \\
&= i\hbar \sum_j f_j(S_j - 1),
\end{aligned}
\tag{4.46}
$$

since for collisions occurring at positive times the integral from $-\infty$ to $0$
vanishes. The final version follows because the exponential is the usual $S$-
matrix for a collision occurring at time zero. This is the one-state version of
the impact theory (Baranger 1958 a,b, 1962, Kolb and Griem 1958, Griem
1974). It corresponds to (4.39) and (4.40) above and, in case only diagonal
$\mathscr{L}(\omega)$ matrix elements are important in (4.38), to Lorentzian profiles. If off-
diagonal matrix elements are also important, one obtains a superposition
of Lorentz profiles and interference terms (Kolb and Griem 1958). This
paper uses an impact or collision operator $\phi$ corresponding to $-i\mathscr{L}/\hbar$ or
$-i\mathscr{H}/\hbar$ in Baranger's notation and contains a detailed discussion of the
conditions for $\phi$, etc., to be time- or frequency-independent. [See also
Sholin et al. (1974), but note that these authors average over the times of
closest approach, thus obtaining different results for the interference terms
than Kolb and Griem (1958), in particular on the line wings.] However,
these profiles must be narrow enough for $\mathscr{L}(\omega)$ calculated according to
(4.44) not to deviate significantly from its value at $\omega = \omega_0$.

Returning to (4.46), its generalization to lines for which also perturbations of the final state are important involves instead of the $S$-matrix for

scattering on radiators in the initial states the product of this $S$-matrix with the complex conjugate of the $S$-matrix for scattering on radiators in the final states (Baranger 1962). We also note that in the impact theory it is not necessary to assume that the average interaction is zero.

## 4.2    Electron scattering theory of line broadening

Since the work of Lorentz (1906) it has been known that the FWHM line widths were equal to the rates of effective collisions, provided their duration was short compared, in modern terms, to the decay time of the autocorrelation function $C(s)$ in (4.14). This condition tends to be well met for most lines broadened by electrons, with the especially broad lines of hydrogen, ionized helium, etc., being important exceptions. These exceptions occur when the product of line width and decay time is of order 1, making the restriction on the duration of collisions particularly severe for broad lines.

The problem of formulating expressions for the effective ("optical") cross section was solved by Baranger, and we will begin with his result (1962) for a matrix element of $\mathscr{L}(\omega)$ in the impact limit, namely

$$
\begin{aligned}
\langle\langle if^* | \mathscr{H} | i'f^{*\prime} \rangle\rangle & \\
\equiv\ & \langle\langle if^* | \mathscr{L}(\omega_0) | i'f^{*\prime} \rangle\rangle \\
=\ & [-(2\pi\hbar^2 N_e/m_e)(\delta_{ff'}\langle i|f(0)|i'\rangle - \delta_{ii'}\langle f|f(0)|f'\rangle^*) \\
& + i\hbar N_e v \int\int \langle f|f(\theta,\phi)|f'\rangle^* \langle i|f(\theta,\phi)|i'\rangle \sin\theta\, d\theta\, d\phi]_{av}.
\end{aligned}
\tag{4.47}
$$

This expression could have been inferred from the two-state generalization of (4.46) by expressing the $S$-matrices through the scattering amplitudes $f(\theta,\phi)$. The remaining average is over the initial directions of the colliding electrons with velocity $v$ and over the distribution of these velocities. Also, $N_e$ is the electron density and $f(0)$ the forward scattering amplitude. The matrix elements are in the "line-space", "doubled atom" notation, in which an initial state is associated with the complex conjugate of a final state. This notation is analogous to the tetradic notation used in connection with the Liouville operator formalism (Fano 1963) employed in much of the more recent theoretical developments discussed in the preceding section.

Most of the quantum mechanical calculations of electron impact broadening have been concerned with isolated lines, in which case only diagonal matrix elements of $\mathscr{H}$ are needed. As shown by Baranger (1958b), the width and shifts in (4.39) and (4.40) can then be expressed through

$$
w = \left\{ \frac{1}{2} N_e v [\sigma_i + \sigma_f + \int d\Omega |f_i(\Omega) - f_f(\Omega)|^2] \right\}_{av}
\tag{4.48}
$$

and

$$d = -\left\{ \frac{2\pi\hbar}{m_e} N_e \text{Re}[f_i(0) - f_f(0)] \right.$$
$$\left. + N_e v \text{Im} \int d\Omega[f_f^*(\Omega)f_i(\Omega)] \right\}_{av}, \quad (4.49)$$

if the optical theorem is used to simplify the width expression. The cross sections $\sigma_i$ and $\sigma_f$ are for inelastic collisions with radiators in initial and final states, while the scattering amplitudes are for elastic scattering, and $f(\Omega)$ corresponds to $f(\theta, \phi)$ above. One advantage of this formulation is the rather explicit demonstration of the complete cancellation of pure Coulomb (Rutherford) scattering, or of the almost complete cancellation of elastic scattering contributions to the broadening of radio-frequency $n - \alpha$ lines of hydrogen (Griem 1967, 1974, see also section 10.1). The former cancellation often justifies the so-called one-state versions of (4.48) and (4.49), for situations in which final-state (lower) perturbations can be neglected, provided $f(0)$, etc., are defined relative to pure Coulomb scattering. These one-state versions are simply

$$w = \left\{ \frac{1}{2} N_e v \sigma \right\}_{av} \quad (4.50)$$

and

$$d = -\left\{ \frac{2\pi\hbar}{m_e} N_e \text{Re} f(0) \right\}_{av}, \quad (4.51)$$

in which $\sigma$ stands for the total, non-Coulomb, cross section – elastic and inelastic.

Because of the usual degeneracy with respect to magnetic quantum numbers, one might question the utility of these isolated line formulas. However, in the absence of ion-produced microfields, the electron-induced perturbations are, of course, isotropic and cannot induce rotations of the radiators. The appropriately averaged $\mathscr{L}(\omega_0)$ or $\mathscr{H}$ matrix elements were introduced into the semiclassical theory by Anderson (1949) and into the quantum-mechanical theory by Baranger (1958 a,b), see also Griem (1974). For overlapping lines, which are very sensitive to ion microfields, e.g., lines of hydrogen, the situation is more complicated. One usually uses $\mathscr{H}$ or $\mathscr{L}$ as calculated for the field-free case in terms of spherical coordinate wave functions for the atom. The $\mathscr{H}$-matrix is then indeed diagonal and is simply transformed to the parabolic quantum number representation. Depending on the Stark splitting in the ion field, elastic collision effects thus can become actually inelastic processes, and some ambiguity may arise, in particular with respect to the upper-lower state interference terms from $|f_i(\Omega) - f_f(\Omega)|^2$ (see also section 4.7).

The most important aspect of (4.48)–(4.51) is the establishment of a link with the large topic of atomic collision theory. Corresponding original (see Griem 1974) and recent calculations (see, e.g., Seaton 1988) are normally not making explicit use of scattering amplitudes, but rather involve $S$-matrices or related quantities from, e.g., close-coupling or $R$-matrix theory for the electron-atom or -ion scattering. They also benefit considerably from the simplifications brought by angular momentum theory, but still require averaging over scattering resonances and extrapolations to high partial waves of the colliding electron, which are not included in the electron scattering calculations aimed primarily toward accurate inelastic cross sections for use in collisional-radiative models (chapter 6).

### 4.3　Ion microfields

Equally important as cross sections and related quantities for line broadening calculations are the ion-microfield distributions needed in (4.11) or its generalizations discussed in section 4.5. The archetype of such distributions was derived by Holtsmark (1919) in the ideal gas limit for perturbing ions. Because of isotropy, we can write

$$W(F) = 4\pi F^2 P(\mathbf{F}) dF, \tag{4.52}$$

if $P(\mathbf{F})$ describes the probability of finding the field vector in $dF_x dF_y dF_z$ and $W(F)$ is the distribution of fieldstrength magnitudes.

The general expression for $P(\mathbf{F})$ is

$$P(\mathbf{F}) = \int \cdots \int \delta(\mathbf{F} - \sum_{j=1}^{n} \mathbf{F}_j) p(\mathbf{r}_1, \mathbf{r}_2, \cdots, \mathbf{r}_n)$$
$$\times\, d\mathbf{r}_1 d\mathbf{r}_2 \cdots d\mathbf{r}_n \tag{4.53}$$

in terms of the fields produced by $n$ ions $j$ which are at positions $\mathbf{r}_j$. Holtsmark assumed unshielded Coulomb fields and uniform distributions $p(\mathbf{r}_1, \ldots)$ of the positions, i.e., $p = V^{-n}$ for a normalization volume $V$. In this limit, it is easy to calculate the Fourier transform $A(\mathbf{k}) = A(k)$ of $P(\mathbf{F})$, which is obtained by multiplication with $\exp(i\mathbf{k} \cdot \mathbf{F}) d\mathbf{F}$ and integration over the field. Actually, $A(\mathbf{k})$ is only a function of $k$, and the 3-$n$ dimensional integral in position space becomes the $n$-th power of the integral for, say, ion 1. Except for the final inverse transform, all calculations can be done analytically, leading to

$$W_H(F) = H_0(\beta)/F_0 \tag{4.54}$$

with the reduced fieldstrength

$$\beta = F/F_0 \tag{4.55}$$

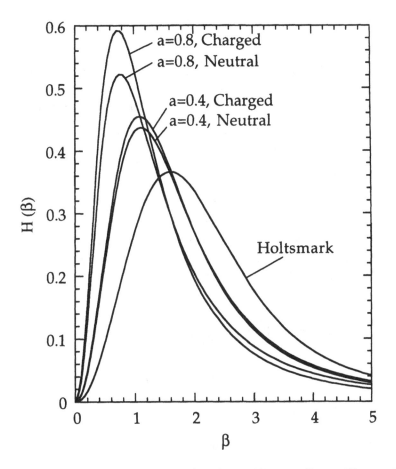

Fig. 4.2. Ion-microfield distribution function $H(\beta)$ according to Hooper (1968) as a function of the reduced fieldstrength $\beta$ (see text) for neutral and singly charged radiators. The parameter $a$ is the ratio of mean ion-ion separation and electron Debye length. The Holtsmark distribution corresponds to $a = 0$.

and Holtsmark's normal fieldstrength

$$
\begin{aligned}
F_0 &= 2\pi \left(\frac{4}{15}\right)^{2/3} \frac{z_p e}{4\pi\epsilon_0} N_p^{2/3} \\
&\approx 2.603 \frac{z_p e}{4\pi\epsilon_0} N_p^{2/3}
\end{aligned}
\tag{4.56}
$$

in terms of ion charge $z_p$ and density $N_p$. Figure 4.2 contains a plot of the Holtsmark function $H_0(\beta)$, together with distribution functions more appropriate for dense plasmas to be discussed now.

The other fieldstrength distributions shown on figure 4.2 include two improvements of the physical model. The configuration space distribution functions can be corrected for correlations between ions by introducing

(Mozer and Baranger 1960) two particle correlation functions $g(|\mathbf{r}_i - \mathbf{r}_k|) = g(r_{ik})$ into

$$p(\mathbf{r}_1, \mathbf{r}_2, \cdots, \mathbf{r}_n) = V^{-n}[1 + \sum_{j<k} g(r_{jk}) + \cdots] \qquad (4.57)$$

and by using Debye-screened fields in place of the Coulomb fields, i.e.,

$$\mathbf{F}_j = \frac{z_p e}{4\pi\epsilon_0 r_j^2} \left(1 + \frac{r_j}{\rho_D}\right) \exp\left(-\frac{r_j}{\rho_D}\right). \qquad (4.58)$$

The Debye radius,

$$\rho_D = \left(\frac{\epsilon_0 kT}{N_e e^2}\right)^{1/2}, \qquad (4.59)$$

here accounts for screening by electrons only, because we require fields that are more or less instantaneous on the ion timescale. In the ion-ion correlation term, on the other hand, it makes sense to also include ion screening effects by using

$$\begin{aligned}
g(r_{jk}) &= \exp\left[\left(\frac{-z_p^2 e^2}{4\pi\epsilon_0 r_{jk} kT}\right) \exp\left(-\frac{r_{jk}}{\rho_D'}\right)\right] - 1 \\
&\approx \frac{-z_p^2 e^2}{4\pi\epsilon_0 r_{jk} kT} \exp\left(-\frac{r_{jk}}{\rho_D'}\right), \qquad (4.60)
\end{aligned}$$

with $\rho_D'$ given by (3.38).

The corresponding fieldstrength distributions, if we assume $z_p = 1$, now depend on the dimensionless parameter

$$a = r_1/\rho_D, \qquad (4.61)$$

$r_1$ being the mean ion-ion separation defined by $4\pi N_1 r_1^3/3 = 1$. Hooper (1966, 1968) improved the model described so far by considering the coordinate space distribution for singly-charged perturbers,

$$\begin{aligned}
p(\mathbf{r}_1, \mathbf{r}_2, \cdots, \mathbf{r}_n) &= \frac{1}{Z_n} \exp\left(-\frac{\Omega}{kT}\right) \\
&= \frac{1}{Z_n} \exp\left[-\frac{e^2}{4\pi\epsilon_0 kT} \sum_{i<k}^{n} \frac{1}{r_{ik}} \exp\left(-\frac{r_{ik}}{\rho_D}\right)\right],
\end{aligned}$$

$$(4.62)$$

with $Z_n$ being the configurational partition function. The interaction potentials are screened by electrons only, the sum beginning with $i = 0$ for neutral and $i = 1$ for charged radiators. The cluster expansion corresponding to (4.57) is avoided. Instead the interaction energy is split into a Debye-like part depending on the ion separations $r_{0i}$ from the radiator at

$r_0$ and a remainder $\Omega_0$, which depends according to (4.62) in a complex way on all $r_{ik}$,

$$\Omega = \Omega_0 + \sum_i \frac{e^2}{4\pi\epsilon_0 r_{0i}} \exp\left(-\frac{\alpha r_{0i}}{\rho_D}\right). \qquad (4.63)$$

Here $\alpha$ is a positive and real constant, which can be varied. The actual calculations involve complicated integrations made possible by the use of collective coordinates. The results are insensitive to the value of $\alpha$ over a wide range, which suggests that errors are rather small as long as $r_1 \lesssim \rho_D$. This high accuracy is supported by comparison with Monte Carlo simulations of the ion fields from statically screened quasiparticles (Hooper 1968, Bailey and Hooper 1972), see figure 2 in chapter II.2a of Griem (1974). For neutral radiators, results obtained by Pfennig and Trefftz (1966) using Mozer and Baranger's (1960) method are practically indistinguishable from Hooper's (1968) distributions at neutral points.

At higher temperatures, multiply-charged radiators and perturbers are of increasing importance. Stark broadening is generally only of interest for such ion lines if densities of perturbers are very high. It is nevertheless customary to express fields, etc., in terms of the electron density, i.e., to retain $F_0$ as given by (4.56) with $N_p = N_e$ and $z_p = 1$ as reference fieldstrength. This is done, although broadening by ions may well be more important than broadening by electrons. O'Brien and Hooper (1972) extended Hooper's (1968) calculations to mixtures of singly- and doubly-charged perturbing ions, and the theoretical development also to plasmas with different ion and electron temperatures. Their functions $P(\epsilon)$, corresponding to $H(\beta)$ here, are for equilibrium plasmas not only characterized by the parameter $a$, which is still calculated using the electron density in (4.61), but also by $R = N^{++}/N^+$, the ratio of the two perturbing ion densities. Some APEX theory results (see below) corresponding closely to the results of O'Brien and Hooper (1972) are shown in figures 4.3(a) and 4.3(b), all calculated assuming the radiator to be singly charged. They exhibit the strong dependence of the reduced field distributions on the correlation parameter $a$ for $R = 1$ and $\infty$, i.e., equal mixtures of ions, or only doubly-charged ions. The variation with $R$ is quite weak, which can be seen better from figures 1-4 of the original paper. This is particularly true for $a$-values approaching 1, indicating the importance of radiator-perturber correlations.

Analogous calculations by Tighe and Hooper (1976) emphasize the last point for multiply-charged ions as radiators and singly-charged perturbers. This work was generalized by the same authors (1977) also to mixtures containing one multiply-ionized species in addition, e.g., to protons, etc., and to plasmas with different ion and electron temperatures. The temper-

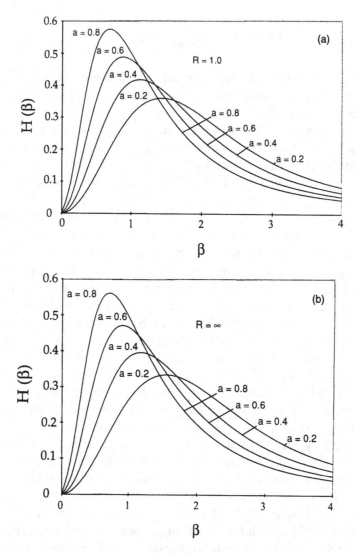

Fig. 4.3. Microfield distribution functions H($\beta$) corresponding to those of O'Brien and Hooper (1972) for singly-charged radiators and various $a$-parameter values (a) for equal densities of doubly- and singly-charged perturbing ions ($R = 1$) and (b) for doubly-charged perturbers only ($R = \infty$). The reduced fieldstrength $\beta = F/F_0$ is defined in terms of the normal fieldstrength $F_0$ at an assumed common electron density. The results presented here were obtained by Haynes and Hooper (1995) using the APEX method (see text).

ature ratio effects can be quite pronounced (figure 4.4). For highly charged radiating ions, this is probably mostly due to radiator-perturber correlations which tend to suppress ion configurations corresponding to large fieldstrengths. However, at given electron density and $a = 0.2$, multiply-charged ions nevertheless produce significantly broader distributions than, e.g., singly-charged perturbers, see figures 7 and 9 of Tighe and Hooper (1977). At $a = 0.4$, this effect is much less pronounced or even inverted, see figures 8 and 10 of the latter paper.

The methods of Mozer and Baranger (1960) and of Hooper (1966 and 1968) are not suitable for very dense, strongly coupled plasmas. Instead of the dimensionless parameter $a$, it is customary to characterize these plasmas by the ion-ion coupling parameter $\Gamma_{ii}$ defined for a single ion species $i$ by

$$\Gamma_{ii} = \frac{z_i^2 e^2}{4\pi\epsilon_0 r_{ik}T},$$ (4.64)

with the mean ion-ion separation defined by $4\pi N_i r_i^3/3 = 1$ as discussed above in connection with the parameter $a$, see (4.61). This parameter $\Gamma_{ii}$, or its generalization for multispecies plasmas, is of order 1 or larger in some high density experiments. It is related to $a$ through

$$\Gamma_{ii} = \frac{z_i}{3}\left(\frac{r_i}{\rho_D}\right)^2 = \frac{z_i^{5/3}}{3}\left(\frac{r_e}{\rho_D}\right)^2 = \frac{z_i^{5/3}}{3}a^2.$$ (4.65)

This can be seen from (4.59) and (4.61), from $N_i = N_e/z_i$, and by using the electron-electron separation $r_e$. Corresponding electron-ion and electron-electron coupling parameters are smaller by factors of $z_i^{2/3}$ and $z_i^{5/3}$, respectively, because they involve $r_e$ instead of $r_i$ in the equivalents of (4.64), and instead of $z_i^2$ only $z_i$ or 1, respectively. There is therefore a substantial range of plasma conditions corresponding to $\Gamma_{ii} > 1$ but $\Gamma_{ei} < 1$, not to mention $\Gamma_{ee}$.

In this situation, it is still justified to calculate the low frequency microfield by considering electron Debye-shielded ions as quasiparticle sources of the fields, as in (4.58). However, if the electron-ion coupling should also be strong or if the electrons must be described quantum-mechanically, the Debye-Hückel (1923) theory is no longer valid, and more involved screening models (Dharma-wardana, Perrot and Aers 1983, Rogers 1984) should be used. At high densities, degeneracy of the plasma electrons becomes very important. Because degenerate electrons act more like a uniform neutralizing background, they are less effective in screening than suggested by the classical Debye-Hückel theory. The condition for

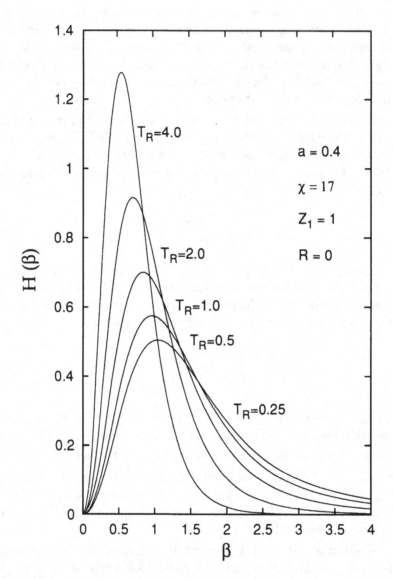

Fig. 4.4.   Microfield distribution functions $H(\beta)$ corresponding to those of Tighe and Hooper (1977) for radiators of charge $\chi = 17$ (e.g., hydrogen-like argon) and singly-charged ($Z_1 = 1$) perturbing ions for different electron-to-ion temperature ratios $T_R$. The correlation parameter is $a = 0.4$ in all cases. These results were also obtained by Haynes and Hooper (1995) using the APEX method.

this reduction in screening not to occur is that the Fermi energy,

$$E_F = (2\pi^2\hbar^2/m_e)[(3/8\pi)N_e]^{2/3}$$
$$= (18\pi)^{2/3}E_H a_0^2/4r_e^2 \qquad (4.66)$$

remains a small fraction of $kT$ or, equivalently, that

$$\Gamma_{ee} \ll \frac{8}{(18\pi)^{2/3}}\frac{r_e}{a_0} \approx 0.54\frac{r_e}{a_0}. \qquad (4.67)$$

Here $\Gamma_{ee}$ is as in (4.64) without the factor $z_i^2$ and substituting $r_e$ for $r_i$.

Normally these conditions on $\Gamma_{ei}$ and $\Gamma_{ee}$ are met, and one can proceed with (4.53), (4.58) and (4.62). Most successful for $\Gamma_{ii} > 1$ has been the adjustable-parameter exponential approximation (APEX) method proposed by Iglesias, Lebowitz and McGowan (1983). For fields at highly charged radiators, it is best understood (Dufty, Boercker and Iglesias 1985) in terms of the Fourier transform $A(\mathbf{k})$ of the field distribution $P(\mathbf{F})$, which was mentioned below (4.53) and is called $F(\lambda)$ by Dufty et al. (1985). These authors recall the two transformations which resulted in the Baranger-Mozer (1959) formulation,

$$A(k) = \langle\exp(\mathbf{k}\cdot\sum\mathbf{F}_i)\rangle_{av}$$
$$= \langle\Pi_i\exp(\mathbf{k}\cdot\mathbf{F}_i)\rangle_{av}$$
$$\equiv \langle\Pi_i(\phi_i+1)\rangle_{av} \qquad (4.68)$$

and

$$A(k) \equiv \exp[G(\phi_1,\phi_2,\ldots)]. \qquad (4.69)$$

By expanding the product in the last version of (4.68) into powers in $\phi_i$, etc., and dropping the zero-order term, which would not contribute to the inverse Fourier transform, and by using a theorem of equilibrium statistical mechanics (see, e.g., Munster 1969, appendix 7), the corresponding expansion for the $G$-function is seen to be

$$G(\phi) = \sum_{p=1}^{\infty}\frac{1}{p!}\int d\mathbf{r}_1\cdots d\mathbf{r}_p h_p(\mathbf{r}_1,\cdots,\mathbf{r}_p|0)\Pi_{i=1}^p\phi_i. \qquad (4.70)$$

In this expression, the functions $h_p$ are Ursell cluster functions, e.g.,

$$h_1(\mathbf{r}_1|0) = f_1(\mathbf{r}_1|0) \qquad (4.71)$$

and

$$h_2(\mathbf{r}_1,\mathbf{r}_2|0) = f_2(\mathbf{r}_1,\mathbf{r}_2|0) - f_1(\mathbf{r}_1|0)f_2(\mathbf{r}_2|0) \qquad (4.72)$$

in terms of the usual correlation functions $f_p$ for a test particle (radiator) at the origin and $p$ perturbing ions at $\mathbf{r}_1,\mathbf{r}_2,\cdots$. These $f_p$ correspond to the functions $p(\mathbf{r}_1,\mathbf{r}_2,\cdots)$ in (4.53), with the index $p$ equal to $n$.

The point of these transformations is to reduce the fraction of the 3-$n$ dimensional volume contributing significantly to the integrations. The cluster functions restrict the ranges in the $p \geq 2$ terms to volumes corresponding to the ion-ion correlation length, i.e., to $\rho'_D$ for $\Gamma_{ii} \lesssim 1$. For small $k$, i.e., large fields, the ranges are further restricted by the $\phi$-functions. If interaction between different perturbing ions can be neglected, only $p = 1$ contributes in (4.70), giving with (4.71)

$$G^{(0)}(\phi) = \int d\mathbf{r}_1 f_1^{(0)}(\mathbf{r}_1|0)\phi_1. \tag{4.73}$$

In the APEX method this independent particle form is retained, but the distribution function $f_1^{(0)}$ is replaced by an effective distribution function

$$F_1^* = f_1^{(0)} R(r_1) \tag{4.74}$$

and the Coulomb field $\mathbf{F}_1$ and $\phi_1$ function by

$$\mathbf{F}_1^* = \mathbf{F}_1/R(r_1) \tag{4.75}$$

and

$$\phi_1^* = \exp(i\mathbf{k} \cdot \mathbf{F}_1^*) - 1, \tag{4.76}$$

with a Debye-like form for $R(r)$, namely

$$R(r) = e^{-\alpha r}/(1 + \alpha r). \tag{4.77}$$

The choice of the effective distribution function in (4.74) ensures that the quasiparticles in a given volume produce the same field at the test particle as the real particles with charge density $f_1^{(0)}$. This pair distribution must be calculated separately, say, from the hypernetted-chain integral equation (Hanson and McDonald 1976). The APEX parameter $\alpha$, the inverse of the effective screening radius, is finally determined by requiring that the second moment of the fieldstrength distribution function agrees with the exact second moment (Iglesias et al. 1983) which is, for quasiparticles as perturbers, also dependent on the pair-distribution function.

The second moment condition would appear useless for neutral radiators, for which the corresponding integral diverges as $F^{1/2}$ at large fields. However, the parameter $\alpha$ in (4.77) is still well defined in the limit of vanishing test charge (Alastuey et al. 1984). Moreover, Dufty, Boercker and Iglesias (1985) show that the second moment condition is equivalent to ensuring that all corrections to APEX vanish to order $k^2$, i.e., for large and intermediate fields. These authors also succeeded in calculating the corrections to the APEX model from pair correlations between perturbing ions in the vicinity of a neutral radiator. Comparisons with Monte Carlo

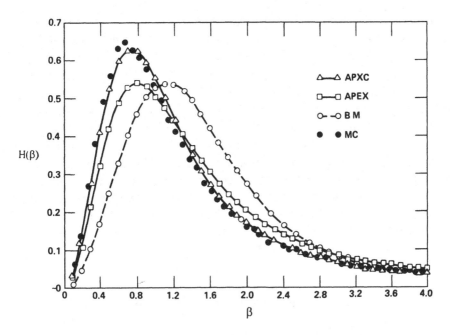

Fig. 4.5. Comparisons between microfield distribution functions at a neutral point for a $\Gamma = 10$ strongly-coupled plasma obtained from one-component plasma Monte-Carlo (MC) simulations with theoretical results of Dufty, Boercker and Iglesias (1985) using the APEX model, a corrected version APEXC of this model, and the Baranger and Mozer method. See text for a description of the models.

simulations indicate that such corrected neutral point field distributions remain accurate up to $\Gamma_{ii} \approx 10$ (figure 4.5). A generalization of the APEX method for plasmas with different electron and ion temperatures has been proposed by Kilcrease (1994).

One additional remark on this very important subject of fieldstrength distribution functions concerns the different normalizations of the fieldstrengths, either in terms of the Holtsmark normal fieldstrength, (4.56), the corresponding quantity for electrons, i.e., $z_p = 1$ and $N_p \to N_e$, or in terms of fieldstrengths defined in terms of ion or electron sphere radii,

$$
\begin{aligned}
F_0^* &= \frac{z_p e}{4\pi\epsilon_0 r_p^2} \\
&= \frac{(4\pi/3)^{2/3}}{4\pi\epsilon_0} z_p e N_p^{2/3} \\
&\approx 2.599 \frac{z_p e}{4\pi\epsilon_0} N_p^{2/3}.
\end{aligned} \tag{4.78}
$$

Fortunately, the numerical factors are practically equal, but care must be exercised for $z_p > 1$ whether or not ion or electron densities are used, and whether or not the $z_p$ factor is included.

Finally, in case of strongly perturbed lines in dense plasmas, not only the quasistatic microfields but also their gradients at the location of the radiators may be required. One then needs to calculate corresponding joint distribution functions for the microfield **F** and its gradient **F′**. Such joint distribution functions can be obtained exactly in the independent particle, Holtsmark, limit (Demura and Sholin 1975) and, approximately, also for dense plasmas either by a generalization (Halenka 1990) of the fieldstrength distribution functions of Mozer and Baranger (1960) or using an extension of the APEX method (Kilcrease, Mancini and Hooper 1993). Closely related to corresponding line profile calculations is the two-center model of Salzmann et al. (1991).

## 4.4    Plasma screening of electron collisions

In most calculations of electron collisional effects on spectral line shapes, correlations between perturbing electrons, and between electrons and ions, are neglected. However, it is also clear that these assumptions can lead to unphysical results. The most striking examples are hydrogen, ionized-helium, etc., lines for which semiclassical calculations (Griem, Kolb and Shen 1959, Griem and Shen 1961) give collisional widths that increase logarithmically with the maximum impact parameter. It was argued on physical grounds that this upper cutoff impact parameter should come at the electron Debye radius $\rho_D$, see (4.59), with possible factors of order 1 arising from various refinements (Griem et al. 1962, Chappell, Cooper and Smith 1969). Before discussing these effects using a combination of the general line shape theory described in section 4.1 and of plasma kinetic theory, the reader is reminded of the static screening of ions by electrons invoked in the calculations of fieldstrength distribution functions. Dynamical corrections to this static screening assumption can be taken up in the present context, but the separation of quasistatic and dynamical effects retains model character, i.e., is not obtained deductively from first principles (see, however, Dufty 1969).

We now return to the general line broadening theory in the form of (4.38) and (4.42). Following the physical arguments just mentioned, we assume $G(\mathbf{r}; s, 0) = 1$ in (4.42), i.e., neglect higher than second-order terms in the electron-radiator interaction. We also make the dipole approximation as in (4.41), because other multipole interactions are only important for close collisions for which screening is usually negligible. A matrix element of the line shape operator can then be written as

$$\langle i|\mathscr{L}(\omega)|i'\rangle \approx -\frac{ie^2}{3\hbar}\sum_{i''}\langle i|\mathbf{r}|i''\rangle \cdot \langle i''|\mathbf{r}|i'\rangle$$

$$\times \int_0^\infty \exp[i(\omega - \omega_{i''f})t]\{\mathbf{F}(t)\cdot\mathbf{F}(0)\}_{av}dt, \quad (4.79)$$

if we assume the electron-produced fields to be isotropic and use the coordinate operator $\mathbf{r}$ instead of the dipole operator $\mathbf{d}$. The Laplace transform of the autocorrelation function of the dynamical fields is thus seen to be the key quantity. Another, implicit, assumption is usually that electron-produced fields can be treated classically. (See, however, the discussion of line shifts in case of degenerate interacting levels at the end of this section and of far wings of hydrogen lines at the end of section 4.7.)

To make contact with plasma kinetic theory, it is convenient to write the Laplace transform of the autocorrelation function of the electron-produced fields in terms of density fluctuations,

$$\int_0^\infty e^{i\Delta\omega t}\{\mathbf{F}(t)\cdot\mathbf{F}(0)\}_{av}dt = \frac{2e^2}{\pi\epsilon_0^2}\int S^+(K, i\Delta\omega)dK \quad (4.80)$$

in Rosenbluth and Rostoker's (1962) notation, but using $K$ for the plasma wave number and multiplying by $(e/\epsilon_0 K)^2$ to change from density to field fluctuations. The former are important in the theory of scattering of electromagnetic waves by plasmas and are therefore emphasized in the plasma literature. Most important for line profiles is the imaginary part of $\mathscr{L}(\omega)$ and therefore according to (4.79) the real part of (4.80),

$$\mathrm{Re}\int_0^\infty e^{i\Delta\omega t}\{\mathbf{F}(t)\cdot\mathbf{F}(0)\}_{av}\,dt = \frac{e^2}{4\pi^2\epsilon_0^2}\int S(K, \Delta\omega)dK$$

$$\approx \frac{e^2}{2\pi^2\epsilon_0^2}\int \frac{N_e}{K}\left(\frac{\pi m}{2kT}\right)^{1/2}\left[1 + \left(\frac{1}{K\rho_D}\right)^2\right]^{-2}$$

$$\times \left\{1 + \left(\frac{M}{m}\right)^{1/2}\left(\frac{1}{K\rho_D}\right)^4 \exp\left[-\frac{1}{2}\frac{M}{m}\left(\frac{\Delta\omega}{\omega_p K\rho_D}\right)^2\right]\right\}dK,$$

$$(4.81)$$

using an approximation to the general result (Rosenbluth and Rostoker 1962) which is reasonably accurate for $\Delta\omega$ intermediate to ion and electron $(\omega_p)$ plasma frequencies. The term involving the ion-to-electron mass ratio accounts, approximately, for the fluctuations in the screening cloud of ions and implies equal ion and electron temperatures. The leading term "1" represents the contribution of (statically) screened electrons. Its accuracy can be improved by replacing $[1 + (1/K\rho_D)^2] = \epsilon(K, 0)$ with the frequency dependent dielectric constant $\epsilon(K, \Delta\omega)$. After integration over $K$ to some maximum value $K_{max} \gg \rho_D^{-1}$, (4.79) and (4.81) give for the "width" operator matrix element

$$\text{Im}\langle i|\mathscr{L}(\omega)|i'\rangle \approx -\frac{8}{3\hbar}\left(\frac{e^2}{4\pi\epsilon_0}\right)^2\left(\frac{\pi m}{2kT}\right)^{1/2}N_e\sum_{i''}\langle i|\mathbf{r}|i''\rangle\cdot\langle i''|\mathbf{r}|i'\rangle$$

$$\times\left[\ell n(K_{max}\rho_D)-\frac{1}{2}+\left(\frac{M}{m}\right)^{1/2}f(x_{ii''})\right] \tag{4.82}$$

with

$$x_{ii''}=\frac{1}{2}\left(\frac{M}{m}\right)\left(\frac{\omega-\omega_{i''f}}{\omega_p}\right)^2 \tag{4.83}$$

and

$$f(x) = \frac{1}{2}[(1+x)e^x\int_x^\infty e^{-y}dy-1]$$

$$\approx \frac{1}{2x^2}\left(1-\frac{4}{x}+\cdots\right). \tag{4.84}$$

Before discussing these expressions (Griem 1970, 1974), the reader is reminded of their region of validity, $|\Delta\omega| \equiv |\omega-\omega_{i''f}| \lesssim \omega_p$, which is generally only marginally fulfilled for so-called overlapping lines with spacings between levels $i$, $i'$, $i''$, ... of the order of the line width. In case of isolated lines with $i, i'$ level spacings larger than the line width and, usually, also larger than the electron plasma frequency, one would have to consider the full dielectric constant, but then screening is usually not important. The leading, logarithmic, term in (4.82) is as expected from the earlier physical arguments, while the $-1/2$ term (double-shielding correction) was obtained by Zaidi (1968), Dufty (1969), and, more explicitly, by Chappell, Cooper and Smith (1969). The ion cloud term is negligible for $\Delta\omega$ values approaching the electron plasma frequency, i.e., for $x \gg 1$. However, for $\Delta\omega$ values close to the ion plasma frequency $\omega_{pi}$, i.e., for relatively narrow lines, $x \approx 1/2$ is possible, corresponding to $f(x) \approx 0.2$. In that case and in view of $(M/m)^{1/2} \gtrsim 43$ and typical values of $K_{max}\rho_D$ $\lesssim 10^3$, the ion-cloud term would actually be somewhat larger than the logarithmic term. If, instead, we assume $x \approx 2$, i.e., $f(x) \approx 0.042$, the ion-cloud term is already less than 20% of the leading term. Still, care should be used for hydrogen, etc., lines whose widths are not much larger than the ion plasma frequency. An estimate for the ion-cloud contribution to $\text{Re}\mathscr{L}(\omega)$ can be made similarly and can be used to estimate line broadening from the corresponding field fluctuations (Griem 1978, Griem, Blaha and Kepple 1979). This additional broadening could be of some importance in dense plasmas, together with the contribution to $\text{Im}\mathscr{L}(\omega)$, but both have been ignored in almost all line shape calculations.

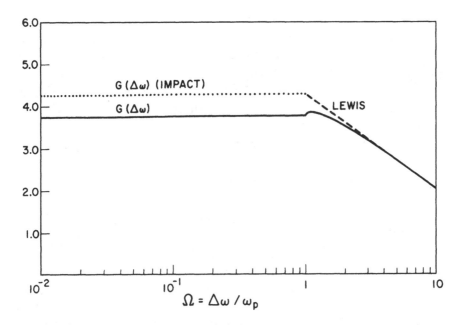

Fig. 4.6. Frequency-dependent factor $G(\Delta\omega)$ of the unified theory width operator Im$\mathscr{L}(\omega)$ calculated by Chappell, Cooper and Smith (1970) using the full plasma dielectric constant as a function of the ratio of frequency detuning $\Delta\omega$ and electron plasma frequency $\omega_p$ (solid curve). The dotted line corresponds to the impact approximation with a simple Debye cutoff, the dashed line to the impact approximation with the Lewis (1961) cutoff. See text and note that the separation between dotted and solid lines is due to the "$-1/2$" in the last factor of (4.82).

The reader may wonder why the $|\Delta\omega| \lesssim \omega_{pi}$ case is not discussed in the present context. The answer is that this region is of practical importance only at very low densities and that also the effects of perturbing ions can now be described using the impact approximation. Since the impact broadening for hydrogen, etc., lines is inversely proportional to the relative radiator-perturber velocity, this broadening by ions would be dominant. It has been calculated, e.g., by Lisitza (1977), Stehlé (1990) and by Ispolatov and Oks (1994). The ion impact broadening is the high-temperature, low-density limit of ion-dynamical theories of line broadening. However, this limit is not necessarily reached by all of the kinetic theory models discussed in section 4.5. In any event, the relatively large broadening by ions renders the discussion of corrections to the electron broadening somewhat academic.

Returning to the electron contribution to Im$\mathscr{L}(\omega)$, it has been calculated using the full dielectric $\epsilon(K,\omega)$ by Chappell, Cooper and Smith (1970). They find that (4.82), with $f(x) = 0$, is practically exact for $\Delta\omega \lesssim \omega_p$ (figure 4.6), and that for $\Delta\omega > \omega_p$ there is a rapid transition to the

$\epsilon = 1$ result first obtained by Lewis (1961) and then by the relaxation or unified theories of electron broadening (section 4.1). The matrix elements of $\mathrm{Im}\mathscr{L}(\omega)$ decrease with increasing $\Delta\omega$ because only collisions with durations $\lesssim |\Delta\omega|^{-1}$ can contribute.

Corresponding calculations of $\mathrm{Re}\mathscr{L}(\omega)$ for overlapping lines (Smith 1968, Dufty 1969, Chappell, Cooper and Smith 1970) involve a principal value integral which must be evaluated with care (Boercker and Iglesias 1984). Instead of (4.81), we now require the corresponding imaginary part of the transform of the field autocorrelation function, i.e., in analogy to our (2.99) or, rather, to equation (V-8) of Griem et al. (1962)

$$- \mathrm{Im} \int_0^\infty e^{i\Delta\omega t}\{\mathbf{F}(t)\cdot\mathbf{F}(0)\}_{av} dt = \frac{e^2}{4\pi^3\epsilon_0^2} P \int\int \frac{S(K,\omega')}{\Delta\omega - \omega'} d\omega' dK, \quad (4.85)$$

with $S(K,\omega)$ corresponding to $S(k,-ku)$ in Rosenbluth and Rostoker (1962), except for a small, but important, quantum correction. In the classical theory, $S(K,\omega)$ is an even function of $\omega$, so that in the impact limit for overlapping lines, i.e., for $\Delta\omega' = 0$, this version of the autocorrelation function results in $\mathrm{Re}\mathscr{L}(0) = 0$, consistent with the original semiclassical impact parameter method results. Actually, $S(K,\omega)$ cannot be exactly even and, at the same time, be consistent with detailed balancing. Rather, we must have (Boercker and Iglesias 1984)

$$S(K,-\omega) = \exp(\hbar\omega/kT)S(K,\omega). \quad (4.86)$$

Hence, for $\hbar\omega/kT << 1$ and $\Delta\omega = 0$, (4.85) becomes

$$\mathrm{Im} \int_0^\infty \{\mathbf{F}(t)\cdot\mathbf{F}(0)\}_{av} dt$$

$$= \frac{e^2}{4\pi^3\epsilon_0^2} \int dK P \int_0^\infty \left[-\exp\left(\frac{\hbar\omega'}{kT}\right) + 1\right] S(K,\omega')\frac{d\omega'}{\omega'}$$

$$\approx -\frac{e^2\hbar}{4\pi^3\epsilon_0^2 kT} \int dK P \int_0^\infty S(K,\omega')d\omega'$$

$$\approx -\frac{e^2\hbar}{8\pi^3\epsilon_0^2 kT} \int dK \int_{-\infty}^\infty S(K,\omega')d\omega', \quad (4.87)$$

if we replace $\omega'$ in the negative frequency range by $-\omega'$. The remaining integral over $\omega'$ gives

$$\int_{-\infty}^\infty S(K,\omega')d\omega' = 2\pi N_e K^2 \rho_D^2 [1 - \mathrm{Re}\epsilon_0^{-1}(K,0)]$$

$$= 2\pi N_e K^2 \rho_D^2 [1 + (K\rho_D)^2]^{-1}. \quad (4.88)$$

For $\Delta\omega = \omega - \omega_{inf} \to 0$ the relation analogous to the logarithmic term in

(4.82) is thus from (4.79), (4.87) and (4.88)

$$\text{Re}\langle i|\mathscr{L}(\omega_{i''f}|i'\rangle \approx -\frac{4}{3}\left(\frac{e^2}{4\pi\epsilon_0}\right)^2\frac{N_e}{kT}K_{\max}\sum_{i''}\langle i|\mathbf{r}|i''\rangle\cdot\langle i''|\mathbf{r}|i'\rangle, \qquad (4.89)$$

which depends strongly on $K_{\max}\approx\rho_{\min}^{-1}$, the minimum impact parameter. (The same holds for other contributions to the shifts (Iglesias, Boercker and Lee 1985, Griem 1988a). This result can also be obtained directly from Baranger's (1958b) result, our (4.51), as shown by Griem, Iglesias and Boercker (1991b). It is automatically included in fully quantum-mechanical calculations for hydrogen and hydrogenic ion lines.

Static shielding corrections to (4.89) result in an approximate correction factor $1-\pi/2K_m\rho_D$ (Griem 1988a), and for the hydrogen Lyman-$\alpha$ line the estimates of corresponding line shifts have been verified by higher-order calculations (Günter 1993). In case of broad lines reaching toward the electron plasma frequency, the $\Delta\omega=0$ limit discussed so far is inappropriate also for $\text{Re}\mathscr{L}(\omega)$. Calculations based on a Green's-function approach (Hitzschke and Röpke 1988) for the hydrogen Balmer-$\alpha$ line at high densities (Böddeker et al. 1993) include the full frequency dependence of the plasma field. They show a less than linear increase with density of the electron-produced line shift, as had been expected on physical grounds in the papers cited at the beginning of this section.

## 4.5 Kinetic theory models of dynamical ion effects

The general theory of spectral line broadening in plasmas discussed in sections 4.1, 4.2 and 4.4 is very satisfactory for the quantitative description of effects that are primarily due to interactions between radiators and plasma electrons. The practical reason for this is that the dominant contributions to the line shape operator $\mathscr{L}(\omega)$ are usually only of first (for shifts of ion lines) and second order in the interaction energy. However, even for electrons as perturbers, this is of course no longer true on the far line wings (Lewis 1961), which correspond to short times in the time evolution and very close electron-radiator encounters. In principle, one must therefore do the perturbation calculation of $\mathscr{L}(\omega)$ to all orders to recover the quasistatic limit (Smith, Cooper and Vidal 1969) which, however, may not agree with the semiclassical result (see section 4.7).

Except for the last proviso, the same is true for ions as perturbers, although the transition to the quasistatic limit occurs much closer to the line center. This is equivalent to requiring $\mathscr{L}(\omega)$ calculations to high orders in the interaction almost everywhere, except for hydrogen line cores at low densities where unified theory calculations (Greene 1982a, Oza et al. 1988b) give a very satisfactory account of ion-dynamical effects

(Kelleher et al. 1993). Although a formal theory for avoiding higher-order dynamical calculations by introducing microfield distributions and systematic corrections to the quasistatic (all order) profiles has been formulated (Dufty 1969), theoretical models and comparisons with experiments and computer simulations have generally been more useful.

A fairly successful model due to Brissaud and Frisch (1971, 1974) involves stochastic fields being constant most of the time, with sudden jumps from one value to the next at random times. Solutions of the time-independent Schrödinger equation were connected at these times to obtain solutions for various field sequences, etc. The field sequences were determined such that the amplitudes were consistent with the appropriate distribution functions (section 4.3) and jump frequencies such that the autocorrelation function of the model field agreed with the theory described in section 4.4. Although not unambiguous, this procedure ensured that the profiles obtained were correct in the impact and quasistatic limits. Computer simulations (Stamm, Smith and Talin 1984, Stamm et al. 1986, Seidel 1985, 1987, Hegerfeldt and Kesting 1988, Cardeñoso and Gigosos 1989) of the broadening by ions, which calculate ion motions and fields, and from the fields time-dependent wave functions, etc., may be viewed as a test of this model microfield method. Deviations between the two predictions are typically less than 20%. A detailed analysis of fields from interacting quasiparticles at a neutral radiator by Berkovsky et al. (1995) suggests that these deviations are inherent in the field model (kangaroo process) used in Brissaud and Frisch's (1971, 1974) model microfield method.

Other kinetic theory models were developed by Boercker, Iglesias and Dufty (1987), see also Dufty, Boercker and Iglesias (1990) and Boercker (1989, 1993). These authors discuss general time correlation functions and their Laplace transforms in plasmas, in which interactions with the radiator involve only its monopole and dipole moments and, therefore, also only the electric fields produced by the plasma ions and electrons. This approach provides useful relationships between line broadening and transport properties. Using Fano's (1963) Liouville operator method various self-consistent kinetic theory models were formulated, beginning with an equation for the dynamics of the plasma average of the time-dependent dipole moment operator $\langle \mathbf{d}(t) \rangle_p$, whose matrix elements determine the line spectrum. This dynamical equation is according to the time-dependent Schrödinger equation, i.e., (4.21),

$$\frac{\partial}{\partial t}\langle \mathbf{d}(t)\rangle_p + \frac{i}{\hbar}[(H+\mathbf{F}\cdot\mathbf{d}), \langle \mathbf{d}(t)\rangle_p] + \frac{i}{\hbar}\mathscr{L}\langle \mathbf{d}(t)\rangle_p = \frac{d}{dt}\langle \mathbf{d}(t)\rangle_p\bigg|_{\text{ions}} \quad (4.90)$$

The left-hand side is the same as for quasistatic broadening by ions of electron-broadened (and shifted) profiles described by the relaxation or unified theory line shape operator $\mathscr{L}(\omega)$ introduced in section 4.1. In the commutator, $H$ is the unperturbed radiator Hamiltonian and $\mathbf{F} \cdot \mathbf{d}$ the ion microfield, dipole interaction with the radiator. (The relevant $\mathbf{d}$ matrix elements here are between states of the same principal quantum number.) If ion dynamics is negligible, the right-hand side can be set to zero. Otherwise, one encounters complex operator expressions (Boercker, Iglesias and Dufty 1987), which so far have only been used after drastic simplifications.

To expose the effects due to changes in the ion microfield, it is useful to separate the plasma average into two steps: first an average for a fixed value of the field at time $t$, then a sum over all field values, i.e.,

$$
\langle \mathbf{d}(t) \rangle_p = \int d\mathbf{F} \langle \mathbf{d}(t) \delta(\mathbf{F} - \mathbf{F}(t)) \rangle_p
$$

$$
\equiv \int d\mathbf{F} P(\mathbf{F}) \mathbf{D}(\mathbf{F}, t). \tag{4.91}
$$

The first step is the formal way to introduce the fieldstrength distribution function $P(\mathbf{F})$, which was discussed in section 4.3. From the second step follows

$$
\mathbf{D}(\mathbf{F}, t) = P(\mathbf{F})^{-1} \langle \mathbf{d}(t) \delta(\mathbf{F} - \mathbf{F}(t)) \rangle_p, \tag{4.92}
$$

not to be confused with the quantity $D$ defined in (4.20). The new dipole operator $\mathbf{D}(\mathbf{F}, t)$ is therefore the conditional average of $\mathbf{d}(t)$ at a specified ion fieldstrength $\mathbf{F}$. The kinetic equation for $\mathbf{D}(\mathbf{F}, t)$ is the same as (4.90) for $\langle \mathbf{d}(t) \rangle_p$, except that the right-hand side may now, in analogy to the collision term in the Boltzmann equation, be written more explicitly as

$$
\frac{d}{dt} \mathbf{D}(\mathbf{F}, t) \bigg|_{\text{ions}} = \int d\mathbf{F}' [W(\mathbf{F}' \to \mathbf{F}) \mathbf{D}(\mathbf{F}', t) - W(\mathbf{F} \to \mathbf{F}') \mathbf{D}(\mathbf{F}, t)]
$$

$$
= \int d\mathbf{F}' W(\mathbf{F}' \to \mathbf{F}) [\mathbf{D}(\mathbf{F}', t) - \mathbf{D}(\mathbf{F}, t)], \tag{4.93}
$$

using the symmetry relation

$$
W(\mathbf{F}' \to \mathbf{F}) = W(\mathbf{F} \to \mathbf{F}'). \tag{4.94}
$$

The quantity $W(\mathbf{F}' \to \mathbf{F})$ is the transition rate from microfield state $\mathbf{F}'$ to $\mathbf{F}$. It should in a more general theory be replaced by an operator with respect to the radiator states, e.g., also describe ion collision-induced transitions between radiator states. The resulting ion-dynamical effect would be analogous to the difference between dynamical and quasistatic Stark effects near forbidden components discussed in section 4.11, i.e., to the difference between $S_+ + S_-$ from (4.128) and $S_f$ from (4.131). A model accounting for such effects will be discussed at the end of this section.

The simplest choice for the fieldstrength transition rate consistent with detailed balancing and an equilibrium distribution is

$$W(\mathbf{F}' \to \mathbf{F}) = v P(\mathbf{F}'), \tag{4.95}$$

$P(\mathbf{F}')$ playing the role of a population density of (field) states. Substitution of (4.93) and (4.95) into the equivalent of (4.90) then yields

$$\left[ \frac{\partial}{\partial t} + \frac{i}{\hbar}\mathcal{L} + \frac{i}{\hbar}L(\mathbf{F}) \right] \mathbf{D}(\mathbf{F}, t)$$

$$= -v[\mathbf{D}(\mathbf{F}, t) - \int d\mathbf{F}' P(F')\mathbf{D}(\mathbf{F}', t)], \tag{4.96}$$

if we replace $[H + \mathbf{F} \cdot \mathbf{d}, \cdots]$ by its Liouville operator equivalent $L(\mathbf{F})$ and use $\int P(\mathbf{F}')d\mathbf{F}' = 1$. Needless to say, $L(\mathbf{F})$ must not be confused with the line shape function $L(\omega)$ in (4.9), etc.

Since $\mathbf{D}(\mathbf{F}, t)$ is essentially the dipole correlation function, we require its Laplace transform,

$$\hat{\mathbf{D}}(\mathbf{F}', \omega) = \int_0^\infty dt e^{i\omega t}\mathbf{D}(\mathbf{F}', t), \tag{4.97}$$

which follows from (4.96) to

$$\hat{\mathbf{D}}(\mathbf{F}', \omega) = [\mathbf{d} + v\mathbf{I}(\omega)][-i\omega + v + \frac{i}{\hbar}\mathcal{L}(\omega) + \frac{i}{\hbar}L(F)]^{-1}, \tag{4.98}$$

with

$$\mathbf{I}(\omega) = \int d\mathbf{F}' P(\mathbf{F}')\hat{\mathbf{D}}(\mathbf{F}', \omega) \tag{4.99}$$

accounting for the field state population term, i.e., for the first term on the right-hand side of (4.93). It can, after substitution of (4.98) into (4.99), be determined from

$$\mathbf{I}(\omega) = \int d\mathbf{F}' P(\mathbf{F}')[-i\omega + v + \frac{i}{\hbar}\mathcal{L}(\omega) + \frac{i}{\hbar}L(\mathbf{F}')]^{-1}[\mathbf{d} + v\mathbf{I}(\omega)]$$

$$\equiv G(\Delta\omega)[\mathbf{d} + v\mathbf{I}(\Delta\omega)], \tag{4.100}$$

with the profile function

$$G(\Delta\omega) = i \int d\mathbf{F} P(\mathbf{F})[\Delta\omega + iv - \frac{1}{\hbar}\mathcal{L}(\Delta\omega) - \Delta\omega(\mathbf{F})]^{-1}. \tag{4.101}$$

In this expression, $\Delta\omega(\mathbf{F})$ represents the quasistatic level shifts generated by $L(\mathbf{F})$, with the constant term being combined with $\omega$ into the frequency detuning $\Delta\omega$. (This $G$ is different from that in figure 4.6.)

Substitution of (4.100) into (4.98) finally yields the Laplace transform

$$\hat{\mathbf{D}}(\mathbf{F}, \Delta\omega) = i[\Delta\omega + iv - \frac{1}{\hbar}\mathcal{L}(\Delta\omega) - \Delta\omega(\mathbf{F})]^{-1}[1 - vG(\Delta\omega)]^{-1}\mathbf{d}, \tag{4.102}$$

which gives according to (4.15), (4.91) and (4.101) for the line shape

$$L(\Delta\omega) = \frac{\hbar}{\pi} \mathrm{Re}\, Tr' \mathbf{d} \cdot [1 - \nu G(\Delta\omega)]^{-1} G(\Delta\omega)\mathbf{d} \qquad (4.103)$$

in this simple form of the kinetic theory. (Note that the trace over radiator states now includes the integral over fieldstrengths.) This version is equivalent to using a Poisson step process in the model microfield method (Boercker 1993). An important generalization is to use effective ion frequencies, $\nu(\Delta\omega)$, in (4.101) and (4.103). In that case the line shape formula derived here is seen to agree with equation (5.2) of Boercker, Iglesias and Dufty (1987), except for a normalization factor and a factor $f(a)$ describing the relative level populations because the relative intensities of the components given by $\mathbf{d} \cdot \mathbf{d}$ were assumed to be independent of the low frequency microfield. The sign of $\nu(\Delta\omega)$ in (4.101) is consistent with the discussion following the original equations of Boercker, Iglesias and Dufty (1987), as is an overall sign factor in this equation.

The last factor in (4.103) corresponds to the last factor in (4.38), while the factor $[1 - \nu(\Delta\omega)G(\Delta\omega)]^{-1}$ is required to retain detailed balancing. For $\nu = 0$, this factor is 1, while the real part of (4.101) generates, for fixed **F**, the relaxation or unified theory electron-broadened profiles. These are then averaged according to the microfield distribution function as in the standard model of hydrogen, etc., line broadening (Griem, Kolb and Shen 1959, Griem 1974). The proposed generalization of this model therefore involves the replacement of $\Delta\omega$ by $\Delta\omega + i\nu$ and the insertion of the detailed balancing factor. However, the additional broadening due to ion dynamics cannot be estimated by simply adding $\nu$ to the line width from a standard model calculation, because the detailed balancing factor actually steepens the core of the line profile. A proper reduction to the ion impact limit has been discussed by Boercker (1989).

For their numerical calculations, Boercker, Iglesias and Dufty (1987) assumed $\nu(\Delta\omega)$ to be independent of the internal radiator state and related it to the momentum autocorrelation function of an ion. Low and high frequency limits were shown to be well represented by

$$\nu(\Delta\omega) = \nu/(1 + i\Delta\omega\tau), \qquad (4.104)$$

with the parameters $\nu$ and $\tau$ determined, e.g., from one-component plasma simulations. Results of this model, and of its refined versions, agreed reasonably well with computer simulations (Stamm et al. 1986) for Lyman-$\alpha$ and -$\beta$ lines of hydrogen-like aluminum and argon in dense plasmas. This can be seen on figures 2-7 in Boercker, Iglesias and Dufty (1987). These authors also emphasize the importance of field changes from the motion of charged radiators and possible correlations between Stark and Doppler effects. Another important generalization is the inclusion of corrections to the allowed and forbidden component mixing effect mentioned below (4.94). A corresponding kinetic theory model has been developed by Talin et al. (1995). It is based on the partition of profiles obtained from the quasistatic approximation into a number of frequency bins or "radiative channels" (Godbert et al. 1994a) and then to assume exchanges between the frequency groups at a frequency obtained from calculated ion field correlation times. As discussed by Talin et al. (1995), this theoretical frequency fluctuation model does not reduce to the ion impact limit for high fluctuation rates. Rather, the profile structure collapses in that case to a single line (velocity narrowing). Such models should therefore probably be used only reasonably close to the quasistatic regime, unless care is taken to find the appropriate limit of the ion collision frequency $v(\Delta\omega)$ (Boercker 1989). However, considering jumps between spectral structures rather than between microfield values may yield more realistic profiles if ion-dynamical effects are important.

A third method for assessing ion-dynamical effects is based on a small time expansion of the perturbing fields (Kogan 1960, Wimmel 1961, Sholin et al. 1971, Lisitsa 1977, Griem 1974). Properties of the underlying special functions have been further developed by Hey (1976, 1989), but this method has not been particularly successful, presumably because it mostly accounts for changes in the magnitude of the field and is not suitable for unshifted components. For lines with strong unshifted Stark components, field rotation is the dominant cause of dynamical effects (Seidel 1981) and simple model calculations (Griem 1979, Griem and Tsakiris 1982) indeed removed practically all deviations from measured profiles in the central regions of Ly-$\alpha$ and Ly-$\beta$.

### 4.6  Collisional narrowing and correlations between Doppler and Stark broadening

Normally, Doppler and Stark broadening, or any other kind of pressure broadening, can be considered to be statistically independent processes, and the corresponding profiles be convolved to obtain the resultant line shape. (For the case of Lorentzian profiles, e.g., due to electron collisions with radiators having widely spaced levels and Maxwellian radiator velocity distributions, this convolution was demonstrated in figure 4.1 above.) However, as first pointed out by Dicke (1953), collisional effects and motional Doppler shifts can actually act in a coherent manner, resulting under some conditions even in a collisional narrowing of the usual Doppler profiles.

To discuss these interesting correlations quantitatively, the factors $\exp(\pm i\mathbf{k}\cdot\mathbf{r})$ omitted in the relations following (2.40) must be reinstated to account for the radiator's center-of-mass motion. In particular, (4.18) should be replaced by

$$C(s) = \sum_\alpha Tr'\{x_\alpha(0)x_\alpha(s)\exp[i\mathbf{k}\cdot\mathbf{r}(s) - i\mathbf{k}\cdot\mathbf{r}(0)]\rho(0)\}_{av}, \qquad (4.105)$$

if we take $\mathbf{r}(s)$ as the classical coordinate vector of the radiator, i.e., neglect the electron's displacement in the exponents, which is of course consistent with the dipole approximation. Another simplification made here conceals an actual difference between the autocorrelation functions for different polarizations (Nienhus 1973) which turns out to be very small after averaging over the directions of the radiator velocities (Seidel 1985). Before this averaging process, profiles calculated in the radiator's frame also depend on the direction of emission; and this anisotropy is especially important if the broadening by heavy ions can be described in the impact limit (Seidel 1979, Derevianko and Oks 1995). In the following discussions, we can set $\mathbf{r}(0) = 0$ and assume that $\mathbf{r}(s)$ is affected only by ion collisions.

As discussed, e.g., by Rautian and Sobel'man (1967), the Laplace transform of $\exp[i\mathbf{k}\cdot\mathbf{r}(s)]$ multiplied with the Maxwell distribution leads for free-streaming, i.e., $\mathbf{r}(s) = \mathbf{v}s$, to the normal thermal Doppler profile given in (4.1). Since the Doppler width in (4.2) is $\sim kv_{av}$, it takes $\Delta s \approx 1/k\bar{v}$ for the correlation function to become small compared to its initial value. Therefore the radiator's collision frequency $v$ must be smaller than $k\bar{v} = 2\pi\bar{v}/\lambda$ for the usual Doppler profile to apply, still neglecting collisional effects on the internal states of the radiator. If the condition $k\bar{v} > v$ is not fulfilled, the correlation function decays more slowly, resulting in a collisional narrowing by a factor of the order $k\bar{v}/v = \Delta\omega_D/v$, which is consistent with Dicke's (1953) original model and detailed calculations by Podgoretskii and Stepanov (1961). In the limit of high collision frequency, the radiator motion can be described as a diffusion process, $r \sim \sqrt{t}$, and the central portion of the Doppler profiles acquires a Lorentzian shape with widths given by $\Delta\omega_D^2/2v_d$ (Dicke 1953, Rautian and Sobel'man 1967). Here the effective collision frequency is defined by

$$v_d = \bar{v}^2/2D \qquad (4.106)$$

in terms of the diffusion coefficient $D$ for the radiating species. The collisional narrowing is now determined by a factor $\Delta\omega_D/2v_d$, and accounting for deceleration by dynamic friction (Gel'Medova and Shapiro 1991) enhances the narrowing even more.

For conditions between the free-streaming and diffusion regimes, collisionally narrowed Doppler profiles have been obtained from kinetic theory using expansion methods (Rautian and Sobel'man 1967) for both weak and strong collision models. Podivilov and Shapiro (1992) emphasized the importance of accounting for the velocity dependence of the Coulomb collision frequency, which is of course particularly small for radiators with large velocities and Doppler shifts, thus reducing the narrowing effect compared to estimates based on average cross sections and velocities. Molecular dynamics simulations (Pollock and London 1993) for highly ionized ions in strongly coupled plasmas showed further reductions in the collisional narrowing compared to previous estimates (Griem 1986), mostly because of ion-ion correlations which lead to larger mean-free paths for the ions than estimated originally. However, ion-ion collision frequencies may be high enough to prevent the rebroadening of saturated laser lines (Koch et al. 1994, see the end of section 8.2).

We now turn to the possible correlations between Stark and Doppler broadening. Even if the free-streaming approximation for radiators is valid, there can still be a relationship between Stark and Doppler broadening invalidating the usual convolution. At least in principle, this problem will arise (Griem 1979, Seidel 1985) when, due to ion-dynamical effects, Stark broadening depends on the radiator velocities, and when Doppler effects and Stark broadening are both significant. This will occur, for a given line of, e.g., hydrogen, at relatively low densities. At high densities, the free-streaming approximation becomes questionable, and the same collisions affect the radiator's center-of-mass motion and its internal state. The kinetic theory approach (Boercker, Iglesias and Dufty 1987) discussed in section 4.5 can, before making the simplifications leading to (4.90) and (4.91), be used to treat such joint Doppler-Stark broadening, as can computer simulations (Seidel 1985, Stamm et al. 1986). In a more general context, the use of a classical description for the radiator motion has been questioned (Berman and Lamb 1971, Smith et al. 1971), but for plasmas these quantum effects (Zaidi 1972) have so far not been noticed.

## 4.7  Stark broadening calculations

It is no longer practical to provide immediate access to the results of calculations of Stark widths, shifts and profiles through tables or graphs of these and related quantities (Griem 1974). The emphasis here is instead on giving an overview of suitable methods of calculation and of available computer codes. Experimental verifications of the very numerous calculated data have also advanced considerably. Such experiments will be discussed in sections 4.9 and 4.10, but have not gone far enough so far to provide critical tests of theory and calculations for extreme conditions in regard to density, temperature and charge state.

One generally separates calculations according to whether or not lines are isolated or overlapping, and according to elements and charge state.

The former classification, as mentioned before, is based on a comparison between line widths and separations between levels involved in the radiative transition and perturbing levels, i.e., levels having large dipole (or other important multipole) matrix elements with, usually, the upper level of the line. If line widths are small against these relevant level separations, Stark broadening is usually dominated by electron collisions, at least for neutral atoms and for ions in low and intermediate charge states and except for very low densities, where ion impact broadening may also be important. The profiles are Lorentzian in the absence of Doppler broadening, etc., with widths and shifts according to (4.48) and (4.49) or their semiclassical equivalents. The semiclassical calculations mainly have second-order terms in the dipole interaction which are obtained according to the procedures developed by Griem et al. (1962), Bréchot and Van Regemorter (1964), and Sahal-Bréchot (1969 a,b). The various calculations differ through the estimates used to allow for close collisions, including elastic scattering, as discussed recently by Van Regemorter and Hoang-Bin (1993). For details of these calculations, the reader is referred to the original literature or to sections II.3.c and II.3.d of Griem (1974) and to some recent papers (Alexiou 1994 a,b, Alexiou and Ralchenko 1994a, Alexiou 1995). Note, however, that the last paper reports on a nonperturbative calculation and that probably equally effective optimization procedures than those discussed by Alexiou (1994b) and by Alexiou and Maron (1995) were used to generate the electron-produced widths and shifts tabulated in Griem (1974).

It is usually assumed that fine structure need not be allowed for in the calculations, i.e., that averaged level splittings and matrix elements, e.g., from the Coulomb approximation (section 3.2), can be used. The total line profile then becomes a sum of equally broad but suitably shifted and weighted Lorentzians, usually assuming statistical populations of the initial levels. In cases where deviations from the implied LS-coupling are particularly large, line widths and shifts should probably be calculated separately (Hey 1977, 1985, Hey and Blaha 1978) to achieve the same accuracy as has become the rule after the original calculations for neutral helium lines (Griem et al. 1962). Another simplification frequently made is the neglect of lower state perturbations. This generally only causes small additional errors in line widths, high $n_\alpha$ lines of hydrogen, etc., being a notable exception (Griem 1974), but tends to affect shifts more seriously.

Calculated electron-impact widths and shifts of many neutral and singly-charged ion lines can be found in tables IV and V, respectively, of Griem (1974), for electron densities $N_e = 10^{16}\,\text{cm}^{-3}$ (table IV) and $N_e = 10^{17}\,\text{cm}^{-3}$ (table V). Except for the Debye shielding correction estimated on page 321 of Griem (1974), both widths and shifts are linear in density until widths approach relevant level spacings. (The numerical factor

$10^{-8}$ cm/Å must be replaced by $10^{-10}$ m/Å if S.I. units are used.) For neutral atom lines, table IV also contains an ion broadening parameter, $\alpha$, which can be used to estimate quadratic, quasistatic, Stark effects from ion microfields according to equation (224) of that monograph. In case quadrupole interactions between plasma ions and radiators are more important, a possible Gaussian contribution to the profile of ion lines can be estimated following Sahal-Bréchot (1991). Also listed in these tables are calculated values of some quantities indicative of errors. The fractional contribution from "strong" collisions is usually not at all small, causing the results to be sensitive to cutoff impact parameters and associated estimates of correction terms. However, since some optimization was attempted based on dispersion relations (Griem and Shen 1962), residual errors of calculated widths appear to amount typically only to a fraction of this strong collision term. Other estimated errors are associated with the neglect of higher multipole than quadrupole interactions and of possible perturbing levels. Again, the resulting errors are probably only a fraction of the corresponding entries. We finally note that the calculated widths for He I, and for C I, N I, and O I can be fitted to very simple semiempirical formulas (Seaton 1989) to within a factor $\sim 2$.

To supplement the detailed calculations of electron impact widths of ion lines, a semiempirical method involving effective Gaunt factors had been proposed (Griem 1968a) which has been extended since then to multiply-ionized ions. As can be seen from the last column in table V (Griem 1974), effective Gaunt factors inferred from the semiclassical calculations for singly-charged ions and from equation (459) of this reference typically range between $\sim 0.2$ and $\sim 1$, and increase with temperature. This compensates some of the explicit $T^{-1/2}$ dependence on temperature. Direct calculations of effective Gaunt factors were provided by Hey and Breger (1980 a,b, 1982, 1989). (See below for an extension of the semiempirical method.) For relatively high electron energies the semiempirical method agrees with a second-order Born approximation calculation (Bassalo and Cattani 1993). At lower energies, calculations of semiclassical width functions and convenient approximations to these functions are available (Alexiou 1994a, Alexiou and Maron 1995). Even simpler is equation (526) of Griem (1974) which gives results, e.g., for Si XII and Mg X lines in reasonable agreement with close-coupling calculations (Fill and Schöning 1994).

Many more semiclassical calculations have been published in the more than two decades since the tabulations just discussed were made. Most of the newer results were obtained by Dimitrijević and Sahal-Bréchot (1990b, 1992 a,b,c,d). [See Fuhr and Lesage (1993) for further references to extended tabulations for many spectra, including those of multiply-ionized atoms.] These authors also give collisional widths from proton and helium

(He$^+$) ion collisions, together with an estimate of the maximum density for using the ion impact approximation. They provide electron collisional widths and shifts as functions of both temperature and density of the electrons in order to include Debye screening explicitly. As to the ion impact broadening calculations, they only account for elastic scattering, an assumption which is not valid at very high temperatures.

Especially for resonance lines and other lines involving not very highly excited levels, the classical path assumption per se becomes questionable, because in quantum scattering evaluations of Baranger's equations (4.48) and (4.49) partial wave contributions of small angular momentum are quite important. For isolated lines, such strong-coupling, $R$-matrix calculations (Bely and Griem 1970, Barnes and Peach 1970, Barnes 1971, Sanchez, Blaha and Jones 1973, Dimitrijević, Feautrier and Sahal-Bréchot 1981, see also chapter II.3.e of Griem 1974) remained rare exceptions until the advent of the Opacity Project (Seaton 1987a). In this context such calculations were first made for C III lines (Seaton 1987b) connecting states of the $2s^2$, $2s2p$, and $2p^2$ configurations, with careful attention to the elastic scattering resonances and errors associated with distant collisions, which are estimated to be $\lesssim 15\%$ for the widths. Much larger errors, by factors $\lesssim 2$, would be incurred if the widths were set equal to the sum of all inelastic scattering rates, as has been the practice in some applications. In other words, the non-Coulomb elastic scattering rate corresponding, approximately, to the last term in (4.48), is quite important. Comparison of the close-coupling, quantum-mechanical results with semiclassical calculations (Sahal-Bréchot and Segre 1971) would suggest that the latter overestimated widths of these lines by factors 1.5–2 (see table 7 in Seaton 1987b). However, experimental data seem to suggest otherwise (see section 4.9). Note also that the semiclassical calculations almost always omit the $\Lambda = 0$, penetrating monopole interaction term.

In an extended close-coupling, $R$-matrix calculation (Seaton 1988, Burke 1992), electron impact widths and shifts were obtained for 42 transitions from $n \leq 3$ upper levels in Li- and Be-like ions up to Ne VIII and Ne VII, respectively. The upper and lower state $S$-matrices were calculated on a common energy mesh to save computer time. In this way, corrections could be made for multipole (especially dipole) interactions in the outer region, which were found to be quite large for 3p and 3d. To reduce cancellation errors, the $S$-matrices were kept unitary to ensure appropriate corrections for their diagonal elements. Convergence with regard to the angular momentum $L$ of the total system, target ion plus colliding electron, was found to be slow. Also, since, e.g., for Li-like ions, only $n = 2$ and 3 states were included in the close coupling expansion, calculations were only extended to energies just beyond the highest inelastic threshold, assuming a constant collision strength for higher energies.

Results for the Be II resonance line width are about a factor 1.5 below earlier close-coupling calculations by Sanchez, Blaha and Jones (1973). Comparisons with semiclassical calculations (Griem 1974) for the widths of 6 Be II lines are typically within a factor $\sim 1.5$, except for the resonance line, for which the semiclassical result is larger by a factor 3.5. The two sets of shift values are very different from each other, with those from the semiclassical calculations being much larger. This could be related to the resonances, which are ignored in semiclassical calculations, or could be due to cancellations of contributions of various levels.

In two papers, Seaton (1987b, 1988) also compares the close-coupling results with a refined version of the semiempirical, effective Gaunt factor, method (Griem 1968a). Instead of extending the approximate inelastic collision rates below threshold down to zero energy, this extrapolation is only done to the energy of the lowest effective resonance. This estimate of elastic scattering contributions leads to more nearly constant effective Gaunt factors, $g_{eff} = (3^{3/2}/8\pi)G$, if the lowest resonance is taken to correspond to an effective quantum number near 3.5. Best fit values of $G$ increase from $\sim 3$ to 6, or of $g_{eff}$ from $\sim 0.6$ to 1.2, as the ionic charge increases, and $G$ still tends to increase slowly with temperature. Agreement with detailed calculations for $n \leq 3$ is within $\pm 20\%$ in over half of the cases, and always better than a factor of 2. Judging by the last column in table V of Griem (1974), a slightly larger spread would be expected for lines with $n \geq 4$. Nevertheless, such estimates are useful, or even essential, for applications involving a large number of lines. A change of effective Gaunt factors with ionic charge had been invoked before, especially also by Dimitrijević and Konjević (1981), and agreement of corresponding line widths for the B III and C IV resonance lines and for two other C IV lines with the close-coupling results is very satisfactory (see table 12 of Seaton 1988).

For lines consisting of overlapping components, no analytic presentation analogous to the Lorentzian shapes for isolated lines is possible, although a relatively simple parameterization of calculated widths is possible and useful (Lee 1985). Another difference is in the relative importance of ions and electrons as perturbers. Ion-produced fields are usually responsible for most of the broadening of overlapping lines, i.e., of Stark broadening proper. Electron collisions, on the other hand, account for most of the broadening of isolated lines except, perhaps, of those from highly charged ions (section 4.9). Most realistic calculations so far are based on convolutions of quasistatic ion-broadened profiles according, e.g., to (4.11), with electron-broadened profiles (generalized Lorentzians), according to (4.38), i.e., on (4.101) and (4.103) in the limit $v = 0$. (Important exceptions are the computer simulations mentioned in section 4.5.) For hydrogen and ionized helium lines such profiles, obtained using semiclassical results for the electron broadening operator in the impact limit, i.e., $\mathscr{L}(\omega_0)$ (section 4.1) and Hooper's fieldstrength distribution functions, are listed in table I of Griem (1974). Another tabulation for hydrogen lines, based on frequency-dependent electron broadening operators $\mathscr{L}(\omega)$ according to the unified theory (section 4.1) was provided by Vidal, Cooper and Smith (1973), including profiles convolved with thermal Doppler profiles. The unified profiles are consistent with the quasistatic limit also for electrons, which is approached on the wings of broad lines provided the semiclassical and dipole approximations remain valid for the corresponding close encounters (see below). They should generally be more accurate in this region. Another difference between the two tabulations concerns the central regions of lines for which lower state interactions are important, e.g., $H_\alpha$. The difference originates from the upper-lower level interference term, which was originally overestimated through the omission of a complex conjugate sign on one of the dipole operators. However, one can argue on physical grounds (Hey and Griem 1975, Griem and Hey 1976) that the original error may have led to more accurate profiles. This might be due to the ion cloud contributions (section 4.4) or ion dynamical effects omitted in both calculations. Recent calculations by Stehlé (1994a) allow for ion-dynamical effects on He II line profiles but seem to underestimate line widths at high densities (section 4.9). As pointed out by Kelleher et al. (1993), a new calculation and tabulation of hydrogen line profiles including these effects and others, like line shifts, fine structure, etc., would be very desirable and also feasible. It could include electron broadening operators calculated (Sholin, Demura and Lisitsa 1974) as a function of the quasistatic microfield and a proposal by Ispolatov and Oks (1994) and by Oks, Derevianko and Ispolatov (1995) to allow for the dynamic splitting of energy levels during close collisions. One would thus avoid some of the ambiguity associated with strong collisions in semiclassical calculations, but it remains to be shown that second-order perturbation theory gives an accurate answer for the effects of field components perpendicular to the quasistatic field. (See section 4.10 for some quantum mechanical calculations of the electron broadening and shift of hydrogen and ionized helium lines.) Using the model microfield method for both electrons and ions, Stehlé (1994b) has provided tabulated profiles for Lyman- and Balmer-series lines of hydrogen for conditions in stellar envelopes. For applications not requiring an accurate description of the central profile regions, a method of calculation proposed by Seaton (1990) should be very useful. Ion dynamical effects could then be estimated as suggested by Schöning (1994a), except that detailed calculations should be used to obtain the corresponding profile parameters (Stehlé 1994a). For hydrogen lines such parameterized profiles have been obtained by Clausset, Stehlé and Artru (1994).

Because of their great similarity to hydrogen lines, it is appropriate to discuss the interesting and important case of neutral helium lines with nondegenerate but closely spaced perturbing levels. In the quasistatic approximation, one then obtains field-dependent eigenstates of mixed parity so that, e.g., the $2^3P$-$4^3D$ line acquires a $2^3P$-$4^3F$ forbidden component, and at higher fields also $2^3P$-$4^3P$. Analogous effects occur in the dynamical theory picture (section 4.11), and it becomes necessary to include, at least, the $4^3P$, $4^3D$ and $4^3F$ levels as upper states in (4.11) and (4.38) or (4.101) and (4.103), with $v = 0$ in most calculations so far. Further complications arise especially in case of highly ionized atoms for which radiative rates often exceed collisional rates. In that case the assumption concerning the density operator made in (4.18) is invalid, and level populations and the collisional-radiative line broadening must be considered on the same footing (Kosarev and Lisitsa 1994).

Since the early calculations for such neutral helium lines (Griem 1968b, Barnard, Cooper and Shamey 1969, Gieske and Griem 1969, Barnard and Cooper 1970, see also chapter II.3.c of Griem 1974) the calculational methods have been improved to account for the effects of ion dynamics, the importance of which for lines with forbidden components was first suggested by Burgess (1970). One approach consisted of an application of the unified theory (section 4.1 and Griem 1974) to the broadening by ions (Barnard, Cooper and Smith 1975). It led to improved agreement with experiments (section 4.9), but was not entirely satisfactory because of the difficulty of treating simultaneous strong collisions. Another approach was based on the model microfield method (Brissaud and Frisch 1971) discussed in section 4.5, which was used by Adler and Piel (1991) for the He I 2P-4D singlet and triplet lines. Their calculations for the singlet line, in particular for the 2P-4F forbidden component, are shown on figure 4.7 at an assumed electron density of $N_e = 2 \times 10^{14} \, \text{cm}^{-3}$. By varying ion and atom (Doppler) temperatures separately, the relative importance of ion-dynamical and Doppler effects is clearly exhibited. The Doppler broadening is generally more important at a given temperature, except in the "dip" region between forbidden and allowed lines. The smallness of the ion-dynamical effects at these plasma conditions is consistent with the insensitivity of the half width of the $H_\beta$ line of hydrogen to these effects for $N_e \gtrsim 10^{14} \, \text{cm}^{-3}$ (section 4.9). The model microfield method was also used in recent calculations of He I and He II line profiles (Schöning 1994a), and Schöning (1994b) also obtained close-coupling results for the electron impact broadening of some He I lines. Some of them, however, deviate by as much as a factor of 2 from semiclassical results (Griem et al. 1962, Griem 1974, Dimitrijević and Sahal-Bréchot 1990b). This deviation would be inconsistent with measurements (Kelleher 1981) as discussed in section 4.9 and is not understood theoretically.

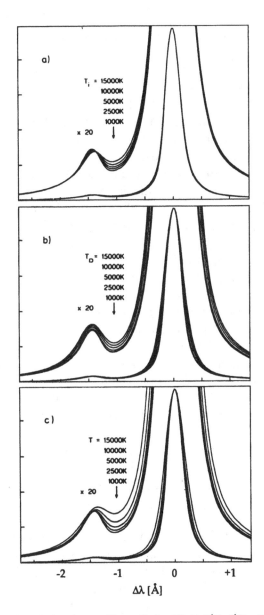

Fig. 4.7. Calculated Stark profiles of the He I $2^1$P-$4^1$D 4922 Å line and its forbidden $2^1$P-$4^1$F component at $N_e = 2 \times 20^{14}$ cm$^{-3}$ according to Adler and Piel (1991), who used the model microfield method for the ion effects and included electron impact and Doppler broadening. In (a) the ion temperature $T_i$ is varied to indicate the ion-dynamical effects, in (b) the gas temperature $T_D$ to show the effect of Doppler broadening, while in (c) the two temperatures are equal. Note that the region of the forbidden component is also shown with the intensity enhanced by a factor of 20.

Numerous calculations have been made mostly of Stark profiles of H-, He-, and Li-like ions, using codes as those described by Lee (1988), by Calisti et al. (1990), or by Mancini et al. (1991, 1994b). Most of these calculations were motivated by experiments on laser-compressed plasmas and have been briefly reviewed by Hauer and Baldis (1988) and Griem (1992). These codes are for overlapping lines, capable even of including levels of different principal quantum number from that of the upper level of the line (Mancini et al. 1991). Calculations for H-like lines of the Lyman-series of C, O, Ne, and Si (Kepple and Griem 1978, Griem, Blaha and Kepple 1979) used diagonal matrix elements of $-\mathrm{Im}\mathscr{L}(\omega_0)$ as obtained from quantum scattering calculations, which were then also used in many later calculations. Corresponding values for matrix elements of width and shift operators, i.e., of $-\mathrm{Im}\mathscr{L}(\omega_0)$ and $\mathrm{Re}\mathscr{L}(\omega_0)$ for H-like ions were calculated by Nguyen et al. (1986) and for He-like ions by Koenig, Malnoult and Nguyen (1988) and, for the $\ell = 1$ levels, by Griem, Blaha and Kepple (1990). Although the predicted shifts from (penetrating) monopole-monopole ($\Lambda = 0$) interactions are quite large, they have so far only been included in one set of profile calculations, namely those for Ar XVII (Griem, Blaha and Kepple 1990). One reason for this omission are the difficulties with the average interaction discussed in section 4.1 following (4.33). Omission of the matrix elements of $\mathrm{Re}\mathscr{L}(\omega)$ as, e.g., in the work of Griem, Blaha and Kepple (1979), Lee (1988), or of Kilcrease, Mancini and Hooper (1993), besides masking possible line shifts, may of course also affect profile shapes and line widths. They had therefore been considered, e.g., by Tighe and Hooper (1976).

Line broadening calculations for dense plasmas have also been made for more complicated spectra, e.g., those of neon-like ions (Keane et al. 1990), and the calculations, e.g., for Li- and Be-like ions by Calisti et al. (1990) have been extended both to ions with more complicated level structure and to include ion-dynamical effects using a kinetic theory method (Godbert et al. 1994a, Talin et al. 1995, see section 4.5). Also the work on dielectronic satellite spectra in dense plasmas (Woltz et al. 1991, Mancini et al. 1992) is based on a multielectron-ion line profile code (Woltz and Hooper 1988), which is being extended to provide evermore realistic descriptions of line broadening in dense plasmas. Jacobs, Cooper and Hahn (1994) have provided a unified description of Stark-broadened dielectronic satellite spectra.

As mentioned before, in cases where the impact approximation is more appropriate than the quasistatic approximation, e.g., for protons as perturbers, careful calculations are needed to supplement the probably more reliable calculations of the electron impact broadening discussed earlier. One such calculation of proton impact broadening has been published recently (Alexiou and Ralchenko 1994b). Another special situation occurs in strongly magnetized plasmas (section II.6 of Griem 1974). For such situations the ion-impact broadening calculations of Derevianko and Oks (1994) may be appropriate. These use the method of Ispolatov and Oks (1994) to treat the effects of ion-produced electric-field components, which are parallel to the magnetic field, to all orders using the exact solution of (4.21) for diagonal $U'(s) = U(s)$. The effects of perpendicular electric field components are again calculated by second-order perturbation theory which converges now also at small impact parameters, although higher-order calculations would be required to check the accuracy of the result.

Especially for opacity calculations, as discussed in chapter 8, and in particular for hydrogen lines, it is important to supplement detailed profile calculations by asymptotic formulas for the extended line wings (Seaton 1995). It had long been taken for granted (Griem 1974) that both ion and electron effects could then be estimated using the nearest-neighbor, quasistatic approximation. However, quantum-mechanical, one-perturber calculations (Tran-Minh and Van Regemorter 1972, Feautrier et al. 1976, Tran-Minh et al. 1976, Stehlé 1986, Seaton 1990, 1995) show that the electron contribution to the far wings of Lyman-$\alpha$ is always smaller than the ion contribution, typically by a factor of $\sim 2$. This can be understood qualitatively by comparing the nearest-neighbor separation $r_p \approx |3\hbar/m\Delta\omega|^{1/2}$ from equation (485) in Griem (1974) both with the de Broglie wavelength $\hbar/mv$ and with the validity condition $r_p > v/|\Delta\omega|$ for the quasistatic approximation, which defines the Weisskopf radius. These considerations require $(1/3)mv^2 < |\hbar\Delta\omega| < 3mv^2$, i.e., for Lyman-$\alpha$ there is, at best, a small region beyond $|\hbar\Delta\omega| \approx mv^2/3 \approx kT$ for which the quasistatic approximation might apply. However, as pointed out by Seaton (1995), this region is, at least on the high frequency side and for typical temperatures, already midway to Lyman-$\beta$ so that the linear Stark effect would not be applicable.

For other hydrogen and hydrogenic-ion lines, the 3's in our estimate for the semiclassical, quasistatic range, are to be replaced by $\sim 3n^2/4Z$, i.e., the above frequency range expands and would begin at $\sim |\hbar\Delta\omega| = (4/3n^2)kT$ for other hydrogen lines. Nevertheless, since line separations scale as $n^{-3}$, this still is well beyond the linear Stark effect regime. For ionized helium lines, etc., one could begin at $|\hbar\Delta\omega| \approx (4Z/3n^2)kT$, which because both line separations and $kT$ scale with $Z^2$, would go even further beyond the linear Stark effect regime. Moreover, for $4Z/n^2 > 3$, the thermal de Broglie wavelength is larger than the Weisskopf radius, invalidating the semiclassical approximation altogether for the wing broadening by electrons.

Although there is no regime in which electron broadening can be attributed to quasistatic linear Stark effects, it appears that corresponding errors for lines other than hydrogen Lyman-$\alpha$ are not large (see figures 6 and 7 of Seaton 1995 and note that the short broken curves are for $kT/Z^2E_H = 0.05$). The higher relative wing intensities for the Lyman-$\delta$ line ($n = 5$) may be attributed to the $n^4$ scaling of collisional widths, rather than the $n^3$ scaling of the asymptotic quasistatic wing intensities. The increases for $Z = 2$ and 4, on the other hand, are probably caused by the long-range Coulomb correlations and by quadrupole, etc., interactions.

The long-range repulsive interaction between ions is equally important, reducing the asymptotic contribution essentially by a Boltzmann factor $\exp[-(Z-1)e^2/4\pi\epsilon_0 rkT]$ for singly-charged perturbing ions. Midway between lines, i.e., for $|\hbar\Delta\omega| \approx Z^2E_H/n^3$, the exponent is approximately $[\cdots] = -(8/3n^5Z)^{1/2}(Z-1)Z^2E_H/kT$, e.g., $-0.2Z^2E_H/kT$ for $Z = 2$, $n = 2$. Since $kT/Z^2E_H \approx 0.2$ is quite typical, the ion contribution is substantially reduced relative to the Holtsmark estimate, and the far wings are dominated by electron effects (see figure 9 of Seaton 1995). For $n \geq 3$, these correlation effects are much smaller.

Not surprisingly, both calculated (Tran-Minh et al. 1976, Feautrier et al. 1976) and measured Lyman-$\alpha$ profiles exhibit asymmetries for a number of reasons (section 4.10) and perhaps also because of $H_2^+$ molecular effects (Boggess 1959, Dunn 1968, Kielkopf and Allard 1995) or because of resonances in the $H^-$ system (Doughty et al. 1966, Macek 1967). Although a description of proton and electron broadening effects in terms of $H_2^+$ and $H^-$ seems physically reasonable, it appears that more work is needed to establish a dynamical theory for the corresponding profile regions containing various satellites (Stewart et al. 1973, Drake 1973). Recent experimental and theoretical research on far Lyman-$\alpha$ line wings has been reported by Kielkopf (1993, 1995).

The new calculations of entire hydrogenic line profiles (Stehlé 1994a and 1994b, Clausset et al. 1994) still depend on the validity of the quasistatic, semiclassical, linear Stark effect limit for electrons on the far wings (Stehlé 1994b). However, the corresponding errors should generally not be very important, hydrogen Lyman-$\alpha$ being a possible and important exception.

## 4.8    Effects of neutral perturbers

In only partially ionized gases, interactions between atoms and molecules and the radiator may compete with the perturbations caused by charged particles. We shall only consider atoms as additional perturbers and generally assume that their density as well as the ensuing line widths are small enough for the impact approximation to be appropriate for the central region of the profile. Any contribution on the line wings would normally have to be estimated using the quasistatic approximation, i.e., the analogue of (4.11), and the appropriate interatomic potentials (Lewis 1980, Peach 1996). However, in the special case of resonance broadening to be discussed now, the collisional line width is independent of the velocity and (4.11) would give $L(\omega) \sim 1/\Delta\omega^2$ for the line wings, suggesting that the line profile remains nearly Lorentzian. However, on the distant wings of the Lyman-$\alpha$ line in hydrogen plasmas satellites occur, some of which can be interpreted in terms of $H_2$ molecules (Kielkopf and Allard 1995).

The effective leading term in the multipole expansion of the atom-radiator interaction Hamiltonian is instead of (4.41) the dipole-dipole interaction

$$U(\mathbf{r}, t) \approx [|\mathbf{r}|^{-3}\mathbf{d}_r \cdot \mathbf{d}_p - 3|\mathbf{r}|^{-5}(\mathbf{d}_r \cdot \mathbf{r})(\mathbf{d}_p \cdot \mathbf{r})]/4\pi\epsilon_0. \qquad (4.107)$$

Here $\mathbf{r} = \mathbf{r}(t)$ is the vector separation of radiator and perturber, and $\mathbf{d}_r$ and $\mathbf{d}_p$ are the dipole moments of radiator and perturber, respectively. In case of charged radiators and perturbers, there are also monopole-monopole and monopole-dipole terms which, however, would in the approximation used here only indirectly affect the line width through their influence on the perturber orbits. Note also that $\mathbf{r}(t)$ is not the same quantity as $\mathbf{r}$ in section 4.1, which stood for the set of initial perturber coordinates and velocities.

The (HWHM) width is according to (4.39) and (4.46)

$$
\begin{aligned}
w &= -\frac{i}{\hbar}\langle i_r g_p | \mathrm{Im}\mathscr{L}(\omega_0) | i_r g_p \rangle \\
&= \mathrm{Re}\sum_j f_j \langle i_r g_p | 1 - S_j | i_r g_p \rangle \\
&= \mathrm{Re}\sum_j f_j \langle i_r g_p | \left[1 - \exp\left(-\frac{i}{\hbar}\int_{-\infty}^{\infty} U'_j dt\right)\right] | i_r g_p \rangle. \quad (4.108)
\end{aligned}
$$

The perturber is in its ground state $g_p$ before and after collision $j$, of collision frequency $f_j$, while $i_r$ is the initial radiator state. In case of resonance broadening, intermediate states corresponding to exchange of excitation energy between radiator and perturbers of the same species are involved. The relevant $U'$-matrix elements are thus between degenerate states of the radiator-perturber system, i.e., the prime can be omitted. For distant (weak) collisions, the U-matrix elements are small, and the exponential in (4.108) can be replaced by a few terms in its Taylor series. The average of the first-order term is zero, as is the third-order term.

The second-order term is

$$\left(\frac{i}{\hbar}\right)^2 \langle i_r g_p| \int_{-\infty}^{\infty} dt_2 U(t_2) \int_{-\infty}^{t_2} dt_1 U(t_1)|i_r g_p\rangle$$

$$= \frac{1}{2}\left(\frac{i}{\hbar}\right)^2 \langle i_r g_p| \int_{-\infty}^{\infty} U(t_2)dt_2 \int_{-\infty}^{\infty} U(t_1)dt_1|i_r g_p\rangle$$

$$= \frac{2/4\pi\epsilon_0}{v^2\rho^4} \langle i_r g_p|\mathbf{d}_r \cdot \mathbf{d}_p - \frac{2}{\rho^2}(\mathbf{d}_r \cdot \boldsymbol{\rho})(\mathbf{d}_p \cdot \boldsymbol{\rho}) - \frac{1}{v^2}(\mathbf{d}_r \cdot \mathbf{v})(\mathbf{d}_p \cdot \mathbf{v})|g_r' i_p'\rangle$$

$$\times \langle g_r' i_p'|\mathbf{d}_r \cdot \mathbf{d}_p - \frac{2}{\rho^2}(\mathbf{d}_r \cdot \boldsymbol{\rho})(\mathbf{d}_p \cdot \boldsymbol{\rho}) - \frac{1}{v^2}(\mathbf{d}_r \cdot \mathbf{v})(\mathbf{d}_p \cdot \mathbf{v})|g_r i_p\rangle$$

$$(4.109)$$

with $\mathbf{v}$ and $\boldsymbol{\rho}$ from the assumed straight path

$$\mathbf{r}(t) = \boldsymbol{\rho} + \mathbf{v}t. \qquad (4.110)$$

Averaged over angles associated with $\boldsymbol{\rho}$ and $\mathbf{v}$, which are correlated, and summed over magnetic quantum numbers of the intermediate states $g_r' i_p'$ this gives (Ali and Griem 1965, 1966)

$$\left(\frac{i}{\hbar}\right)^2 \langle i_r g_p| \int_{-\infty}^{\infty} dt_2 U(t_2) \int_{-\infty}^{t_2} dt_1 U(t_1)|i_r g_r\rangle = -\frac{g_g}{g_i}\left(\frac{e^2 f_{gi}/4\pi\epsilon_0}{m_e \omega_{ig} v \rho^2}\right)^2 .$$

$$(4.111)$$

Here we used (2.48) and (2.51) to express the dipole matrix elements of the two atoms in terms of the absorption oscillator strength $f_{gi}$ for absorption from the ground state to the "initial" state $i$, which could be either the upper or the lower level of the line. The frequency $\omega_{ig}$ is the angular frequency associated with the excitation $g_p \to i_p'$, not necessarily that associated with the line.

Instead of calculating higher-order terms, one may estimate the line width from (4.108), (4.109), (4.111) and $f_j \to 2\pi N_g v \rho d\rho$ as

$$w_r \approx 2\pi N_g v \left[\int_{\rho_0}^{\infty} \rho d\rho \frac{g_g}{g_i}\left(\frac{e^2 f_{gi}/4\pi\epsilon_0}{m_e \omega_{ig} v \rho^2}\right)^2 + \frac{1}{2}\rho_0^2\right]. \qquad (4.112)$$

The last term is an estimate for collisions within the Weisskopf (1933) radius $\rho_0$ which corresponds to a suitable mean value for the matrix elements of $1-S$ for $\rho < \rho_0$, taken by Ali and Griem (1965) as $|\langle|1-S|\rangle| = 1$ or, from (4.111)

$$\frac{g_g}{g_i}\left(\frac{e^2 f_{gi}/4\pi\epsilon_0}{m_e \omega_{ig} v \rho_0^2}\right)^2 = 1. \qquad (4.113)$$

With some corrections (Ali and Griem 1966), this gives for the line width due to resonance broadening

$$w_r \approx 2\pi N_g \left(\frac{g_g}{g_i}\right)^{1/2} \frac{e^2 f_{gi}}{4\pi\epsilon_0 m_e \omega_{ig}}. \tag{4.114}$$

As discussed by Omont and Meunier (1968), the estimate comes very close to more accurate calculations, which allow for higher-order terms in the expansion in powers of $U'$, and to calculations of cross sections for excitation transfer between identical atoms (Watanabe 1965 a,b). Another accurate calculation of resonance broadening was performed by Stacey and Cooper (1969) and a close-coupled calculation by Leo et al. (1995).

Besides the expansion in powers of $U'$, one should of course also check the possibility of higher multipole contributions to $U'$. From (4.113) it follows that contributing impact parameters are of the order

$$\begin{aligned}
\rho_0 &\approx \left(\frac{g_g}{g_i}\right)^{1/4} \left(\frac{e^2 f_{gi}/4\pi\epsilon_0}{m_e\omega_{ig}v}\right)^{1/2} \\
&= \left(\frac{g_g}{g_i}\right)^{1/4} \left(\frac{2E_H f_{gi}}{m_e\omega_{ig}va_0}\right)^{1/2} a_0. \tag{4.115}
\end{aligned}$$

For neutral atoms, one typically finds $\rho_0 \gg a_0$, so that higher multipole than dipole-dipole interactions are indeed negligible in most situations where resonance broadening is important. These estimates must not be used when the groundstate density $N_g$ is so large that one should use a theory accounting for simultaneous collisions.

More universal than resonance broadening is broadening by neutral perturbers of lines of a different species as the radiator, or of lines of the same species as the perturber whose levels are not connected by dipole transitions. One then speaks of Van der Waals broadening, although in general the atom-atom interactions involved do not only correspond to $1/r^6$ forces (see, e.g., Lewis 1980 and Peach 1996). Nevertheless, one can begin the discussion by reconsidering (4.107) and (4.109). Reinstating the exponential factors required for $U'$, a typical second-order matrix element is then in the adiabatic approximation, i.e., after integrating over $t_1$ by parts and neglecting the derivative,

$$\begin{aligned}
\left(\frac{i}{\hbar}\right)^2 &\langle i_r g_p| \int_{-\infty}^{+\infty} dt_2 \exp\left[-\frac{i}{\hbar}(E_{ig} - E_{i'g'})t_2\right] U(t_2)|i'_r g'_p\rangle \\
&\times \langle i'_r g'_p| \int_{-\infty}^{t_2} dt_1 \exp\left[-\frac{i}{\hbar}(E_{i'g'} - E_{ig})t_1\right] U(t_1)|i_r g_p\rangle \\
&\approx -\frac{i}{\hbar} \int_{-\infty}^{\infty} \frac{\langle i_r g_p|U(t)|i'_r g'_p\rangle\langle i'_r g'_p|U(t)|i_r g_p\rangle}{E_{i'g'} - E_{ig}} dt. \tag{4.116}
\end{aligned}$$

The energies $E_{ig}$ and $E_{i'g'}$ are total energies of radiator and perturber in initial (final) and intermediate states of the system. With (4.107) and (4.110) the integral can be evaluated and averaged over directions. An approximate result for the matrix element of $1 - S$ is (Griem 1964)

$$\langle i_r g_p | 1 - S | i_r g_p \rangle \approx \pi \frac{i}{\hbar} \left( \frac{e^2}{4\pi\epsilon_0} \right)^2 \frac{1}{\rho^5 v}$$

$$\times \sum{}' \frac{|\langle i | z_r | i' \rangle|^2 |\langle g | z_p | g' \rangle|^2}{E_{i'g'} - E_{ig}} + \cdots, \quad (4.117)$$

the sum being over intermediate states of the system and $+ \cdots$ standing for terms with $x$- and $y$-matrix elements.

Following Unsöld (1955), one can estimate the sum by assuming that the radiator states $i'$ are quite close to, e.g., the upper state of the line and that therefore the energy denominators are essentially the excitation energies $E_p$ of the perturber. According to (2.48), the $z_p$-matrix elements can be written as

$$|\langle g | z_p | g' \rangle|^2 = \frac{\hbar^2}{2m_e E_p} f_p, \quad (4.118)$$

with $f_p$ being the appropriate absorption oscillator strength. More accurate expressions can be obtained in terms of the polarizability of the perturbing atoms (Al-Saqabi and Peach 1987). The radiator matrix elements, on the other hand, can be summed immediately over $i'$, giving in the Coulomb approximation discussed in section 3.2

$$\sum |\langle i | z_r | i' \rangle|^2 \approx \frac{a_0^2}{6} \frac{E_H}{E_\infty - E_i} \left[ 5 \frac{E_H}{E_\infty - E_i} + 1 - 3\ell_i(\ell_i + 1) \right]$$

$$\equiv \frac{1}{3} R^2 a_0^2, \quad (4.119)$$

where $E_\infty$ is the (neutral) radiator's ionization energy, $E_i$ the energy of the upper level of the line, and $\ell_i$ the orbital angular momentum quantum number of this level. To estimate the Weisskopf radius, consider again $|\langle |1 - S| \rangle| = 1$, which leads with (4.117), (4.118) and (4.119) to

$$\rho_0 \approx \hbar \left( \frac{R^2}{m_e^3 v E_p^2} \right)^{1/5} \quad (4.120)$$

and to a HWHM width

$$w \approx \pi N_g v \rho_0^2$$

$$= \pi N_g \overline{v^{3/5}} \frac{\hbar^2 (\overline{R^2})^{2/5}}{m_e^{6/5} E_p^{4/5}}. \quad (4.121)$$

In this crude approximation, collisions beyond $\rho_0$ only contribute to the shift, and $f_p = 1$ was used to account for all dipole excitations of the perturber. If the perturbing atom has $n_e$ valence shell electrons, the width would be larger by $n_e^{2/5}$; and if lower state broadening is not negligible, one should use the difference of the $R^2$ values.

This simple estimate is in fair agreement with more accurate impact theory calculations (Lindholm 1945, Foley 1946, Anderson 1952) and a generalization covering the transition into the quasistatic regime (Anderson and Talman 1955). Another generalization to densities beyond the validity range of the impact approximation (Royer 1980) allows a systematic calculation of nonlinear terms in a density expansion. Returning to impact approximation calculations, they also give line shifts $d$, to lower frequencies, which approximately amount to 2/3 of the width $w$.

The most serious limitations of these estimates and also of accurate calculations based on the Van der Waals force law arise from the fact that $\rho_0$ as estimated in (4.120) is often of the same order as excited state atomic radii so that short-range repulsive forces should be allowed for. Therefore, if good accuracy is required, the reader should turn to the specialized literature, e.g., in the proceedings of the international conferences on spectral line shapes. Other estimates of broadening by neutral atom collisions can be found, e.g., in Traving (1960) and Sobel'man, Vainshtein and Yukov (1981 and 1995). None of these estimates is appropriate for highly excited atoms as perturbers. An error estimate for this contribution may be made (Griem 1962, 1964) by adding the density of Rydberg atoms with Bohr radii larger than the mean electron-electron separation $r_e$ to the electron density. This correction is usually negligible.

## 4.9 Line profile and width measurements

Because of the complexity of line broadening calculations, which is the reason for the many approximations to the general theory made in most calculations, there is a great need for reasonably accurate measurements under well-defined conditions. Most of such measurements have been concerned with isolated lines of neutral atoms and of ions in low and intermediate charge states of a great number of elements. For most of these lines, electron impact broadening should indeed be the predominant mechanism, except for very partially ionized gases. A first experimental check of this are fits to Lorentzian line shapes or, if fits to Voigt profiles, see (4.4), are more satisfactory, the value of the ratio of Lorentzian and Doppler widths. If the fits are poor and the measured line shapes are symmetric with respect to their actual peaks, the emission is very likely from a plasma with strong density gradients along the line of sight. Since low density regions correspond to relatively narrow lines, the observed width can be substantially smaller than the width corresponding to the maximum density in the source. Another test is to determine the electron density dependence of widths (and shifts) which should be essentially linear. Deviations from this linear dependence are to be expected beyond some density, depending on the line, and have indeed been observed by Büscher et al. (1995) for the HeI $2^3P$-$3^3D$ line for $N_e \lesssim 2 \times 10^{18}\,\mathrm{cm}^{-3}$. This may be due both to increased Debye shielding of electron collisions (section 4.4) and to overestimating the broadening by ions (see below and equation 224 of Griem 1974).

The importance of plasma homogeneity and, in case of transient sources, also of their stability, was emphasized and discussed already in the first critical reviews of Stark width and shift measurements (Konjević and Roberts 1976, Konjević and Wiese 1976). These distributional features are always present to some extent and can be investigated by making space-resolved observations on stationary plasmas of known symmetry to extract local emission coefficients, e.g., by Abel inversion, as discussed in section 8.5, or by making time-resolved measurements on transient plasmas. The latter are of course essential for measurements on multiply-ionized atoms. However, the corresponding plasmas may have somewhat flatter density and temperature distributions than would be typical of stationary plasmas.

Besides correcting for any instrumental and Doppler broadening, one may in case of strong lines also have to account for radiative transfer effects, see chapter 8, which tend to broaden the line by self-absorption or, if induced emission exceeds absorption, may even cause gain narrowing (Casperson and Yariv 1972, Koch et al. 1992). Especially for lines of neutral atoms, any resonance or Van der Waals broadening should at least be estimated as discussed in section 4.8. This requires some measurement or estimate of neutral atom densities, besides the usual electron density and temperature measurements.

For isolated lines showing some profile asymmetry, broadening by ions in or near the quasistatic regime is the most likely cause, especially in case of neutral atom lines. Corresponding corrections for HeI lines were calculated by Griem et al. (1962) and characterized by a parameter $\alpha$. For other neutral atom lines, this parameter can be found in table IV of Griem (1974) for $N_e = 10^{16}\,\mathrm{cm}^{-3}$. It scales with $N_e^{1/4}$, see equation (224) of Griem (1974), and can be used to estimate corrections to the electron impact width $w$ and shift $d$, see equations (226) and (227) of the same reference. Measurements of asymmetries of CI and ArI line profiles, which will be discussed in section 4.10, give good agreements with calculated profiles corresponding to a convolution of Lorentz and quasistatic profiles, but the best-fit $\alpha$-values are larger than calculated values by as much as a factor of $\sim 1.5$. This could be due to errors in the calculated quadratic Stark coefficients and electron impact widths, but may also indicate deviations from the quasistatic approximation (Kelleher 1981, Kobilarov, Konjević and Popovich 1989). Such deviations could be inferred, e.g., from unified theory calculations (see section II.4.a of Griem 1974). The width comparison in one of the papers on profile asymmetries (Hahn and Woltz 1990) misses a factor of 2 in the calculated widths from the conversion to FWHM widths.

At relatively low densities but high temperatures the impact approximation may be more appropriate also for broadening by ions, especially for light ions, e.g., $H^+$ or $He^+$ as perturbers. Because of their large abundances in stellar atmospheres, calculated values for these contributions to the Lorentzian width are therefore included in tabulations for astrophysical applications, see, e.g., Dimitrijević and Sahal-Bréchot (1992 a,b,c). These ion broadening calculations, as already mentioned in section 4.7, account for elastic scattering only, i.e., would be underestimates for very high temperatures or systems with close perturbing levels. The reader should also keep in mind that there are no telltale signs of ion impact contributions on a given line profile like the asymmetry in case of quasistatic effects on isolated lines.

A most critical task in Stark broadening experiments is the measurement of electron density, preferably by a method independent of Stark broadening. At moderate densities, laser interferometry is the preferred method (Konjević and Wiese 1976), but Thomson scattering has also been very useful (Gawron et al. 1988, DeSilva et al. 1992, Wrubel et al. 1996a). Then there is a number of purely spectroscopic methods, which will be discussed in chapter 10, including comparison with already verified Stark widths, e.g., of the hydrogen $H_\beta$ line. Temperature measurements, see chapter 11, are also important but less critical, because Stark profile parameters are relatively weakly dependent on temperature.

As far as the numerous results for isolated lines are concerned, the reader is referred mostly to the various critical reviews (Konjević and Roberts 1976, Konjević and Wiese 1976, Konjević, Dimitrijević and Wiese 1984 a,b, Konjević and Wiese 1990). The tabulated widths in these reviews are FWHM widths, of various estimated accuracies. Their ratio to available calculated widths, $w_{th}$, is also given. It is close to 1, with typical deviations of $\pm 20\%$ for neutral atom lines, much smaller deviations for He I lines (Kelleher 1981, Pérez et al. 1995, Mijatović et al. 1995) and only slightly larger typical deviations for lines from singly-ionized atoms. Somewhat surprisingly, deviations from close-coupling, fully quantum-mechanical calculations tend to be larger, e.g., for the resonance 2s-2p doublets of Be II (Sanchez, Blaha and Jones 1973) and of B III (Glenzer and Kunze 1996), by as much as a factor of 2. This can probably be attributed only partially to broadening by ions. On the other hand, measurements of the AlII $(3p)^2$ $^1$D-3s4p$^1$P line agreed quite well with distorted wave calculations of the electron broadening (Allen et al. 1975) and measured 2s-2p near-threshold excitation cross sections (Bell et al. 1994) for O VI also agree well with close-coupling calculations. This agreement, however, may not carry over to line broadening calculations, which not only require other inelastic cross sections, but also elastic scattering data. For multiply-charged ions, for which fewer data are available, typical deviations are also often large, measured widths again tending to be larger than calculated electron impact widths. Before pursuing this point, the interested reader is referred also to a recent bibliography on atomic line shapes and shifts (Fuhr and Lesage 1993), which continues previous bibliographies.

The scarcity of experimental data for line widths, etc., for multiply-ionized atoms from well-diagnosed, nearly homogeneous and optically thin plasmas has been eased considerably by a series of experiments in a gas liner pinch (Kunze 1987) and other pinch discharges (see, e.g., Purić et al. 1988 and Djeniže et al. 1988), and by a laser plasma experiment (Moreno et al. 1993). Most of these experiments were concerned with transitions involving relatively low-lying levels, for which semiclassical calculations per se could become less reliable as the charge state is increased along an isoelectronic sequence (section 4.7). An early gas liner experiment (Böttcher et al. 1988) gave widths for the 3s-3p transitions of Li-like ions which were all larger than calculated values and, in frequency or energy units, decreased only as $\sim Z^{-1}$, rather than the expected $\sim Z^{-2}$ scaling, $Z$ being 4 for C IV, etc. This excess over the predicted electron broadening was not confirmed for C IV in experiments with improved diagnostics but was again found for higher ions up to Ne VIII (Glenzer, Uzelac and Kunze 1992, Glenzer 1995) with the suggestion that it might arise from ion broadening via quadrupole interactions. An analogous study of 3s-3p and 3p-3d transitions in B-like ions (Glenzer, Hey and Kunze 1994a) led to a similar conclusion. Moreover, the excess broadening also seems to hold relative to fully quantum mechanical calculations and to some semiclassical calculations (Alexiou and Ralchenko 1994a, Alexiou 1994b). It is much larger than estimated possible Zeeman effects from internal magnetic fields (Davara *et al.* 1995) or than uncertainties in the temperature scaling. This scaling is important because the various ions are observed at substantially different temperatures. The temperature scaling has therefore been investigated experimentally by Glenzer et al. (1992, 1994a) for Li- and B-like ions, respectively, and by Blagojević et al. (1994, 1996) for 3s-3p and 3p-3d transitions in B-like ions. If one uses a theoretical upper limit for the strong collision term (Alexiou 1994b, Alexiou and Maron 1995), marginal agreement is obtained between calculated electron impact broadening and measured widths, e.g., of the Ne VIII 3s-3p lines. This upper limit may seem to be too high, judging by the fact that the measured Stark width corresponds to an effective Gaunt factor $\bar{g} \approx 2.4$. This is well above the values discussed in section 4.7 but still below a $\bar{g} \approx 3$ value inferred from fully quantum-mechanical calculations (Koenig et al. 1988) for analogous lines of helium-like ions. Such calculations (Griem, Blaha and Kepple 1979) account for the full Coulomb interactions and contain, besides the usual dipole interactions, significant contributions from $\Lambda = 0$ and 2 multipole interactions. As to other fully quantum-mechanical calculations, especially those based on the close-coupling method (section 4.7), these tend to give smaller electron impact widths than semiclassical calculations by a factor of about 1.5 (Dimitrijević and Sahal-Bréchot 1990a). Returning to the semiclassical calculations, which are valid only marginally for 3s-3p, $Z \approx 7$ lines, the large error associated with the strong collision term can be traced to serious violations of the unitarity condition during, rather than only after, the collisions (Alexiou 1995). As a matter of fact, a recent measurement of 3s-3p line widths of Be-like Ne VII (Wrubel et al. 1996b) agrees rather well with such calculations. It is noteworthy that there is now a fairly large set of measured width data for low principal quantum number transitions in Li-, Be- and B-like ions.

For a discussion of experiments devoted to the widths of 3s-3p and 3p-3d transitions of successive ions of a given species, the reader is referred to a paper on Ne II-Ne VI and F IV and F V lines (Uzelac et al. 1993). Also in this case, ratios of measured to calculated widths increase with charge state. Until the mechanism responsible for this deviation is cleared up, caution should be exercised in the use of empirical fits to measured widths (see, e.g., Purić et al. 1988 or numerous references in Fuhr and Lesage 1993) for extrapolations to higher charge states. As already mentioned, proton and other light perturbing ion collisions should for highly charged radiators and high temperatures begin to compete with electron impact broadening also for isolated lines (see, e.g., Griem 1974, section II.3.f, and Griem 1993). For overlapping lines of multiply-charged ions, ion broadening effects are, of course, well known to be dominant over electron impact broadening.

Before turning to overlapping lines and reemphasizing the difficulties in comparisons between results obtained at different temperatures, the importance of the nearly linear density dependence of electron impact widths, except for the Debye shielding effects discussed in section 4.4, for such comparisons must also be stressed. Measurements by Vitel and Skowronek (1987 a,b) indicated a stronger nonlinearity, but these were not confirmed by later experiments under similar conditions (Uzelac and Konjević 1989, Konjević and Uzelac 1990, Hutcherson and DeSilva 1997).

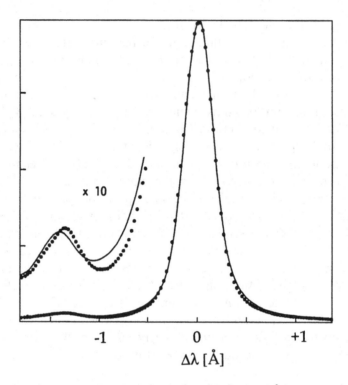

Fig. 4.8.   Comparison of measured He I $2^1$P-$4^1$D, $^1$F 4922 Å line profiles measured by Adler and Piel (1991) at $N_e = 2 \times 10^{14}$ cm$^{-3}$, $T = T_i = T_D = 7,500$ K, $T_e = 20,000$ K with a calculated profile as shown in figure 4.7. The region of the forbidden component is enhanced in intensity by a factor of 10.

For overlapping lines it is usually not practical to characterize the profiles by a small number of parameters, He I lines with their forbidden Stark components being an important exception (Adler and Piel 1991). Although these profiles are no longer single Lorentzians (figures 4.7 and 4.8) the relative peak intensities of the two components, the intensity minimum between the components and their peak separations, in addition to the widths of the components, are all quantities useful for checking profile calculations. He I lines measured at substantially higher electron densities (Hey and Griem 1975, Suemitsu et al. 1990, Uzelac et al. 1991) are also found to agree very well with calculations up to $N_e \approx 2 \times 10^{17}$ cm$^{-3}$. At higher densities systematic differences, e.g., in the separation of the peaks of the two components of the lines, are observed. This could be caused by interactions with higher levels.

Hydrogen lines are the prototype of overlapping lines, except at very low densities, where fine-structure and proton impact broadening must

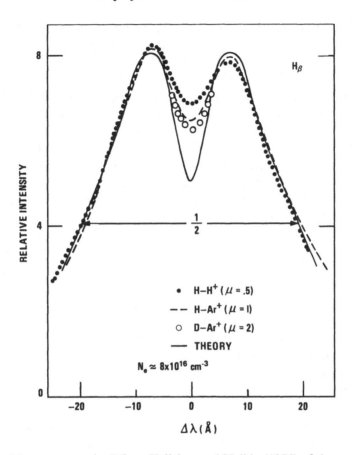

Fig. 4.9. Measurements by Wiese, Kelleher and Helbig (1975) of the central part of $H_\beta$ and $D_\beta$ line profiles at $N_e = 8 \times 10^{16}$ cm$^{-3}$ for different perturbing ions, resulting in various reduced masses $\mu$ (in proton mass units). The theoretical profile is according to Vidal, Cooper and Smith (1973), i.e., independent of $\mu$.

be considered (Weber and Humpert 1981, Stehlé 1990). In applications, this low density regime, which requires Doppler-free spectroscopy (Danzmann, Grützmacher and Wende 1986), is not too important except for its extensions to highly charged radiators. As reviewed recently (Kelleher et al. 1993), experiments for the most studied Stark broadened line, namely $H_\beta$, can be said to result in widths agreeing with theory down to $N_e \approx 10^{14}$ cm$^{-3}$ (Helbig 1991, Thomsen and Helbig 1991). Since iondynamical effects on the $H_\beta$ width are small in this density regime and the line is quite sensitive to Stark broadening, it continues to be very useful as a comparison standard for plasma density. The central minimum in the $H_\beta$ line profile, see figure 4.9, on the other hand, is quite sensitive to ion dynamics, but these deviations are well described by calculations accounting for radiator-ion motions (Kelleher et al. 1993).

For lines that either possess unshifted Stark components, such as $L_\alpha$, $H_\alpha$, $H_\gamma$, etc., or are relatively insensitive to Stark effects, like $L_\beta$, ion-dynamical effects are much more important. This was found in a series of experiments beginning with work by Kelleher and Wiese (1973), Wiese, Kelleher and Helbig (1975), and Grützmacher and Wende (1977), and is now quite well understood (Kelleher et al. 1993). Nevertheless, these lines are less suitable than $H_\beta$ for density measurements, because they are narrower or weaker and more dependent on plasma composition and temperature. Moreover, the $\alpha$-lines tend to be optically thick in hydrogen-rich plasmas. These lines are therefore best measured in gas mixtures containing hydrogen as a minority, say, in helium (Elton and Griem 1964). A recent experiment on $H_\alpha$ of this kind has been performed in a gas liner pinch (Böddeker et al. 1993) for densities up to $10^{19}$ cm$^{-3}$, corresponding to a temperature near 10 eV. At lower densities, also the temperature was lower so that the perturbing ions are both He$^+$ and He$^{++}$ in varying proportions. The measured FWHM range up to $\sim 250$ Å and are in better agreement with calculations (Kepple and Griem 1968) which only include elastic contributions to the upper-lower state interference term (section 4.7) at lower densities, and at higher densities with calculations including the full interference term. This interesting trend can probably not be explained by ion-dynamical effects.

Corresponding measurements of ionized helium lines (Musielok et al. 1987, Gawron et al. 1988) using the same apparatus, but with helium as an $\sim 1\%$ admixture to the hydrogen test gas and pure hydrogen as driver gas, were performed on the VUV Balmer $\alpha$ and $\gamma$ transitions and on the Paschen $\alpha$ transition. Densities were again rather high, up to 0.8 and $2.2 \times 10^{18}$ cm$^{-3}$, and temperatures varied around 10 eV. For the He II Balmer $\alpha$ line, widths were found to agree with Kepple's (1972) calculations. Ion-dynamical corrections calculated analogously to Oza et al. (1988 a,b) of 10-20% were not included, because they were within the experimental errors. The $\gamma$-line, on the other hand, was found to be $\sim 30\%$ narrower than calculated, consistent with two earlier experiments. This could perhaps be explained (Greene 1982b) by the omission in Kepple's (1972) calculations of inelastic contributions to the interference term in the electron broadening, but ion-dynamical effects would counteract these corrections. Further results, up to $N_e \approx 10^{19}$ cm$^{-3}$, have been reported by Böddeker (1995) for the He II Balmer $\alpha$ line, and additional experiment-theory comparisons were made by Günter et al. (1995) and by Oks et al. (1995). The most frequently studied He II line is Paschen $\alpha$ at 4686 Å. As shown in figure 4.10, which is taken from Büscher et al. 1996, its FWHM width ranges from $\sim 4$ Å just above $N_e = 10^{17}$ cm$^{-3}$

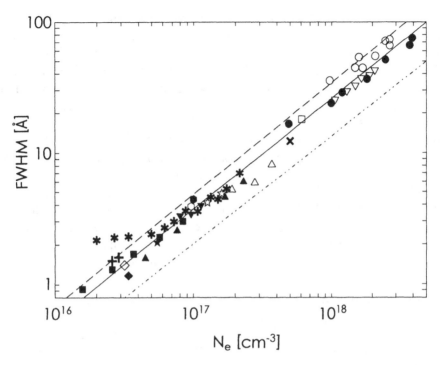

Fig. 4.10. Comparison of measurements of the full half width of the He II Paschen-α line at 4686Å according to Büscher et al. (1996) (solid circles) with earlier measurements and calculations by Greene (1976) (dash, double dots), Kepple (1972) (long dashes), and Griem and Shen (1961) (solid line). The earlier measurements were by Eberhagen and Wunderlich (1970) (solid hexagon), Einfeld and Sauerbrey (1976) (solid tilted squares), Jones et al. (1971) (open hexagons), Jones et al. (1977) (open circles), Piel and Slupek (1984) (stars), Pittman et al. (1980) (solid squares), Soltwisch and Kusch (1979) (inverted solid triangles), Stefanović et al. (1995) (pluses), Wulff (1958) (tilted open squares), Pittman and Fleurier (1986) (asterisks), Gawron et al. (1988) (inverted open triangles), Bernard et al. (1981) (open squares), Oda and Kiriyama (1980) (open triangles), Bacon et al. (1977) (solid triangles), Bogen (1970) (times), and by Berg et al. (1962) (open stars). See Büscher et al. (1996) for some corrections to electron densities measured by the original authors and note the differences between widths obtained by Pittman et al. (1980) and other authors below $N_e = 5 \times 10^{16} \, \mathrm{cm}^{-3}$ and those of Pittman and Fleurier (1986). These differences are not understood.

to $\sim 80 \, \text{Å}$ near $N_e = 4 \times 10^{18} \, \mathrm{cm}^{-3}$. Surprisingly, best overall agreement was found with the first modern calculation (Griem and Shen 1961) for this line, which effectively omitted the entire interference term (by omitting complex conjugate signs on the operators $\mathbf{R}_b$) and neglected any ion dynamics. The largest deviations, by a factor $\sim 2$, are from a calculation including the entire interference term (Greene 1976), while

Kepple's (1972) calculation exceeds most measurements by $\sim 20\%$. Any temperature dependence is neglected in these comparisons, and clearly more experimental and theoretical work, including ion dynamics and the interference term, is needed before accuracies approaching those for hydrogen $H_\beta$ might be achieved. Büscher et al. (1996) also measured the He II $P_\beta$ line and made comparisons with earlier results. Recent measurements (Stefanović, Ivković and Konjević 1995) of the He II $P_\alpha$ and $P_\beta$ lines at $N_e \lesssim 3 \times 10^{16}\,\text{cm}^{-3}$ demonstrate the importance of ion dynamics, which may also have contributed to the relatively large widths measured by Pittman and Fleurier (1986) for $N_e \lesssim 4 \times 10^{16}\text{cm}^{-3}$; whereas the summary of width measurements of the $\alpha$-line up to $N_e \approx 4 \times 10^{18}\,\text{cm}^{-3}$ shows that recent calculations (Stehlé 1994a) underestimate the width by factors of 2-3 at high densities. This could again be attributed to an overestimate of the interference term in the electron impact broadening or also to an underestimate of the effects of strong electron collisions (Oks et al. 1995).

For overlapping lines from more highly-charged ions, only a few experiments have been reported, in which in particular the density of perturbers was measured by an independent method. Examples are measurements on the principal quantum number $n = 7$ to 6 transition in C VI (Iglesias and Griem 1988a, Olivares and Kunze 1993), of this transition and the 6 to 5 transition in B V blended with N V (Huang et al. 1990a) and also of 7 to 6 transitions in C V and N V (Olivares and Kunze 1993). Earlier measurements on similar transitions were made by Irons (1973), who inferred the densities from measured line intensities. Although C V and N V are helium- and lithium-like, respectively, most of the $n\ell$ levels remain unresolved, and comparisons with calculations for the 7 to 6 transition in hydrogen-like C VI (Kepple and Griem 1982) seem reasonable, as does the scaling to the 6 to 5 transition and to various perturbing ions (Huang et al. 1990a). As predicted, the profiles are close to Lorentzians and the measured widths scale with $N_e^{2/3}$. The proportionality factor agrees with calculations within $\pm 20\%$ in case of the C VI measurements, after the radial density distribution is allowed for in the laser plasma experiments. For B V and N V, a similar accuracy is indicated by comparisons between measured profiles and calculated profiles including the radial distribution (see figures 5 and 6 of Huang et al. 1990a). Without this correction, widths differ by larger amounts, equivalent to $\lesssim 50\%$ error in the electron density. The C V and N V line widths measured by Olivares and Kunze (1993), on the other hand, are almost twice the predicted values, but also show the $N_e^{2/3}$ scaling. This discrepancy is not understood, and there are similar unresolved differences between the original calculations (Kepple and Griem 1982) and the results of current line broadening codes (Kepple

1995, Schöning 1995). However, the gas liner experiment supports the finding in laser-produced plasmas (Iglesias and Griem 1988a, Wang et al. 1989) of the anomalous broadening of C V $n = 2$, $\Delta n = 0$ lines, presumably due to Doppler broadening associated with turbulent motions in laser-produced plasmas.

Other laser plasma experiments (Iglesias and Griem 1988b, Wang et al. 1992) were concerned with the $H_\alpha$, $H_\beta$, and $L_\delta$ lines of C VI and with the $H_\alpha$ and $L_\delta$ lines of B V, respectively. The $H_\beta$ and $L_\delta$ lines were used to determine electron densities, near $10^{20}\,cm^{-3}$, which were then used to calculate $H_\alpha$ line profiles or, rather, line widths, according to, e.g., Oza, Greene and Kelleher (1986) or Lee (1985), i.e., with and without ion-dynamical corrections and allowing for fine-structure splitting. As discussed by Wang et al. (1992), measured $H_\alpha$ widths for these hydrogen-like ions do agree with some calculations within experimental errors but are up to a factor 2 larger than earlier theoretical results. It was further concluded that ion-dynamical corrections should not be important in x-ray laser experiments involving higher members of this isoelectronic sequence.

Both gas-liner and laser-plasma experiments have been made on Li-like ion lines from nearly degenerate $n\ell$ levels, including forbidden components. Böttcher, Musielok and Kunze (1987) measured line shapes and widths of 3p-4d, 3d-4f, 4d-5f, and 4f-5g transitions in C IV and N V. They found that the line widths were larger than estimated electron impact widths by factors of 3 to 4, but smaller than quasistatic, linear Stark effect ion-produced widths by factors of 1.2 to 3. (Detailed calculations were not available.) This discrepancy has meanwhile been removed by a comparison (Godbert et al. 1994b) between improved measurements for C IV and N V at higher densities with calculations using the quasistatic ion approximation. Further improvements for the C IV, N V and O VI $n = 4$ to 5 transitions, by 10 to 20%, are obtained (Glenzer et al. 1994c) if ion-dynamical effects are included according to Godbert et al. (1994a). Moreno et al. (1993) measured widths and profiles of 3d-5f transitions for higher members of the lithium-like series and also of the isoelectronic series of some 3d-4f transitions in helium-like ions up to Cl XV. Comparison with detailed profile calculations suggested agreement for an electron density $\sim 10^{20}\,cm^{-3}$ in case of Mg X, and also for Al XI if gain narrowing (Moreno et al. 1989) is considered. For the two more highly charged ions observed, there is either some excessive broadening by factors $\lesssim$ 1.5, or a corresponding increase in the electron density in the emitting region. Experiment-theory comparisons have also been made by Schöning (1993 a,b), who used close-coupling and model-microfield methods, respectively, for the electron and ion broadening. Calculations by Loboda et al. (1994) of Mg XI and Al XII lines as measured by Moreno et al. (1993)

indicate the sensitivity of these line profiles to the level structure of the unperturbed ions.

Discussions of measurements of lines from more highly charged ions in laser compressed plasmas at substantially higher densities and temperatures than encountered in the experiments discussed in the present section are more appropriate for chapter 10. The principal reason for these profile measurements is the determination of density in the absence of more reliable methods. This is not to imply that there are no other density diagnostics. For example, in one of the first compression experiments involving Stark broadening of Ne X lines (Mitchell et al. 1979) the corresponding density was compared with that inferred from the compression ratio as measured by x-ray imaging. Another case in point is an experiment (Hooper et al. 1990, see figure 10.5b) in which He-like argon and Li- and Be-like krypton lines were found to have consistent Stark profiles at $N_e = 1.5 \times 10^{23}\,\mathrm{cm}^{-3}$. The unexplained $Z$-scaling mentioned above therefore is not universal.

In work on highly charged ions, fine-structure, e.g., of $L_\alpha$ lines is often well resolved or, at least, is a cause of pronounced asymmetries (Akhmedov et al. 1985, Zhidkov, Tkachev and Yakovlenko 1986). Recently even hyperfine splitting of neon-like ion lines has become important in x-ray laser experiments (Nilsen et al. 1993) at densities small enough for Doppler broadening to be the dominant broadening mechanism.

## 4.10   Line shift and asymmetry measurements

Experiments concerned with deviations from symmetrical line shapes, e.g., from Lorentzians or from symmetrical hydrogen line profiles calculated using the procedures discussed in section 4.7, are quite important in checking the validity of the usual approximations which yield symmetrical line shapes. Only if observed line shapes are indeed symmetrical can shifts be defined easily. Generally one would then speak of the shift of the profile maximum or, e.g., in case of $H_\beta$, of the shift of a central minimum. A more robust definition is generally in terms of the mean wavelength, or frequency, etc., of opposite points on the profile having equal fractional intensities. For broad lines, it does matter also whether wavelengths or frequencies are used, because then the transformation leading, e.g., to (4.5) and (4.6) must be carried to higher order. Only if the shifts measured in this way are the same for different fractional intensities within experimental errors, including errors from continuum background subtraction, can the line profile be taken as symmetrical and its shift be characterized, e.g., by that of its peak.

Isolated lines from neutral atoms with small quasistatic broadening by ions and isolated lines from ions in low charge states are good candidates for such simple shift representation. As can be seen from the critical reviews mentioned in section 4.9, the number of shift measurements is substantially smaller than that of width measurements, and the agreement with calculations as discussed in section 4.7 tends to be poorer. This remains true even for the much studied He I lines (Mijatović et al. 1995, Djeniže et al. 1995). For the semiclassical calculations this is as expected theoretically (Griem and Shen 1962), and not enough quantum mechanical, *R*-matrix calculations are available for lines with measurable shifts to judge their accuracy against measurements.

Shifts are usually but not always to longer wavelengths, because close interacting levels tend to be above the upper level of the line, rather than below. If the shifts are relatively small in terms of line widths, they also may be strongly temperature dependent. This effect can be traced to cancellations between the effects of different perturbing levels. It may even lead to a change in the sign of the shift, as observed by Heading, Marangos and Burgess (1992) and Büscher et al. (1995) in case of the He I line at $\lambda = 5876\,\text{Å}$. In the first experiment, which was at lower temperature, also an asymmetry was measured and found to change from one side of the profile to the other.

Similar to studies for the widths of isolated lines, systematic studies of measured shifts also reveal regularities and trends. Shift studies have been summarized by Wiese and Konjević (1992) with the following conclusions:

(a) Measured Stark shifts of lines belonging to the same multiplet generally agree within $\pm 10\%$.

(b) Shifts of lines belonging to the same supermultiplets and transition arrays vary typically only by $\pm 25\%$ if the shifts are relatively large, say, about 50% of the FWHM width. For smaller shifts, the variations are much larger, again as expected theoretically (Griem and Shen 1962).

(c) Shifts of lines of the same spectral series show pronounced increases with principal quantum number. Interpolation and limited extrapolation along the series should yield fairly accurate shift values.

(d) Related transitions in homologous atoms and ions exhibit systematic trends in the measured Stark shifts, again suggesting inter- and extrapolations.

(e) Experiments regarding shifts along isoelectronic sequences are scarce, but the only study involving three ions (Kobilarov and Konjević 1990), S II, Cl III, and Ar IV, yielded a regular trend.

(f) Pronounced irregularities have been observed as well. They appear
    to be due to changes in the relative positions of perturbing levels.

As already mentioned in section 4.9, isolated lines are not exactly
symmetrical either. This asymmetry was originally attributed (Griem et
al. 1962) to quadratic Stark effects from quasistatic, ion-produced fields.
Measurements as shown on figures 4.11 and 4.12 for C I (Jones and Wiese
1984) and Ar I (Jones, Wiese and Woltz 1986) confirmed the predicted
asymmetry quite well, but yielded numerical values of asymmetry, i.e.,
ion broadening parameters which were only in fair agreement with predicted values (Griem 1974). These conclusions were supported by further
measurements by Hahn and Woltz (1990). Particularly interesting are the
measurements of Jones, Pichler and Wiese (1987) on N I and Ar I lines,
including lines with blue shifts and asymmetries favoring the blue wings.
Measured and calculated ion broadening parameters compared quite well
in these cases.

Asymmetries and shifts of hydrogen lines in plasmas with densities
near $10^{17}\,\mathrm{cm}^{-3}$ have been observed for a long time (Griem 1954). The
red shift of $\sim 1\,\text{Å}$ of $H_\gamma$ was attributed to quadratic Stark shifts of its
central component, while the $\lesssim 5\%$ excess intensity of the blue maximum
of $H_\beta$ was explained invoking quadratic Stark effect corrections to the
quasistatic theory and including the $\omega^4$ prefactor in (4.7), not to mention
the changes in matrix elements and the transformation from frequencies to
wavelengths. However, variations in Boltzmann factors or their equivalents
in dynamical theories (see, e.g., Smith and Hooper 1967) were neglected,
as were the very important quadrupole contributions to the ion-atom
interaction (Demura and Sholin 1975).

Much experimental and theoretical work on the asymmetry of hydrogen line profiles has been done since. It has been used by Demura and
Stehlé (1995) to check the accuracy of corrections to the usual quasistatic,
linear Stark effect calculations and to discuss the density and temperature dependence. These authors also discuss the influence of electron
impact broadening on the asymmetry, but still assume that electron impacts do not result in net shifts of the lines. As shown by Green's
function method calculations of shifts and asymmetries due to electrons
(Günter and Könies 1994b, Könies and Günter 1994), this assumption is
insufficient for quantitative calculations of profile asymmetries.

A number of experiments have revealed small red shifts of some Lyman
and Balmer series lines that persist after asymmetries are accounted for.
Most of these results are compared in Griem (1988a) with semiclassical
calculations of shifts due to electron impacts, involving interactions with
perturbing levels of both the same (section 4.4) and different principal
quantum number as that of the upper level of the line. For $L_\alpha$ and $L_\beta$,

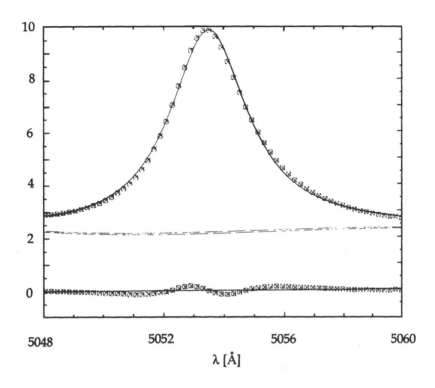

Fig. 4.11. Two profile scans (crosses and open squares) of the C I 5052 Å line (Jones and Wiese 1984). The solid line (actually two completely overlapping lines) is a least-squares fit synthetic spectrum composed of a Lorentzian profile and the background continuum (two broken lines from two scans of the spectrum). At the bottom of the figure the residual deviations are plotted, which are obtained by subtracting the fitted symmetric spectrum from the data points. These small deviations show characteristic oscillations from the base line, as predicted theoretically (see text).

measured shifts were smaller than calculated by a factor $\sim 2$, for $H_\alpha$, $H_\beta$, and $H_\gamma$ by $\sim 20\%$. The discrepancy for $L_\alpha$ and $L_\beta$ has since been reduced by the inclusion of electron-produced asymmetries and by removing the need for the impact parameter cutoff (Könies and Günter 1994, 1995), which allow to use center-of-gravity shifts also in theoretical predictions. A recent experiment (Böddeker et al. 1993) on $H_\alpha$ at densities from $\sim 10^{18}$ to $10^{19}\,cm^{-3}$ and corresponding temperatures from 5.6 to 10.4 eV gave shifts of up to $\sim 35$ Å, which is $\sim 15\%$ of the measured FWHM width. The measured shifts again seem to be somewhat smaller than calculated using both the Green's function method of Günter, Hitzschke and Röpke (1991) and the semiclassical impact parameter method (Griem 1988a). They increase less than linearly with density, which may be due

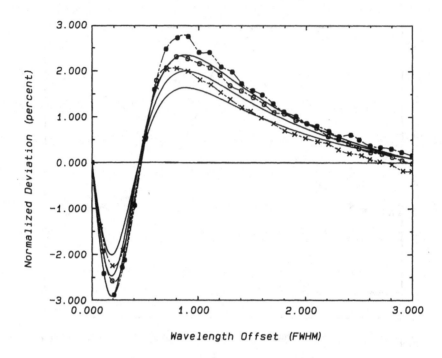

Fig. 4.12. Comparison of the antisymmetric part of the normalized deviation function, basically the ratio of differences of intensities at $\pm\Delta\lambda$ to the mean intensities, from a Lorentzian profile for various Ar I lines (Jones, Wiese and Woltz 1986) as a function of the wavelength offset in units of the Lorentzian full width. The measured lines were Ar I 4511 Å (solid circles), 4272 Å (open circles) and 4259 Å (crosses). The solid lines are calculated for ion broadening parameter values of 0.150, 0.125 and 0.100, whereas the predicted parameter values (Griem 1974) were 0.151, 0.133 and 0.114, respectively.

to dynamical Debye shielding (section 4.4), but in part could also come from the increase in temperature which diminishes predicted shifts. More experimental and theoretical work is needed to verify the significance of these trends, and of the symmetry of the profiles, against experimental and theoretical errors.

Shifts have also been measured for the He II $H_\alpha$, $H_\beta$, $P_\alpha$, $P_\beta$, and $P_\gamma$ lines and have been compared with the same calculations as for hydrogen (Griem 1988a). This comparison already included the He II $H_\alpha$ and $H_\beta$ line measurements of Marangos, Burgess and Baldwin (1988) at $N_e \approx 2 \times 10^{18}\,\mathrm{cm}^{-3}$ and earlier measurements of the He II $H_\alpha$-line and the various Paschen lines. For the He II Paschen-$\alpha$ line, shifts have been measured since up to $N_e = 2.4 \times 10^{18}\,\mathrm{cm}^{-3}$ (Gawron et al. 1989) and $N_e = 3.9 \times 10^{18}\,\mathrm{cm}^{-3}$ (Büscher et al. 1996), respectively. They were found to be typically $\sim 20\%$ smaller than the predicted values (Griem

1988a) and in better agreement with specific calculations presented by these authors. All these measured shifts, of fairly symmetrical profiles, are red shifts, as in the case of hydrogen and in contrast to, e.g., a blue shift reported for He II $P_\alpha$ (Berg et al. 1962). The latter could later be attributed to originally unresolved impurity lines (Berg 1966).

The red shifts measured in the more recent experiments for the Paschen lines mostly agree with calculations (Griem 1988a) within estimated errors of $\sim \pm 25\%$, while the He II $H_\alpha$ measured shift may be about twice the calculated value. The measured values of He II $H_\beta$ line shifts (Marangos et al. 1988) are also larger than calculated values by about the same factor. However, suggestions that this might be due to some additional shift mechanism not allowed for in the shift calculation, e.g., due to the plasma polarization shift proposed by Berg et al. (1962), are probably without foundation relative to almost all more recent shift calculations. This is indicated by the reasonable agreement of the semiclassical results (Griem 1988a) with quantum scattering calculations of the dominant electron impact shifts for ionized helium (Unnikrishnan, Callaway and Oza 1990), and for hydrogen (Callaway and Unnikrishnan 1991, see also Könies and Günter 1995), and by the fact that the quantum calculations account for $\Lambda = 0$, penetrating monopole interactions. They therefore allow for the deviation from charge neutrality in the vicinity of radiating ions, which had been invoked as the reason for plasma polarization shifts relative to the semiclassical, straight perturber path, dipole interactions only, calculations available in 1962. The quantum calculations give, however, a distinctly different temperature dependence of the hydrogen line shifts.

For overlapping lines from more highly charged ions, for which quantum scattering calculations (sections 4.1, 4.2 and 4.7) predict electron-produced red shifts, sufficiently accurate measurements from appropriate dense plasmas turned out to be very difficult. Measurements at densities below $\sim 5 \times 10^{21}$ cm$^{-3}$ for multiply-ionized lines on the whole yielded smaller shifts than predicted by various theoretical models. Moreover, blue shifts reported for C VI resonance lines (Adcock and Griem 1983) were found (Goldsmith, Griem and Cohen 1984) to have been due to errors in the wavelength calibration which, in laser-produced plasmas, is particularly difficult because of Doppler shifts from plasma flows. A recent experiment (Leng et al. 1995) at $N_e \lesssim 10^{22}$ cm$^{-3}$ yielded small red shifts of the C VI $L_\beta$ and $L_\gamma$ lines consistent with calculated (Nguyen-Ho et al. 1986) electron collisional effects. A marked asymmetry of the $L_\gamma$ line (see figure 10.3) can be attributed to the different shifts of the line in plasma regions of various densities. There is also some evidence for about 10 eV downshifts of the $1s^2$ $^1$S-$(1s3p)^1$P line of Ar XVII at $N_e \approx 1.5 \times 10^{24}$ cm$^{-3}$ (Hammel et al. 1993). This would be in good agreement with calculated shifts (Nguyen-Ho et al. 1986, Griem, Blaha and Kepple 1990), but could include some effects of unresolved dielectronic satellites. There is equally good agreement with the calculated shifts of Koenig et al. (1988), whose equation (29) and table VIII scaled from $Z = 10$ to $Z = 18$ results in a shift of 11 eV. Finally, one should at extremely high densities consider the various causes of profile asymmetries as evaluated, e.g., for the $L_\alpha$ and $L_\beta$ lines of hydrogen-like argon by Joyce, Woltz and Hooper (1987) and, more completely, by Kilcrease, Mancini and Hooper (1993).

In F IX and Al XIII $L_\beta$ profile measurements (Leboucher-Dalimier, Poqúerusse and Angelo 1993) at electron densities $\gtrsim 10^{23}$ cm$^{-3}$ asymmetries favoring the red wings have been noted. Besides the usual helium-like ion dielectronic satellites, two additional satellites were interpreted as due to transient $F_2{}^{17+}$ molecules. Calculations for such quasi-molecules with one electron and two bare nuclei (Malnoult, d'Etat and Nguyen 1989) are supporting this interpretation, but internuclear separations corresponding to the satellites are of the same order as the mean ion-ion separation. This suggests that the effects of other ions should be evaluated as well, as discussed in section 4.3, and that the two-center model of Salzmann et al. (1991) may not be appropriate (section 7.1.). Other experiments have been concerned with the broadening and relative intensities of Li-like satellites to the He-like resonance line of Al XII (Peyrusse et al. 1993, Mancini et al. 1994a) and with Li III $L_\alpha$ asymmetries (Jamelot et al. 1990).

So far only two experiments (Baker and Burgess 1977, Chiang and Griem 1978), on the Ar II 4806 and 4848 Å lines, have been reported in which the observed asymmetry or shift was attributed to the perturber bremsstrahlung, line radiation interference effect first discussed by Burgess (1968). More measurements and specific calculations should be made to verify these conclusions.

## 4.11    Effects of plasma wave fields

The dynamical Stark effects discussed so far were implicitly assumed to be caused by the almost random fields associated with the more or less independent charged particles in a plasma. Besides these stochastic electric fields, there are also oscillating fields associated with the various plasma waves, i.e., with the collective behavior of the plasma. Most important of these plasma wave types (Stix 1992, Swanson 1989) are the longitudinal electron plasma, or Langmuir, waves near and above the electron plasma frequency, $\omega_{pe} = (Ne^2/\epsilon_0 m_e)^{1/2}$, especially if their amplitudes are enhanced well above the thermal equilibrium level at the electron temperature. Effects of such plasma oscillations on lines with forbidden Stark components were first considered by Baranger and Mozer (1961), who envisaged a 3-level atom as a plasma probe. Transitions between levels 1 and 0 were responsible for the optical photons, $\hbar\omega_{10}$, while plasmons could cause transitions between the upper level "1" and a nearby perturbing level, "2". Consistent with the selection rules for electric dipole matrix elements (sections 3.1, 3.2 and 3.3), there is then no allowed single-photon transition from "2" to "0". However, there will be two-quantum transitions, one photon and one plasmon, from "2" to "0", with photon energies $\hbar(\omega_{20} \pm \Omega)$, if $\Omega$ is the frequency corresponding to the absorbed or emitted plasmon.

To discuss this interesting effect and its generalization to atoms with different level structures, we now assume a multimode field at the location of the atom or ion represented by

$$\mathbf{F}(t) = \sum_{\alpha,\Omega} \mathbf{F}_{\alpha\Omega} \cos(\Omega t - \alpha) \tag{4.122}$$

and use this field to evaluate the line shape operator $\mathcal{L}(\omega)$ introduced in section 4.1. The $\mathbf{F}_{\alpha\Omega}$ are the electric field amplitudes of modes with frequency $\Omega$ and phases $\alpha$, which are assumed to be random. As in the discussion of Debye shielding in section 4.4, it is convenient to first consider the Laplace transform of the autocorrelation function, i.e.,

$$\int_0^\infty e^{i\Delta\omega' t}\{\mathbf{F}(t) \cdot \mathbf{F}(0)\}_{av}dt = \frac{1}{\epsilon_0} \int d\Omega U_\Omega$$
$$\times \left[\frac{i}{\Delta\omega' + \Omega} + \pi\delta(\Delta\omega' + \Omega) + \frac{i}{\Delta\omega' - \Omega} + \pi\delta(\Delta\omega' - \Omega)\right] \tag{4.123}$$

if we introduce a damping factor $e^{-\gamma t}$ in (4.122) and take the limit $\gamma \to 0$. Moreover, the square of the field amplitude was replaced by the spectral density of the wave electric fields via

$$\int U_\Omega d\Omega = \frac{\epsilon_0}{4} \sum_{\alpha,\Omega} |\mathbf{F}_{\alpha\Omega}|^2. \tag{4.124}$$

The frequency separation $\Delta\omega'$ in these relations is in analogy to that occurring in (4.80)

$$\Delta\omega' = \Delta\omega - \omega_{21} \tag{4.125}$$

if $\Delta\omega$ is the detuning with respect to the allowed line. In other words, $\Delta\omega'$ is the frequency separation from the zero field position of the forbidden Stark component. This fact, together with the $\delta$-functions in (4.123), ensures consistency with the physical picture for the Baranger and Mozer plasma satellites in terms of two-quantum transitions involving $\hbar(\omega_{20} \pm \Omega)$ photons.

Of greatest practical interest are lines and plasma conditions for which the plasma satellites are superimposed on the distant line wing due to electron collisions and ion-produced fields, both evaluated under the assumption that particle-produced fields are not important beyond the Debye length (sections 4.3 and 4.4). By expanding the denominator in (4.38) in terms of $\mathscr{L}(\omega)/\Delta\omega$, the line shape in the region of interest can be approximated by

$$L(\omega) \approx [-1/\pi\hbar(\Delta\omega)^2]\,Tr\,\mathrm{Im}D\mathscr{L}(\omega), \qquad (4.126)$$

which, with (4.79) and (4.123), gives

$$\begin{aligned}
\Delta L(\omega) \quad \approx \quad & \frac{1}{3\epsilon_0}\left(\frac{e}{\hbar\Delta\omega}\right)^2 |\langle 1|\mathbf{r}|2\rangle|^2 \\
& \times [U_\Omega(\Omega = \omega_{21} - \Delta\omega) + U_\Omega(\Omega = \Delta\omega - \omega_{21})]
\end{aligned}$$

$$(4.127)$$

for the contribution from oscillating electric fields. Here we assumed that the vectors $\mathbf{F}_{\alpha\Omega}$ had random directions, and $TrD = 1$ was used to maintain profile normalization, at least approximately. As discussed by Lee (1979) and by Nee (1987), the expansion used here, and therefore the omission of the dispersive (shift) terms in (4.123), can lead to significant changes in the profile structure near the plasma satellites. Another change would occur if one also had to deal with growing plasma waves for which the damping factor used in (4.123) would be inappropriate (Nee and Griem 1976, Nee 1987).

Integration of $\Delta L(\omega)$ over the two frequency ranges centered at $\Delta\omega = \omega_{21} \pm \Omega$ gives the intensities, in terms of the rms fieldstrength $F_{rms} = \langle F(t)^2 \rangle_{av}^{1/2}$,

$$S_{+,-} \approx \frac{3}{8}\frac{\ell_>}{2\ell_1 + 1}\left(\frac{4\pi\epsilon_0\hbar n_1}{mez}\right)^2 \frac{n_1^2 - \ell_>^2}{|\omega_{21} \pm \Omega|^2}F_{rms}^2 \qquad (4.128)$$

of the two satellites relative to the integrated intensity of the allowed "1" $\rightarrow$ "0" line. Also, hydrogen or hydrogenic ion matrix elements corresponding to (3.22) were used, which depend on principal and angular momentum quantum numbers of the upper level of the allowed line, the effective nuclear charge, and $\ell_> = \mathrm{Max}(\ell_1, \ell_2)$. Corrections to these matrix elements can be estimated according to (3.20) and (3.21). Gavrilenko (1993) has derived analytic expressions for these satellite intensities for arbitrary polarization of the observed photons in the case of one-dimensional oscillating fields. Finally, note that $4\pi\epsilon_0$ would not occur in (4.128) and related expressions if cgs units were used.

If, in particular, $S_-$ is no longer small, higher-order terms must of course be considered. They lead to corrections (see chapter II.5 of Griem

1974) which can be written as

$$\frac{\Delta S_{+,-}}{S_{+,-}} \approx \pm S_{+,-} \left[ \frac{\omega_{21}}{\Omega} \mp 1 \mp \left( \frac{2\omega_{21}}{\omega_{21} \mp \Omega} \right)^2 \right] + \cdots. \qquad (4.129)$$

Such calculations also predict satellites at $\pm 2\Omega$ from the allowed line with relative intensities estimated by

$$S_{2\pm} \approx \frac{1}{2} S_+ S_- [(\omega_{21} \pm \Omega) / \Omega]^2. \qquad (4.130)$$

Continuing to still higher order, satellites would also appear at $\pm 3\Omega$ from the forbidden component. However, these higher-order plasma satellites are probably not important under conditions for which the Baranger and Mozer satellites have been observed (see chapter III, section 10 of Griem 1974) and are reasonably well described by (4.128). Truong-Bach and Drawin (1982) have generalized these relations for simultaneous plasma wave and radiation fields.

The spectral structure of the satellites is not only governed by the spectral density $U_\Omega$ in (4.127), but of course is also subject to all the other line broadening effects discussed in the preceding sections. This fact should be kept in mind in attempts to infer plasma wave spectra from high resolution measurements of plasma satellites. Another complication arises from the forbidden Stark component due to low frequency microfields (section 4.3). The relative intensity of this component can be estimated by adding $S_+$ and $S_-$ for $\Omega \approx 0$, i.e., from (4.128)

$$S_f \approx \frac{3}{4} \frac{\ell_>}{2\ell_2 + 1} \left( \frac{4\pi\epsilon_0 \hbar n_1}{mez} \right)^2 \frac{n_1^2 - \ell_>^2}{\omega_{21}^2} \langle F^2 \rangle_{av}, \qquad (4.131)$$

although for neutral ($z = 1$) radiators $\langle F^2 \rangle_{av}$ diverges as $F_{max}^{1/2}$, if $F_{max}$ is determined, e.g., so that the quasistatic shift stays well below $\omega_{21}$. For charged radiators, the second moment of the fieldstrength distribution function is well defined, see (5.23) and Iglesias et al. (1983), but should probably be used in (4.131) only as an upper limit. For lines subject to two perturbing levels, relations due to Gaisinskii and Oks (1987) can be used, although their earlier proposal (1986) to only include so-called quasistatic ions is not only inconsistent, e.g., with (4.128) but also with a unified approach by Astapenko, Kukushkin and Lisitsa (1992).

Another complication may arise from the shift of the allowed line which is caused, in parts, also by the wave fields (dynamical Stark effect). This correction of the usual shifts is according to (4.40) given by the matrix elements of $\mathrm{Re}\mathscr{L}(\omega)$, to which wavefields contribute

$$
\begin{aligned}
\Delta d &= \frac{1}{\hbar}\mathrm{Re}\langle 1|\Delta\mathscr{L}(\omega_0)|1\rangle \\
&\approx \frac{3}{4}\frac{\ell_>}{2\ell_2+1}\left(\frac{4\pi\epsilon_0\hbar n_1}{mez}\right)^2\frac{n_1^2-\ell_>^2}{\omega_{12}^2-\Omega^2}\omega_{12}F_{rms}^2 \\
&= 2(S_+S_-)^{1/2}\omega_{12},
\end{aligned}
\tag{4.132}
$$

if we again use (3.22), (4.79) and (4.123). This is just what one would expect from a quadratic Stark effect, i.e., corresponding shifts from low frequency fields could be estimated by multiplying with $(\omega_{12}^2-\Omega^2)\,\langle F^2\rangle_{av}/\omega_{12}^2F_{rms}^2$. Naturally, electron impact and any Doppler shifts should be considered separately.

Before leaving this subject of plasma satellites to more or less isolated lines, one implicit assumption should be mentioned, namely that of statistical populations of levels "1" and "2" and the absence of other interacting levels. Both of these assumptions may have to be avoided in practical cases, which would then require special calculations for comparisons with measurements.

At sufficiently low electron densities for deviations from statistical populations to occur, laser induced fluorescence (LIF) can be used to increase the sensitivity of plasma satellite measurements (Burrell and Kunze 1972). These authors demonstrated in a model experiment, i.e., substituting microwave photons for plasmons and using a tunable dye laser, that best sensitivity is achieved by tuning the laser across the satellite and forbidden component region. This causes additional population in level "2" and, via electron collisional excitation transfer, also in level "1". The corresponding enhancement of the allowed line gives the actual signal, permitting relatively broadband detection of the induced fluorescence. A smaller increase in sensitivity was achieved by optical pumping of the allowed line, followed by collisional transfer to level "2" and subsequent satellite emission. A common advantage of both methods is the spatial resolution from the intersection of laser beam and observation optics. In low density, low temperature plasma, much greater sensitivity to electric fields is achieved by using various molecular transitions (Moore, Davis and Gottscho 1984, Derouard and Sadeghi 1986, Maurmann and Kunze 1993, Maurmann et al. 1996). In all experiments of this kind, the laser intensity should be kept low enough as not to cause significant changes in the satellite structure (Truong-Bach and Drawin 1982), or these changes should be allowed for in the data analysis.

Depending on the experiment, one may also have to include magnetic field effects (Cooper and Hess 1970, Scott et al. 1970). Another deviation could be connected with the presence of a broad plasma wave spectrum instead of the narrow spectra implied so far. An extreme situation would arise for $U_\Omega$ being constant over a range covered by a line. In such a case, wave fields contribute to the Lorentzian width of the allowed line (Sholin 1961). This can again be seen from (4.38), (4.79) and (4.123), and one should make sure that the real part of $\mathscr{L}(\omega)$ is not too important in the actual case.

For hydrogen, ionized helium, etc., lines, and in general for lines involving degenerate or nearly degenerate levels, the level splittings corresponding to $\omega_{12}$ above become functions of the ion-produced microfield, say, $\omega_{ii''} = C_{ii''}F$. Moreover, one now also requires more matrix elements of the line broadening operator. The wave field contribution to these is from (4.79) and (4.123)

$$\frac{1}{\hbar}\langle i|\Delta\mathscr{L}(\Delta\omega)|i'\rangle = \frac{\pi}{12}\left(\frac{e}{\hbar}\right)^2 \sum_{i'',\pm}\langle i|\mathbf{r}|i''\rangle \cdot \langle i''|\mathbf{r}|i'\rangle$$

$$\times \left[-i|\mathbf{F}_{\Omega'}|^2 + \frac{1}{\pi}P\int_0^\infty \frac{|\mathbf{F}_\Omega|^2 d\Omega}{\Delta\omega - \omega_{i''i} \mp \Omega}\right]$$

$$\equiv -i\langle i|w_w(\Delta\omega)|i'\rangle + \langle i|d_w(\Delta\omega)|i'\rangle \qquad (4.133)$$

with

$$\Omega' = |\Delta\omega - \omega_{i''i}|. \qquad (4.134)$$

The width and shift matrix elements defined on the last line of (4.133) are generalizations of the usual electron impact widths and shifts to which they must be added before insertion into (4.38) for the line shape. Also, $\Delta\omega$ is the detuning from the $i \rightarrow f$ transition, $\mathbf{F}_\Omega$ is already summed over modes, and we still neglect perturbations of the lower levels. Had we used

$$\mathbf{F}(t) = \sum \mathbf{F}_\Omega e^{i\Omega t} + c.c. \qquad (4.135)$$

instead of (4.122) to define the wave field, the prefactor $\pi/12$ in (4.133) would have been $\pi/3$ instead.

For quasistatic fields giving resonance between $\Omega$ and the quasistatic shift of a level the width term in (4.133) is large at $\Delta\omega = 0$, i.e., zero detuning. The generalized line shape according to (4.38) then has a local minimum if $w_w(0)$ from (4.133) is comparable to the usual width from electron collisions, ion-dynamical effects, etc. For $\Delta\omega \neq 0$, one or the other shift term from (4.133) will come into play, leading to a pair of local maxima corresponding to $\mathrm{Re}\mathscr{L}(\Delta\omega) = \Delta\omega$ at $\Delta\omega_m$ obeying

$$\Delta\omega_m \approx \frac{1}{12}\left(\frac{e}{\hbar}\right)^2 |\langle i|\mathbf{r}|i''\rangle|^2 \int_0^\infty \frac{|\mathbf{F}_\Omega|^2 d\Omega}{\Delta\omega_m}. \qquad (4.136)$$

Using again the *rms* field strength and approximate dipole matrix elements one estimates very crudely

$$\Delta\omega_m \approx \pm \left(\frac{\sqrt{3}}{2}\right) \frac{n^2 e a_0}{2z\hbar} F_{rms} \qquad (4.137)$$

for the position of these maxima. Clearly, the error in the integral in (4.136) could be quite large, depending on the actual spectrum of the plasma field.

Since in equilibrium plasmas wave fields are typically only a small fraction of particle-produced microfields $F_p$, namely according to equation (56) of Griem (1974)

$$F_{rms} \approx 0.17(r_1/\rho_D)^{1/2}\overline{F_p} \qquad (4.138)$$

for singly-charged ions, and since corresponding linear Stark shifts are larger than $\Delta\omega_m$ by a similar factor, the interesting profile dip structure discussed here will normally be obscured by the usual line broadening. Nevertheless, such dip structure has been observed on the Ly-$\alpha$ line at $N_e \approx 2 \times 10^{18}\,\mathrm{cm}^{-3}$, $kT \approx 12\,\mathrm{eV}$ (Oks, Böddeker and Kunze 1991), and the separations of the local maxima seemed consistent with estimates of the effects of thermal field fluctuations (Griem 1993). However, after correcting a numerical error in this estimate, one actually finds that the 3 Å peak-to-peak separation observed corresponds to about ten times the thermal wave field, at least according to (4.137). In an earlier measurement (Nee and Griem 1976) similar dips and peaks were observed on two hydrogen $n$-$\alpha$ lines and very recently (Riley and Willi 1995) on very broad $n = 4$ He-like aluminum lines from short-pulse laser-produced plasmas, also interpreted as due to nonthermal fields. The electron density in the short-pulse experiment at the time of peak satellite emission was probably well below $N_e = 3 \times 10^{23}\,\mathrm{cm}^{-3}$. This follows from the $n = 3$ and 5 line widths and raises some questions concerning the interpretation of this experiment.

Detailed line profile calculations (Günter and Könies 1994a) based on the Green's function approach for the electron effects (Hitzschke and Röpke 1988, Günter, Hitzschke and Röpke 1991) and the quasistatic approximation for ions only yield dip structures comparable to those observed if a nonthermal electron distribution function is assumed, with a high energy fraction carrying more than half of the total electron energy. This also corresponds to a much enhanced field fluctuation level near the electron plasma frequency, as should have been anticipated by checking the condition that $w_\omega(0)$ be comparable to the collisional width from (4.46). This width is approximately (Griem, Kolb and Shen 1959,

Griem 1974)

$$
\begin{aligned}
w &= -\frac{1}{\hbar}\mathrm{Im}\langle i|\mathscr{L}(0)|i\rangle \\
&= \frac{8}{3}\left(\frac{e^2}{4\pi\epsilon_0\hbar}\right)^2\left(\frac{\pi m}{2kT}\right)^{1/2}N_e \\
&\quad \times \sum_{i''}\langle i|\mathbf{r}|i''\rangle\cdot\langle i''|\mathbf{r}|i\rangle\ell n(K_{\max}\rho_D)
\end{aligned}
\tag{4.139}
$$

in the notation of (4.82) and only using the leading term. (Again, $4\pi\epsilon_0$ would not occur if cgs units had been used, or it would cancel if the matrix elements were expressed in atomic units.) For thermal fluctuations near and above the electron plasma frequency we estimate

$$
\begin{aligned}
|\mathbf{F}_\Omega|^2 &\approx \frac{2F_{\mathrm{rms}}^2}{\omega_p} \\
&\approx \frac{2r_1}{\omega_p\rho_D}\left(\frac{e}{4\pi\epsilon_0}\right)^2 N_e^{4/3} \\
&= 2\left(\frac{3}{4\pi}\right)^{1/3}\left(\frac{m}{kT}\right)^{1/2}\left(\frac{e}{4\pi\epsilon_0}\right)^2 N_e
\end{aligned}
\tag{4.140}
$$

according to (4.56), (4.61), (4.138) and taking $\overline{F}_p \approx 2F_0$. This is a typical value in dense hydrogen plasmas, as can be seen from figure 4.2. For the thermal plasma wave contribution one therefore obtains from (4.133) and (4.140)

$$
\begin{aligned}
w_\omega(0) &\approx \frac{\pi}{6}\left(\frac{3}{4\pi}\right)^{1/3}\left(\frac{e^2}{4\pi\epsilon_0\hbar}\right)^2\left(\frac{m}{kT}\right)^{1/2}N_e \\
&\quad \times \sum_{i''}\langle i|\mathbf{r}|i''\rangle\cdot\langle i''|\mathbf{r}|i\rangle.
\end{aligned}
\tag{4.141}
$$

The ratio of wave- and electron-collision-produced widths in thermal equilibrium plasmas is finally from (4.139) and (4.141)

$$
\frac{w_\omega}{w} \approx \frac{0.10}{\ell n(K_{\max}\rho_D)}.
\tag{4.142}
$$

The argument of the logarithm, corresponding to the ratio of maximum and minimum impact parameters in the original calculations (Griem, Kolb and Shen 1959), is $\sim 10$ for the conditions of the experiment of Oks et al. (1991), and electron and quasistatic ion contributions are comparable in the $L_\alpha$ profile region of interest. From these considerations and assuming nearly asymptotic behavior of the line shape function, one thus infers a dip in the relative intensity of only $\sim 2\%$ slightly above the electron plasma frequency, which could not have been observed, in particular because it would extend over a frequency region of order $\omega_p/2$ for truly thermal fluctuations.

A fluctuation level at least 10 times thermal is therefore required to explain the observed structure of $L_\alpha$ in the experiment of Oks et al. (1991). Such an enhancement is of the same order as that implied (Krall and Trivelpiece 1986) by the calculation of Günter and Könies (1994a). In their case, high phase velocity waves with frequencies within $\sim 10\%$ of the electron plasma frequency are at $\sim 5 \times 10^3$ the thermal level of the bulk plasma, but have only $\sim 1\%$ of the phase space available for thermally excited waves. The spectrally integrated wave energy density is therefore indeed about 10 times the thermal level for the calculations yielding noticeable dip structure, but not 100 times as would be suggested by the measured structure and the estimate used for (4.137). Moreover, the relatively narrow frequency range of the enhanced field fluctuations is important for the contrast in the calculated dip structure. All of this is consistent with earlier calculations by Lee (1979) for plasmas having bi-Maxwellian electron distributions. For other lines with more and partially overlapping Stark components, any dip structure would be further weakened because of the smooth background of components with different quasistatic shifts. Note also that density-dependent dip structures reported from $z$-pinch (not gas-liner pinch) experiments for the hydrogen $H_\alpha$ line (Finken et al. 1980, Oks et al. 1991) were not confirmed by the gas-liner pinch measurements (Böddeker 1995). These measurements instead exhibited a density-independent 20 Å wide, about 5% deep feature centered about 80 Å below the peak of the line (Böddeker, Kunze and Oks 1995). The proposed interpretation in terms of a transient $(H\,He)^{++}$ molecule, while consistent with the wavelength position is, however, difficult to reconcile with electron collisional broadening and ion-dynamical effects. According, e.g., to the condition (39) of Sholin et al. (1971), the latter effect is more important for the dip structure than for the line as a whole by an order of magnitude, because the transient molecule would form at about one tenth of the mean ion-ion separation. Moreover, the spectrally integrated intensity of the deviation from a smooth line profile contains over 1% of the total line intensity, whereas the proposed theory can account only for less than 0.1%.

The complex interplay between Stark effects from quasistatic and oscillating electric fields was, e.g., addressed by Cohn, Bakshi and Kalman (1972), by Gavrilenko and Oks (1987) and by Yakovlev et al. (1989). It is the main subject of a recent monograph (Oks 1995), together with effects from a combination of various oscillating fields as discussed by Lifshitz (1968). For reviews of earlier experiments concerned with plasma wave field effects, see sections III.10 and IV.3 of Griem (1974). The reader should also realize that most of the wave field effects were discussed here within the generalized impact approximation framework. Such essentially linear response theory description may not be valid for strong and low frequency oscillating fields. Their effects are actually better described in terms of the quasistatic picture, as was done, e.g., to interpret both early (Griem and Kunze 1969, Pal and Griem 1979) and recent observations of nonthermal, low frequency, electric fields (Sarid, Maron and Troyansky 1993, Sarfaty et al. 1995). The rms low-frequency fields can in such cases be estimated by comparison with the usual quasistatic profiles. One first infers a fictitious density and from it a fieldstrength via (4.56), multiplying with a factor $\sim 2$. For specific profile calculations, an argument can be made (Oks and Sholin 1977) that instead of the distribution functions $H(\beta)$ in (4.54), etc., the Rayleigh function

$$H(\beta) = 3(6/\pi)^{1/2}\beta^2 \exp(-3\beta^2/2) \qquad (4.143)$$

should be used. Here $\beta$ is the reduced fieldstrength $\beta = F/F_{\rm rms}$ defined in terms of the rms fieldstrength of the low frequency waves, rather than by (4.55) and (4.56). The waves are assumed to have random phases, and their frequencies must be smaller than the resultant line widths. An alternative approach (Alexiou et al. 1995) is to extract autocorrelation functions from the measured line profiles by inverting (4.15), accounting for all other contributions, e.g., Doppler and electron impact broadening, theoretically.

# 5

# Continuous spectra

In a fully ionized plasma, in which all bound electrons have been removed from their original atoms or molecules, there is besides the continuous spectrum no line emission or absorption, except possibly for features related to plasma resonances or waves (Bekefi 1966, Stix 1992, Swanson 1989). These often nonthermal features are usually at such low frequencies that they do not obscure or interfere with atomic radiation, the subject of principal interest here. Since atoms and incompletely stripped ions possess a continuous spectrum, besides the discrete spectrum providing the pairs of states involved in line radiation, continuous emission and absorption spectra underly and accompany the line spectra discussed in the preceding chapters. These processes are not only important as background to line emission, but also because continuum intensities can provide relatively direct measures of electron density and temperature (see sections 10.2 and 11.4, respectively).

With the usual convention for the energies of bound states as being negative relative to those of zero kinetic energy electrons at large distances from the nucleus of any isolated atom or ion, one might infer that all positive energy states belong to the continuous spectrum. In practice, this is an oversimplification, because there are states corresponding to the excitation of two or more bound electrons which are almost discrete. They can usually decay by spontaneous release of an electron, i.e., by autoionization, but also tend to generate line radiation in the vicinity of lines of the preceding ion in the ionization sequence. These dielectronic satellites can be quite useful for density and temperature determinations (chapters 10 and 11, respectively). Another reason for ambiguity in the relation between the sign of the energy of an electron and the discrete-continuous nature of a state arises from the interactions with other ions and electrons. These interactions give a model characteristic to the usual distinction between negative and positive energy states.

132

Absorption of continuous radiation by boundstate atoms or ions causes photoionization, if the photon energy exceeds the corresponding threshold. Expressions for the corresponding cross section will be derived in the following section as an extension of line absorption. From the photoionization rate we can then obtain the rate of spontaneous and induced continuous emission associated with radiative recombination by using the principle of detailed balance between inverse processes (Tolman 1938, 1979). Numerical calculation of these cross sections is very similar to the calculation of oscillator strengths (chapter 3). Results for photoionization cross sections will be discussed further in sections 5.1 and 5.2, and certain bremsstrahlung cross sections of incompletely stripped ions in 5.3.

Bremsstrahlung emission as such usually occurs as the result of transitions of an electron from one continuum state to another, while this free electron scatters off an ion or atom. Such free-free transitions are therefore a further extension from bound-bound (line) transitions via bound-free (recombination) transitions to free-free transitions. This fact will be used in section 5.4 to derive the corresponding contribution to the continuum emission of a plasma consisting only of fully stripped ions and free electrons. From this emission and the principle of detailed balancing one can obtain the cross section, per incident electron, for free-free absorption, often called inverse bremsstrahlung in the plasma physics literature.

There are several additional continuous emission or absorption processes, including electron-electron collisions in relativistic plasmas, which produce quadrupole radiation (Joseph and Rohrlich 1958), and two-photon decays of excited bound states (Spitzer and Greenstein 1951, Seaton 1955a), which are important only at very low densities. Other possible sources are ion-atom collisions (Mihajlov, Ermolaev and Dimitrijević 1993) and dissociation of molecules (Lishiwa et al. 1985, Kwok et al. 1986) in partially ionized gases with very low degrees of ionization. In most plasma experiments, none of these processes is significant compared to electron-ion recombination and bremsstrahlung. However, ion-atom scattering may contribute comparable continuum emission to that of electron-atom bremsstrahlung (see, e.g., figure 2 of Mihajlov et al. 1993) and there is increased interest in bremsstrahlung, e.g., from energetic proton-atom collisions (Tsytovich and Oiringel 1992). Such protons transfer typically an electron/proton mass fraction of their energy to bound electrons. This dynamical polarization is then responsible for emission or absorption.

The continuous spectra discussed in the present chapter should not be confused with the spectrum of radiation emitted by a nearly homogeneous plasma that is sufficiently extended and dense for most photons emitted in the interior to be absorbed and reemitted several times before they escape. Since strong lines are more affected by this than weak lines or continua, the net result of such radiative transfer (chapter 8) is a smoothing of the spectrum. In the limit of a large but homogeneous plasma with atomic level populations, etc., as in complete thermodynamic equilibrium (chapter 7), this smoothed spectrum approaches that of a blackbody according to Planck's law (2.54). Normally laboratory plasmas do not fulfill all of these conditions, especially not for photon energies near the blackbody maximum.

Another reason for a line spectrum to approach a continuous spectrum is the broadening of spectral lines (chapter 4), which causes higher members of a given spectral series to merge. Often the broadening is mostly due to plasma microfields. The corresponding merging of lines then leads to an extension of free-bound continua to photon energies below the photoionization threshold for isolated atoms or ions. This effect was first analyzed by Inglis and Teller (1939) and will be discussed further in section 5.5.

## 5.1  Photoionization cross sections

The starting point for a quantitative discussion of the photoionization is (2.28), which yielded the transition probabilities (rates) for bound-bound transitions. Using also (2.19), (2.21), (2.34) and (2.38), this gives in the dipole approximation

$$W_{nm} = \frac{\pi e^2 \omega}{3\epsilon_0 L^3} \sum_i |\langle n|x_i|m\rangle|^2 \rho_{E_n} \tag{5.1}$$

for the transition rate if one photon is present initially in the normalization volume. As in (2.46), the operator $-i\hbar\nabla/m_e$ was replaced by $\omega r$, and an average taken over the orientations of the atom. The corresponding photon flux is $c/L^3$. Taking the ratio of $W_{nm}$ and this flux gives for the cross section

$$\sigma_{nm} = \frac{\pi e^2 \omega}{3\epsilon_0 c} \sum_i |\langle n|x_i|m\rangle|^2 \rho_{E_n}, \tag{5.2}$$

with the density of final states in this case corresponding to that of the outgoing electron. In analogy to (2.33), but using $dE = \hbar^2 k \, dk/m_e$, and integrating over the solid angle associated with the electron, this density of states is for given spin

$$\rho_{E_n} = \frac{L^3 m_e k}{2\pi^2 \hbar^2}. \tag{5.3}$$

The factor $L^3$ is compensated by the product of the normalization factors implied for the free electron wave functions, e.g., $L^{-3/2} \exp(i\mathbf{k} \cdot \mathbf{r})$. The remaining factor $m_e k/2\pi^2\hbar^2$ can be included as an additional normalization factor by using $(m_e k/2)^{1/2}\,\pi\hbar\exp(i\mathbf{k}\cdot\mathbf{r})$ as the new continuum function, say, $|n_E\rangle$. Integrating over all space, they are seen to be energy normalized according to

$$
\begin{aligned}
\langle n_E'|n_E\rangle &= \frac{m_e k}{2\pi^2\hbar^2}\int\exp[i(\mathbf{k}-\mathbf{k}')\mathbf{r}]dr^3 \\
&= 4\pi\frac{m_e k}{\hbar^2}\delta(\mathbf{k}-\mathbf{k}')=\delta(E-E'),
\end{aligned}
\tag{5.4}
$$

if we use standard relations for delta functions (see, e.g., Schiff 1955). In other words, for such energy-normalized free electron wave functions the total photoionization cross section is simply as in (5.2), but without the density-of-state factor. More details concerning the normalization of (radial) continuum wave functions can be found in several texts (Bethe and Salpeter 1957, Cowan 1981, Sobel'man 1992).

The modified expression for the total photoionization cross section may be expressed as

$$
\sigma_{nm}=\frac{4\pi\alpha\hbar\omega}{3g_m E_H}S\,\pi a_0^2
\tag{5.5}
$$

in terms of the fine-structure constant $\alpha$, Bohr radius $a_0$, statistical weight of the initial state $g_m$, hydrogen ionization energy $E_H$, and the quantity $S$ which is an analog of the line strength in (3.2), but already in atomic units. The dependence of $S$ on the various angular quantum numbers can be obtained from the corresponding relations for oscillator strengths (sections 3.2 and 3.3). For most applications in plasma spectroscopy, it is sufficient to use the fractional parentage approximation. In that case one has (Sobel'man 1992)

$$
\frac{S}{g_m}=(2L'+1)\left\{\begin{array}{ccc}\ell & L & L_1\\ L' & \ell' & 1\end{array}\right\}^2\ell_>\left(\int R_{m\ell}\,r\,R_{n\ell'}dr\right)^2,
\tag{5.6}
$$

after averaging over fine-structure levels. In this expression, $L'=L,L\pm1$ is the total orbital angular momentum of the final state, ion plus free electron, $L$ that of the initial state. The orbital momentum quantum numbers of the active electron are $\ell'$ and $\ell$ after and before the photoionization, with $\ell_>$ being the larger of the two. The $m$ and $n$ in the radial wave functions $R_{m\ell}$ and $R_{n\ell'}$ are not necessarily the usual principal quantum numbers, but rather suitable labels of the bound and continuum states. Also, $R_{n\ell'}$ is often written as $G_{k\ell'}$ to emphasize that it is an energy-normalized function associated with free electron wave number $k$. After the $6-j$ symbol is, with the help of symmetry relations, written as

$$\sim \left\{ \begin{array}{ccc} \ell & \ell' & 1 \\ L' & L & L' \end{array} \right\}$$

and then replaced by the corresponding Racah coefficient $W(\ell\ell'LL';1L_1)$, (5.6) is seen to agree with (5-10) of Griem (1964). The corresponding coefficients of the radial integrals were tabulated by Burgess and Seaton (1960) [see also tables 5.1 and 5.2 of Griem (1964)]. For equivalent electrons in the outer shell of the target atom or ion, additional factors must be inserted into (5.6), which correspond to the oscillator and line strength factors in section 3.3. Such factors have also been tabulated by Burgess and Seaton (1960).

The implicit assumption of having only valence electrons involved in the continuum absorption must be removed for photon energies high enough to excite or eject also innershell electrons. If relaxation of the core of the remaining ion is neglected, this ion would be in an excited state which is normally degenerate with respect to continuum states. In such cases it is essential to include configuration interactions (see, e.g., Cowan 1981) and to treat the corresponding Fano-type resonances in the photoionization cross sections in considerable detail (Yu Yan and Seaton 1987). An example from this paper for the photoionization of the 4f level of C II, compared with the smooth cross section for the same level in He II, is shown in figure 5.1. Note that the cross section for C II has a minimum falling about a factor 10 below the He II value and a sharp maximum rising several orders above this value. Corresponding line-like resonance features are expected to appear in the inverse process, recombination continuum emission (section 5.4). An example are features in neon and oxygen emission spectra (Elton and Palumbo 1974) which were attributed to radiative Auger effects, in the x-ray spectroscopy notation. The interested reader should consult Cowan (1981) and Yu Yan and Seaton (1987) for discussions of the terminology of the various resonance (interference) effects.

At sufficiently high photon energies, photo-absorption may lead to the simultaneous ejection of two electrons (Krause, Carlson and Dismukes 1968). This process and electron shake-off (Carlson and Krause 1965) therefore generate highly charged ions more rapidly than is usually assumed in plasma modeling as discussed in chapter 6.

### 5.2  Approximate calculations of photoionization cross sections

For hydrogen and hydrogen-like ions, the radial wave functions needed to obtain cross sections from (5.5) and (5.6) are, of course, known exactly

(Bethe and Salpeter 1957), and one can take the sum of the $\ell' = \ell \pm 1$ contributions. Considering also $L' = \ell \pm 1$, $L = \ell$ and $L_1 = 0$ and (5.29) of Cowan (1981) for $6 - j$ symbols with one zero argument, this sum is

$$\sum_{\ell'} \frac{S}{g_{\ell'}} = \frac{1}{2\ell + 1} \left[ (\ell + 1) \left( \int R_{n\ell} r R_{k,\ell+1} dr \right)^2 \right.$$
$$\left. + \ell \left( \int R_{n\ell} r R_{k,\ell-1} dr \right)^2 \right]. \tag{5.7}$$

Now $n$ is the principal quantum number of the initial state, corresponding to $m$ in the earlier relations, and $k$ is used to label the continuum functions. It is the wave number of the emerging electron. The radial integrals are special cases of those occurring in a generalization (Burgess and Seaton 1960, Peach 1970) of the Bates and Damgaard (1949) method for estimates of line oscillator strengths (section 3.2). The generalization is based on the quantum defect method, with quantum defects for continuum states extrapolated from those for bound states in the same series of energy levels. Corresponding formulas and tables of numerical coefficients can be found in the original work and in Griem (1964) and Sobel'man (1992). This method of calculation is probably most appropriate for excited target atoms or ions, as long as the resonance and interference effects already mentioned are not important. For target configurations involving $n_e$ equivalent electrons, i.e., especially for many ground state configurations, estimated cross sections are larger by this factor $n_e$, if the radial integrals are insensitive to total angular momentum quantum numbers of the target and product ions (Sobel'man 1992).

It has become customary (Biberman and Norman 1960, 1963) to compare absorption coefficients obtained from photoionization and related cross sections (Norman 1963) with those for hydrogen or hydrogenic ions, or with a close approximation to these cross sections. The exact hydrogen cross sections, averaged over $\ell$, are

$$\sigma_n^H = \frac{64\alpha}{3^{3/2}} \frac{z^4}{n^5} \left( \frac{E_H}{\hbar\omega} \right)^3 \pi a_0^2 \, g_n^z(\omega), \tag{5.8}$$

where the $g_n^z(\omega)$ are Gaunt (1930) factors multiplying Kramers' (1923) approximate, correspondence principle, formula. The Gaunt factors are typically close to one, as discussed in the following section, and have been calculated analytically by Menzel and Pekeris (1935). They have been tabulated by Karzas and Latter (1961). The cross section in (5.8) is exact for nonrelativistic ejected electrons and one-electron target ions of moderate $z$. It can be reproduced very closely by semiclassical calculations (More and Warren 1991, D'yachkov and Pankratov 1994, Adams et al. 1995), which are readily extended to multielectron atoms and are

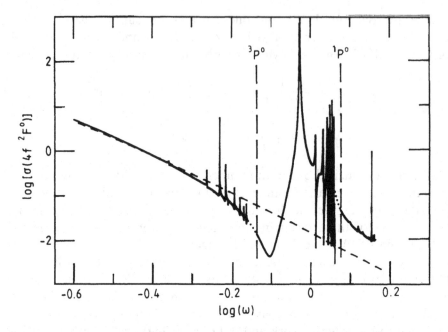

Fig. 5.1. Photoionization cross section of the $2s^2(^1S)4f$ level of C II according to Yu Yan and Seaton (1987) compared with the exact cross section for the 4f level of He II. Note the resonance structure of the C II cross section, which is associated with photoexcitation of the core configuration to $2s2p(^1P^0)$. The angular frequency $\omega$ is in units of $E_H/\hbar$, the cross section in units of $\pi a_0^2$.

computationally very efficient if effective potentials or quantum defects are available. Actually, Kramers (1923) had already used his semiclassical theory to obtain some correction terms in an expansion of the "Gaunt" factor. This is discussed by Hey and Breger (1989), who carried this expansion to higher orders. A detailed discussion of the often surprisingly good accuracy of Kramers' approximation method has been provided by Kogan, Kukushkin and Lisitsa (1992).

In the literature, cross sections for two- and more-electron ions are frequently also expressed in terms of (5.8), except that the Gaunt factor is then replaced by a factor $\zeta$ (the Biberman factor), properly averaged to yield continuum absorption coefficients. As will be discussed in the following section, appropriate weight factors are the statistical weights of the target ion states divided by the statistical weight of the ground state of the resulting ion. Sums of such averaged quantum-defect cross sections for $n = 3$ electrons, in units of the Kramers cross section, i.e., (5.8) with $g_n^z(\omega) = 1$, and multiplied with the statistical weight factor $2n^2$, are shown in figure 5.2 for some light atoms for wavelengths from 2000 to 6500 Å. The exact cross section for hydrogen is also shown, and it

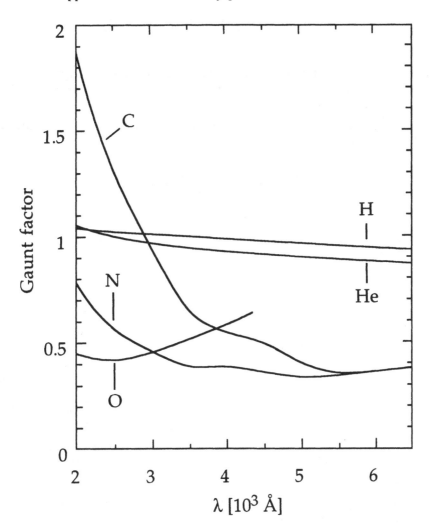

Fig. 5.2. Photoionization cross sections of $n = 3$ levels of light atoms (He I - O I) from quantum-defect calculations in terms of the sum of $n = 3$ Kramers' cross sections, i.e., generalized Gaunt factors. Also shown is the exact cross section for hydrogen; whereas the O I cross section on the long wavelength side of the strong edge at 4350 Å is omitted because it is very close to the N I curve. All other edges were smoothed.

appears that most deviations from hydrogen cross sections are associated with the various thresholds, which affect the results for $\lambda \geq 3000$ Å for C I, N I and O I. The largest deviations at wavelengths below all thresholds occur for C I, by almost a factor of 2. However, the absolute value of the cross section is very small at short wavelengths due to the $1/\omega^3$ factor in (5.8).

### 5.3   Continuum absorption coefficients

The product of target ion density and photoionization cross section, summed over all states which can be ionized at a given photon energy, yields the bound-free contribution to the absorption coefficient $\kappa$, the inverse of the photon's mean-free path due to this process. According to (5.8), this results for one-electron atoms and ions in

$$
\kappa_z(\omega) \;=\; \frac{2^6 \alpha z^4}{3^{3/2}} \pi a_0^2 \left(\frac{E_H}{\hbar\omega}\right)^3 \sum_n \frac{N_n}{n^5} g_n^z(\omega)
$$

$$
\approx 2^9 \alpha \pi \left(\frac{\pi}{3}\right)^{3/2} z^4 a_0^5 \left(\frac{E_H}{kT}\right)^{3/2} \left(\frac{E_H}{\hbar\omega}\right)^3 N_e N_z
$$

$$
\times \sum_n \frac{g_n^z(\omega)}{n^3} \exp\left(\frac{z^2 E_H/n^2 - \Delta E_z}{kT}\right), \tag{5.9}
$$

using the Saha equation (6.24) in the second version to express the target state densities in terms of resulting (completely stripped) ion densities $N_z$, the reduction of the ionization energy,

$$
\Delta E_z \approx z e^2 / 4\pi\epsilon_0 \rho_D' \tag{5.10}
$$

according to (7.30) and the electron density $N_e$. (The validity of this thermodynamic equilibrium relation should be checked against the various criteria discussed in chapter 7.) The second version of (5.9) is particularly suitable for the inclusion of inverse bremsstrahlung, i.e., of absorption involving free-free transitions in the nuclear Coulomb field.

In preparation for this extension, the sum over principal quantum numbers is broken up into

$$
\sum_n (\cdots) = \sum_{n_{min}}^{n_{max}} (\cdots) + \sum_{n_{max}}^{\infty} (\cdots) \approx \sum_{n_{min}}^{n_{max}} (\cdots) + \int_{n_{max}}^{\infty} (\cdots) dn \tag{5.11}
$$

and approximated for high $n$ by an integral. In this manner (5.9) becomes

$$
\kappa_z(\omega) =
$$

$$
2^9 \alpha \pi \left(\frac{\pi}{3}\right)^{3/2} z^4 a_0^5 \left(\frac{E_H}{kT}\right)^{3/2} \left(\frac{E_H}{\hbar\omega}\right)^3 N_e N_z \exp\left(-\frac{\Delta E_z}{kT}\right)
$$

$$
\times \left[ \sum_{n_{min}}^{n_{max}} \frac{g_n^z(\omega)}{n^3} \exp\left(\frac{z^2 E_H}{n^2 kT}\right) + \frac{kT}{2 z^2 E_H} g_f^z(\omega, T) \exp\left(\frac{z^2 E_H}{n_{max}^2 kT}\right) \right],
$$

$$
\tag{5.12}
$$

where $g_f^z(\omega, T)$ is the Maxwell-averaged free-free Gaunt factor. Most importantly, the upper limit of the integral was chosen to correspond to $z^2 E_H/n^2 kT \to -\infty$, i.e., to very large negative binding energies and, therefore, to very large kinetic energies of the free electron.

The choice of $n_{\max}$ is related to the merging of spectral lines, usually because of Stark broadening (Inglis and Teller 1939), and will be discussed in section 5.5. Generally the lowering of the bound-free threshold to energies below the isolated atom (or ion) limits and extrapolation of the bound-free continuum is well justified by the smooth transition, e.g., from line oscillator strengths per boundstate energy interval to the differential oscillator strengths corresponding to the continuum Gaunt factors. The plasma reduction of the ionization energy $\Delta E_z$ in (5.10) (see also section 7.3) is generally different from the effective continuum edge shift $z^2 E_H/n_{\max}^2$, which is usually significantly larger but should never be smaller than $\Delta E_z$. This requirement imposes an upper limit on $n_{\max}$,

$$n_{\max} \lesssim (z^2 E_H/\Delta E_z)^{1/2}. \tag{5.13}$$

Note also that $n_{\min}$ must be large enough for $\hbar\omega$ to exceed the energy difference between levels $n_{\max}$ and $n_{\min}$. For very small photon energies, this condition may not be fulfilled even for $n_{\min} = n_{\max} - 1$. In that case, one has no photoionization except out of levels $n \geq n_{\max}$. Effectively this can be allowed for by using only the inverse bremsstrahlung term in (5.12) with $z^2 E_H/n_{\max}^2$ now determined by $\hbar\omega + \Delta E_z$, such that ionization is possible energetically. (Recall that $z^2 E_H/n_{\max}^2 - \Delta E_z$ is the ionization energy of the level in question.) Corresponding limits for continuum emission will be discussed in the following section.

The Gaunt factors required for accurate calculations of hydrogen or hydrogenic ion continua are shown in figure 5.3. Although the bound-free Gaunt factors $g_n^z(\omega)$ are indeed close to 1, the free-free Gaunt factors $g_f^z(\omega, T)$, which are velocity averages of Sommerfeld's (1931) exact formula, are quite large at relatively high temperatures but small frequencies. The last term in (5.12), corrected for induced emission and dispersion effects (Dawson and Oberman 1962, Basov and Krokhin 1964, Johnston and Dawson 1973) is often called inverse bremsstrahlung in plasma physics. Corresponding Gaunt factors will be discussed in section 5.5. For a discussion of nonlinear effects of intense laser fields, especially in strongly degenerate plasmas, the reader is referred to Polishchuk and Meyer-ter-Vehn (1994a). A classical treatment of inverse bremsstrahlung of intense laser radiation has been provided by Pert (1995); and in highly collisional plasmas with electron collision frequency $v$ a factor of $1/[1 + (v/\omega)^2]$ should be inserted into the velocity integrals for the absorption coefficient (Margenau 1958, Zel'dovich and Raizer 1966, Bekefi 1966, Lee and More 1984, More 1991).

As should be clear from section 5.2, (5.12) cannot be used directly for two- and more-electron targets. One either has to start from *ab initio* calculations of photoionization and (inverse) bremsstrahlung cross sections, or use approximate correction factors, e.g., those shown in figure 5.2. In that case, the sum over $n$ must be broken up again, because generally the correction factors are available only for some lower-lying excited states. If $N_z$ now includes also excited state populations of the resulting ion, a factor $N_i^z/N_z$ should be inserted into the sum over low levels and the Gaunt factors there be replaced by relative cross sections as shown in figure 5.2. Naturally, the actual binding energies must be used and individual thresholds be considered. The sum over higher $n$-states can probably be estimated using the hydrogenic form, assuming that the presence of other bound electrons does not affect the absorption. (In view of the resonance and interference effects discussed in section 5.2, this may be a rather crude approximation.) The use of total rather than groundstate ion densities is also related to the definition of the $\xi$-factor of Biberman and Norman (1960, 1963) which will be discussed in section 5.4.

Not surprisingly, our procedure is very inaccurate for bremsstrahlung absorption at high photon energies. This point was demonstrated recently by Avdonina and Pratt (1993), who presented bremsstrahlung cross sections for various iron and molybdenum ions as function of the initial electron energy, from 5 to 100 keV, in units of the Kramers cross section for the bare nucleus. This quantity therefore corresponds to $g_f^z(\omega, T)$ in (5.12), before the Maxwell average and multiplied with the square of the nuclear charge. At these photon energies, use of the nuclear charge in the last term of (5.12) would probably come to within a factor of 2 of the exact result, even for neutral atoms as targets. (See also Lee et al. 1976 and Gervids and Kogan 1992.) All of these calculations are based on static shielding by bound electrons, an assumption which severely underestimates bremsstrahlung in high energy proton-atom collisions (Ishii and Morita 1984, Tsytovich and Oiringel 1992). Another contribution to bremsstrahlung arises from electron-electron collisions at relativistic energies (Joseph and Rohrlich 1958, von Goeler et al. 1995), for which the usual argument of cancellation of the dipole moments is not valid.

Inverse electron bremsstrahlung on neutral atoms in the thermal range, together with photoionization of negative ions, can give major contributions to continuous absorption coefficients. The best known example is the negative hydrogen ion, $H^-$, with 0.75 eV binding energy for the second electron. The effective absorption coefficient can first be written as

$$\kappa' = \kappa[1 - \exp(-\hbar\omega/kT)]\sigma(\omega)N_{H^-}, \qquad (5.14)$$

where the bracketed factor allows for induced emission (section 8.1). Then one adds inverse bremsstrahlung, which is proportional to the product of

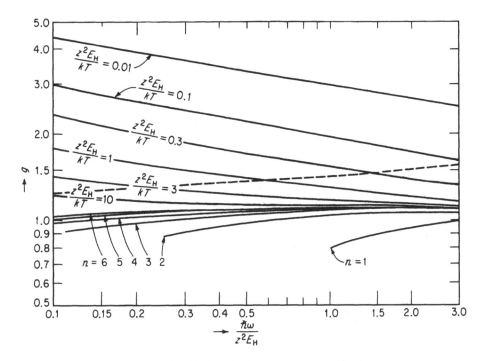

Fig. 5.3. Bound-free and free-free Gaunt factors for hydrogen ($z = 1$) and hydrogenic ions ($z = 2, 3$, etc.). The free-free Gaunt factors are averaged over Maxwellian electron distributions (Karzas and Latter 1961).

atom and electron densities. This product is, according to the appropriate Saha equation,

$$N_H N_e = \frac{1}{2} \left( \frac{kT}{\pi E_H} \right)^{3/2} a_0^{-3} N_{H^-} \exp \left( -\frac{E_\infty}{kT} \right). \qquad (5.15)$$

The total $H^-$ effective absorption coefficient can now be obtained from calculated bound-free (Chandrasekhar 1958, Geltman 1962, Wishart 1979) and free-free (Ohmura and Ohmura 1961, Stilley and Callaway 1970) cross sections. Such effective absorption coefficients per $H^-$ ion, i.e., effective cross sections, are shown in figure 5.4. They are large, so that even relatively small $H^-$ concentrations can be important. In solar and stellar plasmas, inverse bremsstrahlung on He atoms (Sommerville 1967) may also be important. To estimate, especially in case of more-electron ions, relative contributions of bound electrons due to dynamic polarization, one may use equation (11.54) of Kukushkin and Lisitsa (1992), although possible interference between the corresponding processes should be kept in mind.

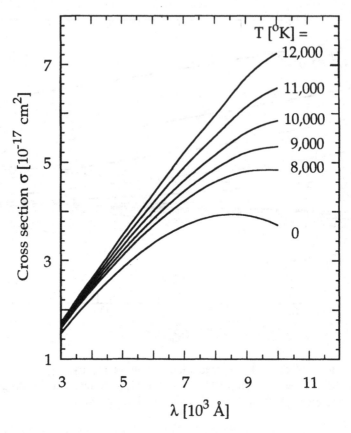

Fig. 5.4. Effective absorption coefficients per H⁻ ion (see text). The zero temperature curve corresponds to the photo-detachment cross section, while the $T = 8,000°$ K, etc., curves also include electron-atom bremsstrahlung and depend on Saha equilibrium according to (5.15).

## 5.4   Continuum emission coefficients

Since we had assumed level populations and ion and electron densities to be according to thermodynamic equilibrium, volume emission coefficients (energy per time, angular frequency, volume and solid angle intervals) accounting for free-bound and free-free emission during electron-ion collisions can be obtained by multiplying the effective absorption coefficient corresponding to (5.12) with the Planck function (2.54). This use of Kirchhoff's law gives

$$\epsilon_z(\omega, T) = \frac{2^5 \alpha^3 z^4}{3(3\pi)^{1/2}} \left(\frac{E_H}{kT}\right)^{3/2} E_H a_0^3 N_e N_z$$

$$\times [\cdots] \exp\left[-\frac{\Delta E_z + \hbar\omega}{kT}\right], \tag{5.16}$$

with the square bracket as in (5.12) or (5.17). Besides being valid only for completely stripped ions, (5.16) requires a Maxwell distribution for the free electrons. For non-Maxwellian distributions, which could, moreover, be anisotropic, one has to start with the basic cross sections, e.g., for bremsstrahlung (Lamoureux, Jacquet and Pratt 1989, Jung 1994). In such cases, the radiation is generally polarized and anisotropic. See also section 12.3.

Returning to Maxwellian plasmas, but now considering target ions that are not completely stripped, it is again necessary to base accurate calculations of continuum emission on cross sections obtained from ab initio atomic structure codes as discussed in the preceding sections and in chapter 3. For approximate calculations, it would be tempting to use (5.16) with the prefactors $z^4$ and $z^2$ of the terms in the square bracket in (5.12),

$$[\cdots] = \sum \frac{g_n^z(\omega)}{n^3} \exp\left(\frac{z^2 E_H}{n^2 k T}\right) + \frac{k T}{2 z^2 E_H} g_f^z(\omega, T) \exp\left(\frac{\Delta E_L}{k T}\right), \quad (5.17)$$

replaced by effective charges and appropriate statistical factors. Also, we introduced

$$\Delta E_L = z^2 E_H / n_{\max}^2 \quad (5.18)$$

for the lowering of the effective photoionization thresholds (advance of the series limits). Following Unsöld (1938), and bearing in mind that the number of edges is larger for nonhydrogenic recombined ions, the first (free-bound) term in (5.17) also may be replaced by an integral, resulting with $g_n^z(\omega) \approx 1$ in

$$[\cdots]_{fb} \approx \frac{k T}{2 z^2 E_H} \left[\exp\left(\frac{E_L}{k T}\right) - \exp\left(\frac{\Delta E_L}{k T}\right)\right]. \quad (5.19)$$

The quantity $E_L$ corresponds to the largest binding energy of contributing recombined electrons, i.e.,

$$E_L = \text{Min}(\hbar\omega + \Delta E_L, E_\infty^{z-1}), \quad (5.20)$$

$E_\infty^{z-1}$ being the ionization energy of the recombined atom (or ion). To allow for deviations from the unit Kramers-Gaunt factor approximation, Biberman, Norman and Ulyanov (1962) introduced a correction factor $\xi_{fb}(\omega, T)$ calculated, e.g., using the quantum defect method discussed in section 5.3 (see also Schlüter 1968). For the corresponding emission coefficients, there is also a statistical, Pauli-blocking factor corresponding to the ratio of actually available final bound states for a given initial ion state to $2n^2$, the degeneracy factor for completely stripped target ions. This statistical factor is especially important, i.e., small, for recombination into groundstates involving equivalent electrons. It naturally also depends on the initial state of the recombining ion.

If one is mostly interested in recombination into excited states, it is reasonable to omit the statistical factor and to use the first version of (5.20). Adding back the bremsstrahlung term, the continuum emission coefficient can now be written as

$$
\epsilon_z(\omega, T) \approx
$$
$$
\frac{16\alpha^3 z^2}{3(3\pi)^{1/2}} \left(\frac{E_H}{kT}\right)^{1/2} E_H a_0^3 N_e N_z \exp\left(\frac{\Delta E_L - \Delta E_z}{kT}\right)
$$
$$
\times \left\{ \xi_{fb}(\omega, T) \left[1 - \exp\left(-\frac{\hbar\omega}{kT}\right)\right] + \xi_{ff}(\omega, T) \exp\left(-\frac{\hbar\omega}{kT}\right) \right\}.
$$

$$(5.21)$$

(This equation must not be used for $\hbar\omega > E_\infty^{z-1} - \Delta E_L$, because this would include recombination into nonexisting states below the ground state.) A relation extending (5.21) to higher frequencies can easily be obtained, using the second version of (5.20) and again (5.16), (5.17) and (5.18). However, such a relation is less useful because of the statistical factors and of the often very different (from Kramers') cross sections. Equation (5.21) must also be modified at small photon energies associated with recombination into merged levels only and with true bremsstrahlung. As discussed following (5.12), the edge shift is then to be replaced by $\hbar\omega + \Delta E_z$, resulting in a cancellation of the exponential factors for the bremsstrahlung term in (5.21). The recombination term vanishes in this case.

Both calculations and experiments (section 5.6) typically give free-bound Biberman factors $\xi_{fb}(\omega, T)$ near one, whereas the free-free Biberman factors $\xi_{ff}$ can be quite large; and this not only for $kT \gg E_\infty^{z-1}$ as in case of the free-free Gaunt factors (figure 5.3). An early example for this are free-free Gaunt (or Biberman) factors calculated by Rozsnyai (1979) for cesium by representing the electron-ion potential in terms of two superimposed Yukawa potentials. His results correspond to $\xi_{ff} \lesssim 4$. Again, the contributions to continuum emission due to dynamically polarized bound electrons may also have to be considered. They can, neglecting interference with ordinary bremsstrahlung, be estimated from equations (1.10) or (1.11) of Amus'ya et al. (1992).

The quantities $\Delta E_L$ and $\Delta E_z$, i.e., the shift of the series limits and the reduction of the ionization energy, are often neglected. Their physical meanings will be discussed next, after remarking that (5.21) and its high and low frequency extensions can be used also to obtain approximate values for effective absorption coefficients, provided Kirchhoff's law is valid. In that case one may reintroduce the target atom or ion densities via the appropriate Saha equation from chapter 7 and will find that other statistical factors appear.

## 5.5   High density effects

There are at least four physical effects to be considered in interpreting continuum emission from dense plasmas. Two of these were already approximately allowed for in the preceding section. First was the reduction of the ionization energy $\Delta E_z$ estimated by (5.10) in terms of the Debye length according to (3.38) and taking the expection value of the second term in the expansion (3.39). Second, we introduced a downshift $\Delta E_L$ of the photoionization thresholds expressed by (5.18) and (5.24a) or (5.24b) below to allow for the merging of spectral lines from line broadening. Before discussing the other effects, it is appropriate to follow Inglis and Teller (1939) in their evaluation of the principal quantum number $n_{max}$ of the state beyond which a continuous spectrum may be assumed even in accurate calculations, i.e., not making Unsöld's (1938) approximation, etc., as discussed above.

The dominant line broadening process for lines near the reduced series limit tends to be Stark broadening by ions; and since the lines are broad, it is usually well estimated by the quasistatic approximation. Also, and this is almost per definition, linear Stark effects are dominant even for two- and more-electron systems, because one is looking for perturbations of the order of $n_{max}$ and $n_{max} \pm 1$ level splittings. Using an estimate corresponding to (4.137) for the linear Stark effect of each of the two "last" lines and assuming hydrogenic binding energies, $n_{max} = n$ is then seen to be defined by

$$
\begin{aligned}
2|\hbar\Delta\omega| \quad &\approx \quad \frac{n^2 e a_0}{z}\overline{F} \\
&\approx \quad \frac{z^2 e^2}{8\pi\epsilon_0 a_0}\left|\frac{1}{n^2} - \frac{1}{(n\pm 1)^2}\right| \\
&\approx \quad \frac{z^2 e^2}{4\pi\epsilon_0 a_0 n^3},
\end{aligned}
\tag{5.22}
$$

where $\overline{F}$ is a suitable value of the ion microfield. For (neutral) atoms, typical fields are about twice the Holtsmark fieldstrength given by (4.78) in terms of the density of perturbing ions, $N_p$, and their charge $z_p$. For radiating ions, of charge $z - 1$ in our notation, ion-ion correlations and screening by electrons can cause considerable deviations from Holtsmark distributions (section 4.3). However, in this case the second moment of the fieldstrength distribution is well defined, and we can use equation (2.5) of Iglesias et al. (1983) to estimate an effective upper limit

$$\overline{F} \approx [N_p k T z_p / \epsilon_0 (z - 1)]^{1/2}. \tag{5.23}$$

From (5.22) and (4.78) or (5.23), respectively, we finally estimate

$$n_{\text{max}}^{\text{atoms}} \approx \frac{z^{3/5} / \sqrt{2}}{z_p^{1/5} (a_0^3 N_p)^{2/15}} \tag{5.24a}$$

for atoms and for ions

$$n_{\text{max}}^{\text{ions}} \approx \frac{z^{3/5}(z - 1)^{1/10}}{[2\pi z_p a_0^3 N_p (k T / E_H)]^{1/10}}. \tag{5.24b}$$

An important proviso is that (5.24a) should be used for ions as well if it corresponds to a larger $n_{\text{max}}$ value than (5.24b) or, for given $n_{\text{max}}$, to a lower ion density. Insertion of these values or, rather, of the next higher integer into (5.18) then gives the effective edge shift. Since broadening by electrons was neglected, these $n_{\text{max}}$ values are probably overestimates, especially for ions whose line wings tend to be dominated by electron effects (Seaton 1995). However, the interference between electron scattering on levels $n$ and $n + 1$ (Baranger 1962, Griem 1974) would tend to deepen the "windows" between the lines, and we are mostly interested in half widths here.

More exact values of $n_{\text{max}}$ are normally not required, because a calculation should include all lines with upper levels below $n_{\text{max}}$ and use realistic line profiles. In principle, one should also use the actual microfield distribution in order to obtain a smooth transition of the continuum emission instead of the sharp edge. However, it is almost certainly sufficient to model this transition by convolving the continuous spectrum obtained, e.g., from (5.12) with the line profile corresponding to $n \approx n_{\text{max}}$ (Iglesias and Griem 1996). If in such model calculation the last line is still well separated, one would simply iterate by increasing $n_{\text{max}}$ by 1, etc. Only in cases where, say, only one or two lines remain, more involved calculations would be desirable (Gurovich and Engel'sht 1977). See also the following section.

The two remaining high density effects are again as for line radiation, namely the refractive index factor as discussed below (2.33), (2.43) and (2.54), and the changes of oscillator strengths or cross sections corresponding to those discussed in section 3.5. For emission, the refractive index factor is $[1 - (\omega_{pe}/\omega)^2]^{1/2}$ if collisions are not important for the damping of transverse waves, and if only free electrons contribute to the refractive index. For absorption, the inverse of this factor must be used in such nearly collisionless plasmas, as discussed near the end of section 2.5.

Any changes of the differential oscillator strengths $df/dE$, where $dE$ is an energy interval in the (free) electron-ion spectrum corresponding to photons between $\hbar\omega$ and $\hbar(\omega + d\omega)$, would have to be calculated analogously to the changes from Debye shielding in bound-bound oscillator strengths estimated in section 3.5. The corresponding reductions were found to be second order in $n^2 a_0 / z \rho_D'$, i.e., in the ratio of orbit size and the Debye length according to (3.38). However, the third term in the expansion (3.39) of the Debye-shielded Coulomb potential clearly causes downshifts of the upper bound levels, which are also of second order in $1/\rho_D'$. The boundstate energies are therefore approximately

$$E_n \approx - \left[ \frac{z^2}{n^2} - n^2 \left( \frac{a_0}{\rho_D'} \right)^2 \right] E_H, \tag{5.25}$$

giving for the energy intervals

$$\Delta E \approx \frac{2z^2 E_H}{n^3} \left[ 1 - \left( \frac{n^2 a_0}{z \rho_D'} \right)^2 \right] \Delta n. \tag{5.26}$$

There is therefore strong cancellation of these two corrections to the differential oscillator strengths. This is consistent with the numerical results of Höhne and Zimmermann (1982) for both the static Debye model and a cutoff Coulomb potential, a more appropriate model for strongly coupled plasmas. Almost complete cancellation seems to occur also in case of other atoms, e.g., for the electron-argon ion bremsstrahlung, as calculated by Lange and Schlüter (1992) using a scaled Thomas-Fermi potential for the ion core combined with a longer range Debye-shielded Coulomb potential.

The insensitivity of differential oscillator strengths and therefore also of the sum of (Stark-broadened) line and continuum intensities to plasma effects as discussed here is further demonstrated by calculating line and continuum emission in a given microfield and then averaging over the field distribution. Such calculations in the appropriate nonspherical potential and including Stark resonances in the continuum were performed by D'yachkov, Kobzev and Pankratov (1990). They show that the total spectrum is essentially as discussed so far, thereby justifying the use of models based on the "principle of spectroscopic stability." However, these authors also imply that the lower levels involved should not be perturbed by plasma screening, etc. This follows because the constancy of differential oscillator strengths is predicated on the fact that the density of oscillator strengths is not affected by effective potential changes outside the region of localization of the lower state. An implicit assumption of this kind was of course also involved in section 3.5 and therefore in (5.25) and (5.26) above. As discussed by More and Warren (1991) and by More (1994), this insensitivity of differential oscillator strengths to perturbations holds both for Debye and other screening models.

To conclude the discussion of oscillator and cross section modifications in dense plasmas, significant reductions in continuum emission due to changes in effective potentials for the active electrons are therefore to be expected only if the principal quantum numbers in (5.24a or b) approach those of groundstate bound electrons. In such cases, the factor $\exp(-\Delta E_z/kT)$ in (5.21) or equivalent expressions may be very important as well. For example, at $N_e = 10^{19}\,\text{cm}^{-3}$, $kT = 1\,\text{eV}$ it gives a factor of $\sim 0.54$ for $z = 1$, corresponding to almost a factor of 2 reduction in continuum emission for given electron and ion densities. The edge shift $\Delta E_L$ in this case is $\Delta E_L = 0.76\,\text{eV}$ from (5.18) and (5.24a). At, say, $N_e = 10^{22}\,\text{cm}^{-3}$ and $kT \approx 100\,\text{eV}$, but $z \approx z_p = 5$, i.e., using (5.24b), we find $\Delta E_L \approx 19\,\text{eV}$ and $\Delta E_z \approx 10\,\text{eV}$. However, at $kT = 1\,\text{eV}$, i.e., $\Delta E_z \approx 30\,\text{eV}$, the relative size of these two quantities is inverted, and the Saha equation and its correction as used here would not be appropriate. The plasma would become strongly coupled, and electron degeneracy may also be important as discussed, e.g., by Cox and Giuli (1968). Probably a description of radiative properties should then follow the example of condensed matter physics applied to disordered materials. Some of these properties will be discussed in chapter 7. Salzmann, Yin and Pratt (1985) have compared photoionization cross sections for dense aluminum plasmas obtained from various models.

For bremsstrahlung from solid and higher density aluminum plasmas, such a calculation of high density effective potentials and effective Gaunt factors by Lamoureux et al. (1987) yields results suggesting reductions by factors $\lesssim 2$ in the continuum emission at $N_e \approx 10^{25}\,\text{cm}^{-3}$. This emission then comes close to bremsstrahlung produced in neutral (cold) aluminum by, say, keV electrons. A consistent computation of the density of states and normalization of continuum wave functions for two models of dense plasmas, ion-sphere and "jellium," resulted in somewhat larger free-free Gaunt factors for aluminum and iron plasmas from the ion-sphere model than from the other model (Rozsnyai and Lamoreux 1990). This jellium model corresponds to the Debye model, except that electron degeneracy is allowed for. Liberman and Albritton (1995) have recently used the local density approximation (Kohn and Sham 1965) for calculations of bound-free absorption by high-density gold plasmas. These calculations show what may be Cooper minima (Fano and Cooper 1968) near 150 eV. A comparison of various approximations for the calculation of 1s electron photoionization cross sections for high density lithium and beryllium plasmas was made by Furukawa (1995), who found large enhancements over hydrogenic values for photon energies above 400 or 600 eV, respectively.

It is necessary to verify that the use of static effective potentials, i.e., neglecting dynamical polarization (Tsytovich and Oiringel 1992), does not lead to changes in the above conclusion concerning the insensitivity of cross sections and the density of oscillator strengths to plasma screening. For long wavelength bremsstrahlung this question was already answered by Dawson and Oberman (1962) whose result suggests a fractional reduction by at most $\sqrt{3}/2\pi g'$. Here $g'$ is the (classical) result for the free-free Gaunt factor (Scheuer 1960, Griem 1964)

$$g' = \frac{\sqrt{3}}{2\pi} \left\{ \ell n \left[ \left( \frac{kT}{\hbar\omega} \right)^2 \frac{kT}{z^2 E_H} \right] - 0.577 \right\}, \qquad (5.27)$$

which is valid if the argument of the logarithm is large. This reduction only occurs close to $\omega_p$ and is in addition to the effect of the factor $[1 - (\omega_p/\omega)^2]^{1/2}$. At $\omega \gtrsim 3\omega_p$, typical reductions in bremsstrahlung emission from the first cause would be only in the percent level, i.e., less important than the refractive index factor. Since Dawson and Oberman (1962) used the frequency dependent dielectric constant for the classical electron gas, it seems safe to infer that dynamical effects are not important except near or even below the electron plasma frequency. Since no effective potentials are used in this calculation, this comparison together with the work of D'yachkov et al. (1990) may also be taken as indications that any plasma effects on line and continuum spectra are not critically dependent on symmetries or manybody effects. However, as pointed out by Barnett, Huttner and Loudon (1992), near the plasma frequency the emission probability is further modified by absorptive effects.

## 5.6  Experiments

Absolute intensity measurements in the visible region (Wiese, Kelleher and Paquette 1972) of the spectrum for pure hydrogen plasmas at temperatures above 1 eV and electron densities $\lesssim 10^{17}\,\text{cm}^{-3}$ agree very well with continuum emission coefficients calculated essentially according to (5.16). This conclusion is true only if $H^-$ and $H_2^+$ continua are added (Roberts and Voigt 1971) as corrections and if line contributions are included, with allowance for extended line wings. Plasma conditions in this experiment were inferred from absolute continuum emission measurements in the near ultraviolet and the visible, from absolute intensities of $H_\beta$ and $H_\gamma$ and from the total pressure, allowing for small deviations from partial local thermodynamic equilibrium (PLTE, chapter 7). Measured spectra obtained from an Abel inversion of the side-on measurements are shown in figure 5.5. They have been compared by D'yachkov, Kobzev and Pankratov (1990) with their calculation of emission coefficients for ions in a given microfield. The agreement was found well within $\sim 10\%$, except between lines where the calculated intensities are $\sim 20\%$ above the measured values. Some or all of this excess calculated intensity may be

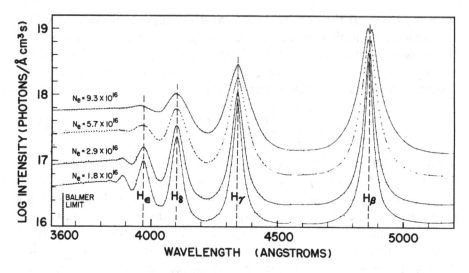

Fig. 5.5. Measured hydrogen Balmer line and continuum spectrum at various electron densities (Wiese, Kelleher and Paquette 1972). The individual data points are connected by curves representing the best fit.

due to an overestimate of the electron impact contribution to the distant line wings beyond the electron plasma frequency (chapter 4). Even better agreement was obtained by Stehlé (1995) using the procedure of Däppen et al. (1987) for the merging of high $n$-levels and line wings calculated using the model-microfield method.

Any remaining doubts concerning the agreement between measured and calculated continuous spectra of hydrogen near and below $N_e \approx 10^{17}\,\mathrm{cm}^{-3}$ because of the use of PLTE relations by Wiese et al. (1972) in the determination of temperatures and densities have been removed since by similar experiments and the use of two-wavelength laser interferometry (Helbig and Nick 1981, Nick, Richter and Helbig 1984). As discussed in section 4.9, these more recent experiments have yielded, e.g., $H_\beta$ line widths as function of electron density consistent with those obtained by Wiese et al. (1972), thus confirming their determination of the electron density from, e.g., line intensity measurements and LTE or PLTE relationships.

At electron densities greater than $N_e = 10^{17}\,\mathrm{cm}^{-3}$, substantial deviations from calculated continuous spectra for hydrogen were reported by Gavrilov, Gavrilova and Fedorova (1985), with reductions in emission by $\lesssim 40\%$ near the Balmer limit at $N_e = 9 \times 10^{17}\,\mathrm{cm}^{-3}$, $T \lesssim 3\,\mathrm{eV}$ in a capillary discharge. On the other hand, a pulsed arc experiment by Radtke and Günther (1986) at very similar conditions, $N_e = 8.4 \times 10^{17}\,\mathrm{cm}^{-3}$, $T \approx 2\,\mathrm{eV}$, appears to agree with calculations (Radtke, Günther and Spanke 1986, D'yachkov, Kobzev and Pankratov 1990). This is particularly interesting because the ion-ion coupling parameter from (4.64) is somewhat larger in the pulsed arc experiment, though still well below unity, namely $\Gamma \approx 0.12$.

In shock-compressed hydrogen-argon mixtures, Fortov et al. (1990), on the other hand, reported a $\lesssim 40\%$ deficiency of the measured continuum absorption near the Balmer limit for similar electron density, but slightly higher temperature, and larger deviations, by factors $\sim 2$, in the extended Balmer continuum at $N_e = 7 \times 10^{18}$, $T \approx 1.3\,\mathrm{eV}$, i.e., $\Gamma \approx 0.34$. Although there could be some doubly-ionized argon in these experiments, which would give somewhat larger $\Gamma$ values, it is not likely that the reduction of the hydrogen continuum absorption is due to any deviations from Debye screening associated with strong coupling effects. More experiments, preferably in pure hydrogen, are needed to settle the question of this "clearing up" at wavelength $\gtrsim 4500\,\text{Å}$ under conditions where only the $H_\alpha$ line remains as a discrete feature. It is interesting to note here that its oscillator strength should be reduced by only $\sim 7\%$ by plasma effects, according to (3.45). This is much less of a reduction than seen experimentally (Fortov et al. 1990), possibly indicating errors in the interpretation of the continuum measurements in terms of absorption coefficients. The plasma frequency in this case corresponds to $\lambda \approx 12\ \mu\mathrm{m}$. Any refractive index effects should therefore be entirely negligible. However, to the extent that the shock-heated plasmas are rapidly ionizing, there may be a depletion of highly excited states below the assumed LTE populations. This effect has been proposed by Hirabayashi et al. (1988a) as a possible explanation for a "transparency window."

Numerous experiments have been concerned with continuum emission of argon plasmas. The measurements for plasma densities up to $1.5 \times 10^{17}\,\mathrm{cm}^{-3}$ have been discussed by Zangers and Meiners (1989) in terms of a simplified version of (5.21), namely

$$\epsilon_z(\omega, T) = \frac{16\alpha^3 z^2}{3(3\pi)^{1/2}} \left(\frac{E_H}{kT}\right)^{1/2} E_H a_0^3 N_e N_z \xi_z(\omega, T), \tag{5.28}$$

i.e., by incorporating all remaining atomic level structure effects into a single Biberman factor $\xi$. The factor $\exp(-\Delta E_z/kT)$ in the usually dominating recombination contribution was omitted, i.e., included in the effective Biberman factor, which typically is $\xi \approx 2$ for the argon experiments. The exponential factor, on the other hand, ranges between 0.95 and 0.90 for most experiments discussed by Zangers and Meiners (1989), including their own measurements. As emphasized by the authors, allowance for line radiation is usually more important. This can also be seen from an earlier experiment by Schnapauff (1968) who measured both argon and neon continuum emission and relative intensities of Ar II lines.

In an experiment with electron densities up to $N_e \approx 1.6 \times 10^{20}\,\mathrm{cm}^{-3}$, Bespalov, Gryaznov and Fortov (1979) observed a decrease in the absorption coefficient below the Kramers ($g = 1$) value by as much as a factor of 2 near 6000 Å at $T \approx 2\,\mathrm{eV}$. The plasma wavelength is near 2500 Å, i.e., refractive effects must be allowed for, and the absorption coefficient must be calculated beginning, e.g., with

$$k(\omega) = \frac{\omega}{c}\mathrm{Im}[n(\omega)], \tag{5.29}$$

which can be inferred from (2.101) and (2.102). Instead of the complex refractive index, it is customary to introduce the dielectric function,

$$\epsilon_d = [n(\omega)]^2, \tag{5.30}$$

for which powerful theoretical methods extending into the strong coupling regime have been developed (Kraeft et al. 1986). Corresponding calculations of inverse bremsstrahlung absorption coefficients by Berkovsky et al. (1993) describe the observed reduction quite well, both for argon and for aluminum (Mostovych, Kearny and Stamper 1990, see also Mostovych et al. 1995). Large decreases in argon continuum emission near the electron plasma frequency were reported by Fortov et al. (1990) at ion-ion coupling parameters $\Gamma = 3.5$.

These authors also measured the continuum emission of xenon plasmas, in this case for a variety of plasma conditions corresponding to $\Gamma$ from 1.2 to 4.6 and $N_e$ from 0.2 to $6 \times 10^{21}\,\mathrm{cm}^{-3}$. The largest decrease in emission occurs at wavelengths slightly above $\lambda = 1\,\mu\mathrm{m}$, i.e., again close to the plasma frequency. However, almost no decrease is seen for $\Gamma \lesssim 3$, presumably because of the relatively large effective collision frequency (Berkovsky et al. 1993). However, not all reported decreases in xenon continuum emission can be explained in terms of effects occurring near the electron plasma frequency, e.g., not the measured decreases by factors $\lesssim 3$ of effective Biberman factors at wavelengths below 1 $\mu\mathrm{m}$ and electron densities $N_e \lesssim 3 \times 10^{17}\,\mathrm{cm}^{-3}$ (Popović and Dordević 1993). The coupling parameter in this case is only $\Gamma \approx 0.15$, so that other than dense plasma effects should be considered to explain the decrease in continuum emission. Turning to hydrogenic ions, e.g., B V Lyman-series lines at $\Gamma \approx 0.07$, a search for a transparency window failed to provide evidence for any substantial deviations from the usual calculations of Stark broadened line and continuum emission (Iglesias and Griem 1996). Actually, as can be seen from figure 3 of this reference, the observed spectra are very reminiscent of the classical hydrogen spectrum shown in figure 5.5.

Another interesting subject is the experimental study of the shift of $K$ photoionization edges in laser-produced, near solid density plasmas (Riley et al. 1989, Schwanda and Eidmann 1992). At lower temperatures, electron degeneracy is important in these plasmas, as is the increase in ion charge at higher temperatures. Also to be considered are the other plasma effects discussed in section 5.5 and any structure near the edges associated with incipient band structure of low energy electrons due to some short-range order of the ions. This effect is analogous to the much-studied EXAFS (extended x-ray absorption fine structure) in solids and has been observed (Hall et al. 1988) and calculated (Gordon 1993, Frey and Meyer-ter Vehn 1995) also in and for plasmas. Finally, in low density regions even of relatively high temperature plasmas, molecular emissions are of considerable importance (Moran 1995).

# 6

# Cross sections and level kinetics

Observable intensities of spectral lines and continua depend just as strongly on the appropriate level populations as on the transition probabilities and related quantities discussed in chapters 2 and 3, and on the radiative transfer processes to be discussed in chapter 8. The level population and transfer problems are not really separable. One must therefore strive for internal consistency, following the example of astrophysicists (see, e.g., Mihalas 1978). Nevertheless, it is reasonable to discuss first the various population and depopulation processes as if these interconnections were not very important. Often only processes affecting discrete states that are not very close to the corresponding continuum limits need to be evaluated explicitly. This important simplification arises because the rates of collisional processes controlling the relative populations of highly excited states tend to be very large, thus facilitating the establishment of so-called partial local thermodynamic equilibrium (PLTE, see sections 7.1 and 7.6). However, the lower excited level populations often need explicit evaluation, especially because these levels tend to give rise to the strongest lines. Note also that electron collisional rates usually, but not always, dominate because of their high velocities and of their energy-dependent cross sections that are similar to those for ions.

The limiting PLTE relations allow us to close the truncated set of equations corresponding to the kinetic models used in practice by invoking LTE relations for relative populations between highly excited levels (figure 6.1), including doubly-excited levels and levels corresponding to the effective series limit, and the population of the appropriate state of the resulting ion times the free electron density. By extension of the assumption for controlling rates between highly excited discrete levels, the energy distributions of free electrons can be assumed to be thermal, i.e., Maxwellian or, if degeneracy is important, to be according to Fermi-Dirac statistics, with even better justification in most situations. Plasmas produced and heated

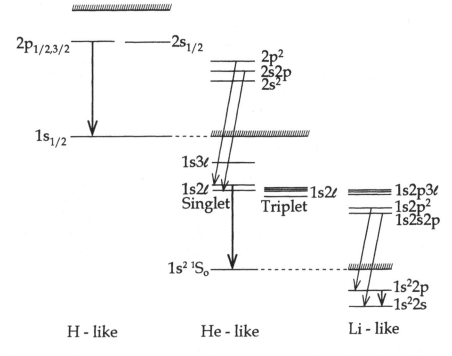

Fig. 6.1. Schematic energy level diagram, adapted from figure 2.1 of Janev, Presnyakov and Shevelko (1985), of one-, two- and three-electron ions. The resonance transitions are indicated by thick vertical arrows, and some of the related dielectronic satellite transitions by slanted thin arrows. The first ionization limits are indicated by the cross hatched lines, but are drawn using compressed energy scales for the He-like and H-like ions.

by intense lasers constitute an interesting exception, as do radio-frequency heated low density plasmas. In such cases, plasma kinetic theory must be used to first determine the electron distribution function (see, e.g., Kruer 1988 or Rubenchick and Witkowski 1991), allowing for its interactions with the ion distribution which may be laser-field dependent (Polishchuk and Meyer-ter-Vehn 1994b). Then one may determine any discrete level populations by using the corresponding nonthermal electron distributions in calculating collisional population and depopulation rates.

An important role in regard to calculating level populations for a given ionization stage of some atomic species, and for relative populations between different stages of ionization, has been played by a variety of theoretical and computer models. Some of these models will be discussed in this and the following chapter. The more the predictions of these models deviate from LTE, the more critically do they depend on the values of collisional and radiative rates. Cross sections for the radiative rates were already derived in chapters 2 and 5, leaving all of the difficulties in the determination of the corresponding photon fluxes or intensities to self-consistent solutions of the radiative transfer problem (chapter 8). For collisional rates, on the other hand, the major difficulty lies in the acquisition of a reasonably complete and accurate set of cross sections. The quantities corresponding to photon fluxes, e.g., the product of electron velocity and density, can usually be treated as given.

The emphasis in the present chapter is therefore on cross sections for important collisional processes (Brown 1959, Shevelko and Vainshtein 1993, Janev et al. 1987, Pal'chikov and Shevelkov 1995, Janev 1995, Drake and Hedgecock 1996) in radiating plasmas. Before attempting to summarize the large body of information concerning these cross sections, some of which are available from the Atomic and Molecular Data Information System of the International Atomic Energy Agency on the Internet (pms@ripcrs0.1.iaea.or.at), it is appropriate to discuss various types of kinetic models and their use in the interpretation of experiments.

## 6.1    Kinetic models

The prototype of a kinetic model for atomic level and charge state distributions is the corona ionization relation originally proposed by Wolley and Allen (1948). Assuming steady state and neglecting any transport of ions, these authors balanced the rate of electron collisional ionization of atoms or ions in charge state $z - 1$ by the rate of radiative recombination with ions in charge state $z$, i.e., per unit volume,

$$SN_a^{z-1}N_e = \alpha N_a^z N_e. \tag{6.1}$$

This results in the corona ionization "equilibrium"

$$N_a^z/N_a^{z-1} = S/\alpha, \tag{6.2}$$

the right-hand side being only a function of the electron temperature, or of the shape of the electron distribution function, and of the structure of the atoms and ions involved, which are all assumed to be in their ground states.

The ionization rate coefficient is in terms of the ionization cross section $\sigma_{ion}$ given by

$$S_a^{z-1} = \langle \sigma_{ion} v_e \rangle, \tag{6.3}$$

$v_e$ being the electron velocity relative to the target atom or ion. The $\langle \cdots \rangle$ indicates the required average over the electron distribution function, the corresponding integrals beginning with the appropriate threshold velocity or energy, i.e., the groundstate ionization energy of ion $z - 1$.

If, as was originally assumed, there is essentially only radiative recombination, then the corresponding rate coefficient $\alpha$ can be obtained by dividing the free-bound continuum emission coefficient in (5.16) by the photon energy $\hbar\omega$ and by the product of recombining electron and ion densities, and by finally integrating over $\omega$. This procedure results for radiative recombination with completely stripped ions of charge $z$ equal to the nuclear charge $Z$ in

$$\alpha_a^z = \left(\frac{\pi}{3}\right)^{1/2} \frac{2^7(\alpha a_0)^3}{3\hbar} z^4 E_H \left(\frac{E_H}{kT}\right)^{3/2}$$
$$\times \sum_{n=1}^{n'} \frac{1}{n^3} \exp\left(\frac{z^2 E_H}{n^2 kT}\right) \int_{\omega_{n\infty}}^{\infty} \frac{d\omega}{\omega} \exp\left(-\frac{\hbar\omega}{kT}\right), \qquad (6.4)$$

if we use the Kramers (Gaunt factor $g_{fb} = 1$) approximation. Normally recombination to the ground state, $n = 1$, dominates in the sum. In general, the threshold frequency in the exponential integral may be expressed in terms of

$$\hbar\omega_{nn'} = z^2 E_H \left(\frac{1}{n^2} - \frac{1}{n'^2}\right) \qquad (6.5)$$

with $n' \gg 1$ chosen to correspond to high Rydberg levels in LTE, i.e., Saha equilibrium, relative to the completely stripped ions and the free electrons as discussed in chapter 7. See the Saha equations (7.24) or (7.25).

For recombination into incompletely stripped ions, the methods discussed in chapter 5 should be used to calculate more accurate radiative recombination coefficients. The dominant effects, however, are Pauli blocking, i.e., the effect of the exclusion principle due to the presence of bound electrons in the various shells, and screening of the nuclear charge by inner electrons. The screening can be accounted for by using for $z$ the ionic charge of the recombining ion and the blocking by extending the sum only over shells that are at least partially open, multiplying each term by a factor $1 - z_n/2n^2$, where $z_n$ is the number of bound electrons in the n-th shell. This factor accounts for the number of vacancies in the partially occupied shells. In any case, it turns out that the radiative recombination coefficients are relatively insensitive to the electron temperature or distribution function. Also, much more important than corrections, including the use of accurate Gaunt factors, to these radiative rates are the inclusion of different recombination processes, especially of dielectronic recombination, which will be discussed in section 6.4, and of three-body recombination. The latter process is the inverse of collisional ionization; it will therefore be discussed together with ionization cross sections in the following section. Note, however, that three-body recombination involves another electron density factor, thus causing the total recombination coefficient to become a function of the electron density.

Returning to (6.2) and anticipating the typical energy dependence of ionization cross sections, it can be said that the major temperature dependence of the corona ionization equilibrium is caused by the essentially exponential dependence, with exponents $-E_a^{z-1}/kT$ for $S_a^{z-1}$ as given by (6.3). Here $E_a^{z-1}$ is the ionization energy of the ground state.

By balancing collisional excitation from the ground state with spontaneous radiative decay, one can also write down corona excitation "equilibrium" relations, namely,

$$\frac{N_m}{N_g} = \frac{X_{mg} N_e}{\sum_{n<m} A_{nm}}, \qquad (6.6)$$

if we now drop superscripts and subscripts $a$. The groundstate density is designated as $N_g$ and excited state densities of the same charge state and species by $N_m$. The $A_{nm}$ are the transition probabilities discussed in chapters 2 and 3, while the collisional excitation coefficients are, in analogy to (6.3),

$$X_{mg} = \langle \sigma_{\mathrm{exc}} v_e \rangle. \qquad (6.7)$$

For excitation energies $E_m > kT$ they are strongly temperature-dependent, similar to the situation for ionization coefficients. In case of corona excitation equilibrium, spectral line intensities are simply proportional to the electron density. However, this model should only be used if the fractional population in excited states is small, i.e.,

$$\sum_{m \neq g}^{m'} N_m/N_g \ll 1. \qquad (6.8)$$

In that case, practically all atoms and ions are in their ground states, as assumed for (6.2). The upper limit $m'$ of the sum in (6.8) essentially corresponds to $n'$ as discussed below (6.5). It tends not to be very critical in the regime appropriate for the corona model.

Two other complications are often much more important, the role of fine-structure splitting of the ground state and that of metastable levels, for which electric dipole radiative decays are forbidden or slow but excitation cross sections are not particularly small. To deal with the former complication, one normally uses a fictitious single ground state with averages according to statistical weights. The underlying assumption is that collisional rates, including ion-ion collisions as discussed in section 6.5, are sufficiently large to establish statistical populations between these closely spaced levels, whose separations are usually much less than $kT$. For the second complication, namely for metastable levels with higher excitation energies, a more physical modification of the corona model is to add electron collisional depopulation rates to the denominator in (6.6).

Early examples for this were models for the $(1s2s)^1S$, $(1s2s)^3S$ and $(1s2p)^3P$ level populations in He-like ions (Elton and Köppendörfer 1967, Kunze, Gabriel and Griem 1968a, Gabriel and Jordan 1972, Fujimoto and Kato 1981, 1984). This modification limits the metastable state populations to the LTE (Boltzmann) fraction, if collisional rates consistent with the principle of detailed balancing are used.

At higher electron densities, but still assuming steady state and spatial homogeneity, the collisional-radiative (CR) model first introduced by Bates, Kingston and McWhirter (1962a) for hydrogen and one-electron ions provides a systematic approach to bridge the gap between the corona model and LTE. These authors considered and solved, for hundreds of levels, the following set of stationary-state rate equations for $m \geq 1$

$$
\begin{aligned}
0 &= \frac{dN_m}{dt} \\
&= \left[ \sum_{m'} X_{mm'} N_{m'} + (\alpha_m + c_m N_e) N_i - \left( \sum_{m'} X_{m'm} + S_m \right) N_m \right] \\
&\quad \times N_e + \sum_{m'>m} A_{mm'} N_{m'} - \sum_{m'<m} A_{m'm} N_m,
\end{aligned}
\tag{6.9}
$$

for excited levels, especially but not exclusively for those not in Saha-Boltzmann equilibrium. The groundstate density $N_1$ and ion density $N_i = N_e/z$ are input parameters, and for high $m$ the system could have been closed according to

$$
\begin{aligned}
N_m &= 2m^2 N_i N_e f(T) \\
&= \frac{2}{z} m^2 N_e^2 f(T),
\end{aligned}
\tag{6.10}
$$

the last factor $f(T)$ being the Saha factor evaluated in section 7.2. This closure and the detailed balancing relations between the (approximate) excitation and deexcitation coefficients, say, $X_{21}$ and $X_{12}$, namely

$$
X_{12} = \frac{g_1}{g_2} \exp \left( \frac{E_{21}}{kT} \right) X_{21}
\tag{6.11}
$$

in terms of statistical weights and excitation energy, ensures the approach to the LTE limit at high densities. The radiative coefficients are the exact hydrogen values, times $z^4$ and statistically averaged over states of given principal quantum number. (Especially for very high $z$ ions, this averaging may be questionable in view of the then appreciable fine-structure splitting of $n = 2$ levels and the relatively small cross sections.) The radiative recombination coefficients are given by the individual terms in (6.4), while $c_m N_e N_i N_e = c_m N_e^3/z$ represents the population rate of level $m$ by three-body recombination. The coefficients $c_m$ are again related to the coefficients $S_m$ for the inverse process, e.g., ionization out of level $m$, by detailed balancing. For one-electron ions, (6.9) should, in principle, be augmented by recombination loss terms and by population terms accounting for ionization of the two-electron ions. Especially important at high densities could be excitation of the one-electron ion associated with dielectronic recombination (Fujimoto and Kato 1982, 1985, 1987).

The system (6.9) is linear in the $N_m$, with the recombination and groundstate excitation and ionization terms providing the inhomogeneous part of the corresponding matrix-vector equation. For given $N_1$ and $N_e$, and therefore also $N_i$, and temperature, the $N_m$ for excited levels can thus be determined by matrix inversion methods. To summarize their results for excited level populations, Bates et al. (1962a) introduce through their equations (22) and (23) effective CR rate coefficients for ionization out of and recombination into $n = 1$ that include excitation followed by ionization, and for recombination an effective rate coefficient that also allows for recombination into $n \geq 2$ followed by radiative and collision-induced decay, etc. These effective ionization and recombination rate coefficients, multiplied with $z^3$ and $z^{-1}$, are shown in figures 6.2 and 6.3, respectively. They are plotted as functions of the scaled electron density $\eta = N_e/z^7$, and with the scaled temperature $T/z^2$ as a parameter. (Note that the ordinate scale on a similar figure 6-4 of Griem 1964 is low by one decade.) The z-scaling originates in the $z^4$ scaling of radiative rates and in an (approximate) $z^{-3}$ scaling of collisional rates at fixed $T/z^2$. This will be discussed further in the following sections. Also plotted are more recent results by Burgess and Summers (1976), who employed different collisional rates. (For hydrogen, such differences are larger, see Johnson and Hinnov 1973.) The important point is that the CR coefficients increase with electron density because of two-step ionization and cascading. As can be gleaned from figure 6.4, even their ratio, which in analogy to (6.2) gives the steady state value of $N_i/N_1$, is a slowly increasing function of $N_e$ for $N_e/z^7 \lesssim 10^{16} \, \text{cm}^{-3}$ before it approaches the LTE (Saha) limit, i.e., $S/\alpha \sim 1/N_e$. Some or all of this overshoot may be due to errors in the collision cross sections used, although more recent calculations with more accurate cross sections support the notion of a significant density dependence of effective rate coefficients, e.g., for neutral carbon (Sasaki et al. 1994). Simple approximate formulas for effective rate coefficients were proposed by Fujimoto (1985) for one-electron ions. They also show the behavior just discussed.

Corresponding calculations of excited state populations under quasi-stationary conditions by McWhirter and Hearn (1963) can, according to equation (20) of Bates et al. (1962a), be expressed as a linear combination of two terms proportional to $N_1$ and $N_i N_e$, respectively, with coefficients $r_1(m)$ and $r_0(m)$. This separation into excitation and recombination contributions is not necessarily transferable to very transient conditions (see

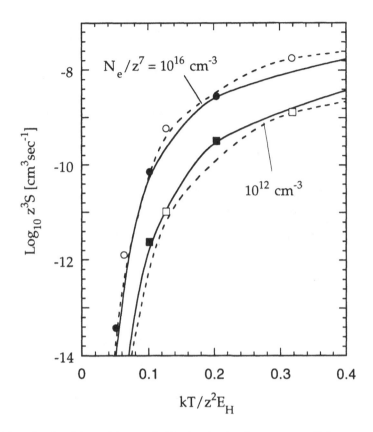

Fig. 6.2. Decimal logarithms of effective ionization rate coefficients, multiplied with the cube of the nuclear charge, for hydrogenic ions according to Bates et al. (1962a) (solid curves), and to Burgess and Summers (1976) (dashed curves). The parameter on the curves is the reduced electron density, $N_e/z^7$, and the independent variable is the reduced or scaled temperature, $kT/z^2E_H$.

Cacciatore, Capitelly and Drawin 1976 and the discussion at the end of this section), and the two coefficients remain functions not only of the temperature but also of the electron density.

In a second paper, Bates, Kingston and McWhirter (1962b) presented results for the same kinetic model, except that transition probabilities, e.g., for all Lyman series lines were set to zero in order to estimate the effects of self-absorption. Normally, these effects were found not to be very important, although they lead to a reduction in effective recombination coefficients.

In all of this work, the sublevels of a given principal quantum number were assumed to be populated according to their statistical weights. At very low electron densities this assumption may not be valid (Hutcheon and McWhirter 1973, Ljepojević, Hutcheon and McWhirter 1984), es-

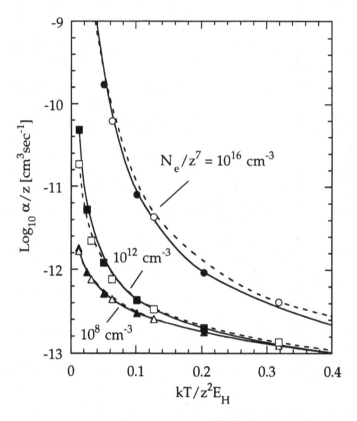

Fig. 6.3. Decimal logarithms of effective recombination coefficients for fully stripped ions of charge $z$, divided by $z$, according to Bates et al. (1962a) (solid curves), and to Burgess and Summers (1976) (dashed curves). The other quantities are as in figure 6.2.

pecially not for the $n = 2$ fine-structure levels. However, the minimum electron density suggested by Hutcheon and McWhirter for collisions to ensure statistical populations is not consistent with relative He II line intensity measurements in a hollow cathode discharge (Kohsiek 1977).

In principle, photo-excitation and -deexcitation and also photo-ionization terms should be added to the system (6.9) if self-absorption is important (Mihalas 1978), together with radiation transport equations for the self-consistent determination of spectrally resolved photon fluxes. These are major complications, except in situations where the radiation field has an essentially continuous spectrum and is prescribed (Abdallah et al. 1993). Two other generalizations of kinetic models are equally important and much more frequently used in laboratory plasma spectroscopy. The first of these generalizations are extensions (Burgess and Summers 1969, Summers 1974, 1977, Fujimoto 1979a, Hess and Burrell 1979, Kawachi, Fujimoto

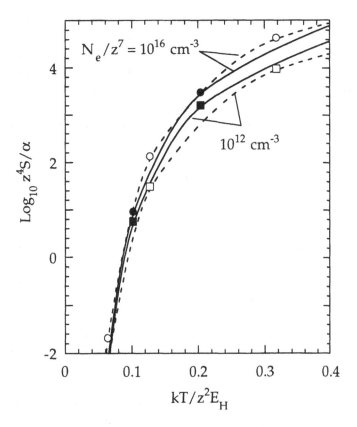

Fig. 6.4. Decimal logarithms of the ratio of ionization and recombination coefficients, multiplied with $z^4$, obtained from the results shown in figures 6.2 and 6.3.

and Csanak 1995, Kawachi and Fujimoto 1995) to two- and more-electron ions or atoms by the inclusion of dielectronic recombination (section 6.4). This extension is actually required already to properly describe hydrogen-like spectra of $Z > 1$ ions, both because the results depend also on the concentration of helium-like, etc., ions and because of unresolved satellite lines involving transitions between doubly-excited states. Such multi-ion kinetic models (see, e.g., Summers et al. 1991 and Summers 1992) require a multitude of atomic data, some of which are described in the following sections, and the use of optimized codes. An important question is how many levels are required for the model to converge reasonably to the appropriate limit. The answer to this question may depend on which spectral features are employed in experiment-model comparisons for the determination of plasma parameters, or which plasma property one wants to describe, e.g., the effects of radiation cooling on x-ray laser gain (Keane and Suckewer 1991). Even for one-electron ions, the questions regarding the number of levels and processes to be included can turn out not to be simple, if fine-structure or $n\ell j$ sublevel populations and ion-ion collisions must be included to make realistic predictions of line spectra (Ljepojević, McWhirter and Volonte 1985, Tallents 1985, Ashbourn and Ljepović 1995). For multielectron ions, such complications often arise already for levels of the ground configuration; and, again, especially proton collisions can be quite effective in populating and depopulating these closely spaced levels (Seaton 1955b, 1964b, Bahcall and Wolf 1968, Landman 1973 a,b, Kastner 1977, Walling and Weisheit 1988). The paper by Tallents (1985) contains results also for transient, namely recombining, plasmas.

This leads us to kinetic models for time-dependent and spatially inhomogeneous plasmas. Instead of system (6.9) or more general steady-state models, one must consider the set of rate equations with $dN_m/dt \neq 0$ and, for inhomogeneous plasmas, add $-\nabla \cdot \mathbf{F}_m$ on the right-hand side. The $\mathbf{F}_m$ are the various particle fluxes, i.e., $\mathbf{F}_m = \langle \mathbf{v}_m N_m \rangle$ involving particle velocities obtained, say, from solutions of hydrodynamic equations describing a given experimental situation. Normally these velocities are assumed to be independent of quantum and charge state and, moreover, distributions over excited states are often assumed to be as in a steady state and homogeneous plasma, so that the full equations only need to be solved for groundstate ion densities, using effective ionization and recombination coefficients plus the corresponding transport terms.

To conclude this section on kinetic models for level populations, some discussion of characteristic time scales for the approach to steady state populations by transient plasmas seems in order. Intuitively, one might assume that this characteristic level lifetime is given by the inverse of the total depopulation rate of a given level, radiative plus collisional, and that the approach to steady state populations is exponential. Especially for highly excited levels, this intuitive picture can be misleading and wrong, because collisional excitation and deexcitation rates are then almost equal. Given the relatively small radiative rates of such highly excited levels, the approach to steady state is more akin to a diffusion process in level space, i.e., slower than one might think and not exponential. (See section 5.4.6 of Sobel'man, Vainshtein and Yukov 1981, 1995 for a Fokker-Planck description of such situations.) These two important points are perhaps best discussed in terms of Ragozin's (1985) transient Green's function $G_{m'}(m, t)$, which gives the fractional population of level $m$ as a function of time for an imposed $\delta$-function excitation of level $m'$ at $t = 0$. The Green's function is a special solution of the time-dependent rate equations, and Ragozin, König and Kuz'micheva (1993) have obtained solutions for hydrogenic models containing up to 560 levels in order to study the dynamics of $m \leq 40$ level populations. Their most important result is that actual relaxation times can exceed the intuitively estimated level life times by an order of magnitude. In a certain sense, this work can be viewed as a generalization of the capture-cascade model of recombination spectra (Pengelly 1964, Pengelly and Seaton 1964). See section 7.6 for further discussions of relaxation effects and note that a Fokker-Planck approach may also be used to model steady-state plasmas (Li and Hahn 1995).

## 6.2 Collisional ionization and three-body recombination

The evolution of a plasma frequently begins with ionization of atoms or molecules by electron collisions. Cross sections and rate coefficients for this process and for further ionization of the initial ions to higher charge states are therefore extremely important for the quantitative analysis of plasmas that are not in LTE. For target atoms or ions in the ground state they can and have been measured by, e.g., crossed-beams or ion-trap experiments, but for ionization of excited atoms or ions, one must rely almost entirely on theory. This is an important point, not only because these processes are the final step in two-step ionization, but also because three-body recombination (see below) is mostly into excited states (D'Angelo 1961) and therefore obtainable from excited state ionization rates by detailed balancing.

Thomson (1912) estimated the ionization cross section by considering the energy exchange between the two electrons involved as if both electrons were free, equating the ionization cross section with the binary collision cross section corresponding to an energy exchange equal to the ionization energy $E_\infty$ of the bound electron. In this manner he obtained, per bound electron,

$$\sigma_{\text{ion}}^{Th} = 4\pi a_0^2 \left(\frac{E_H}{E}\right)^2 \left(\frac{E}{E_\infty} - 1\right) \tag{6.12}$$

for incoming electrons with kinetic energy $E > E_\infty$. This is readily converted into a rate coefficient, assuming a Maxwell distribution, namely

$$f(E) = \frac{2}{\sqrt{\pi}} \left(\frac{E}{kT}\right)^{1/2} \exp\left(-\frac{E}{kT}\right) / kT. \tag{6.13}$$

With $v = (2E/m)^{1/2}$ this gives for the corresponding ionization rate coefficient

$$\begin{aligned} S_{Th} &= \langle \sigma v \rangle \\ &= 8\pi \left(\frac{2E_H}{\pi m}\right)^{1/2} a_0^2 \left(\frac{E_H}{E_\infty}\right)^{3/2} \beta^{-1/2} \\ &\quad \times [\beta e^{-\beta} - \beta^2 E_1(\beta)] \end{aligned} \tag{6.14}$$

in terms of

$$\beta = E_\infty/kT. \tag{6.15}$$

For large $\beta$, the exponential integral can be approximated by

$$E_1(\beta) = \int_\beta^\infty e^{-x}\frac{dx}{x} = e^{-\beta}\left(\frac{1}{\beta} - \frac{1}{\beta^2} + \cdots\right) \tag{6.16}$$

and one obtains (Zel'dovich and Raizer 1966)

$$S_S = 8\pi \left(\frac{2E_H}{\pi m}\right)^{1/2} a_0^2 \left(\frac{E_H}{E_\infty}\right)^{3/2} \beta^{-1/2}e^{-\beta}, \tag{6.17}$$

corresponding to $[\cdots] = e^{-\beta}$ in (6.14). This is twice as large as a semiempirical ionization rate coefficient originally proposed by Seaton (1964a) for near-threshold ionization, but agrees with it in regard to scaling with temperature and ionization energy. For hydrogenic ions and $T \sim z^2$, both (6.14) and (6.17) scale as $z^{-3}$, as already mentioned in the preceding section.

Thomson's (1912) estimate, which was for a weakly bound electron at rest, was refined by Thomas (1927), see also Webster et al. (1933) and Hey (1993), who allowed both for the motion of the bound electron and for the acceleration of the incoming electron by its interaction with, e.g., the hydrogen nucleus. Further improvements of these classical cross sections (Elwert 1952, Gryzinski 1965) not only served to test various quantum-mechanical methods (Seaton 1962), but also suggested semiempirical formulas (Elwert 1952, Drawin 1961, 1963, and Lotz 1967a, 1968, 1969, 1970) that could be adjusted to measured cross sections over a wide energy range, including energies for which the Bethe(1930)-Born approximation is appropriate. These formulas and generalized fitting expressions (Bell et al. 1983, Burgess and Chidichimo 1993, Arnaud and Rothenflug 1985, and Arnaud and Raymond 1992) based on improved theoretical methods (Younger 1981, Kim and Rudd 1994) are now frequently used to predict ionization cross sections for many atoms and ions.

Very useful and rather general expressions for ionization cross sections and rate coefficients are those obtained by Lotz (1967b, 1968) from fits to cross section measurements and extrapolations. Still per electron and writing his expression in the form of (6.17), his semiempirical formula for high-$Z$ one-electron ions is

$$S_L = 8\pi \left(\frac{2E_H}{\pi m}\right)^{1/2} a_0^2 \left(\frac{E_H}{E_\infty}\right)^{3/2} \beta^{-1/2}[0.69e^{-\beta}f(\beta)] \tag{6.18}$$

with

$$f(\beta) = \beta e^\beta E_1(\beta). \tag{6.19}$$

A short table (Sobel'man, Vainshtein and Yukov 1981) for $\beta = 1/4$, 1, 4 and 8, namely $f(\beta) = 0.34, 0.59, 0.83$ and $0.90$, respectively, and comparison with (6.14) and (6.17) suggests that the Thomson formula for the cross section leads to overestimates of ionization rates by factors $\gtrsim 3$ only for $kT > 3E_\infty$, a very unusual situation for ionization of groundstate atoms or ions. Furthermore, Seaton's (1964a) semiempirical formula, i.e., $1/2$ of the rate coefficient given by (6.17), agrees with Lotz's formula (6.18) well within a factor of 2 for all temperatures of interest for ground state ionization. For near threshold ionization of hydrogenic ions, (6.18) also agrees closely with the semiempirical formula proposed by Drawin (1961).

For other than high-$Z$ one-electron ions, (6.18) must be multiplied by a factor $3.0 \times 10^{-3} A \, E_\infty/E_H$, with $A$-factors according to table 3 in Lotz (1967b). These $A$-factors include the number of equivalent electrons, and by introducing averaged ionization energies $\bar{\chi}$, Lotz also allowed for ionization of inner electrons. Further discussions of empirical ionization cross sections and their comparison with measurements and calculations can be found in Lotz (1970), and a collection of ionization energies in Lotz (1973).

As discussed, e.g., by Sobel'man, Vainshtein and Yukov (1981, 1995), the Lotz formula (6.18) is usually closer to the results of *ab initio* quantum mechanical calculation than might be expected. Exceptions are the cross sections from $n > 1$ levels at high energies (Golden, Sampson and Omidvar 1978, Moores, Golden and Sampson 1980). If the Lotz formula is used for innershell ionization, ionization energies must be used which are calculated ignoring the relaxation of outer electrons, corresponding to leaving the resulting ion in a doubly-excited state.

For ionization out of excited states of hydrogenic ions or out of Rydberg levels of any atom or ion, which results in an ion of charge $z$, near threshold ionization rates, e.g., according to (6.17), would scale with $n^4$ because we then have $E_\infty \approx z^2 E_H/n^2$. More often than not this would be an overestimate, because $\beta = z^2 E_H/n^2 kT$ tends to be quite small and $f(\beta)$ according to (6.19) can be approximated by

$$f(\beta) \approx \beta(\ell n\beta^{-1} - 0.577)e^\beta. \tag{6.20}$$

The ionization coefficient therefore becomes, according to (6.18),

$$S_L^{\text{exc}} \approx \frac{8\pi}{z^3} \left(\frac{2E_H}{\pi m}\right)^{1/2} (na_0)^2 \left(\frac{z^2 E_H}{kT}\right)^{1/2} 0.69 \left(\ell n\frac{n^2 kT}{z^2 E_H} - 0.577\right), \tag{6.21}$$

writing the result to exhibit the usual $z$ and $T$ scaling. For high (Rydberg) levels, ionization rates are therefore only increasing as $n^2$, not as $n^4$ as one might infer from semiempirical expressions for near threshold ionization. An even slower increase is predicted by *ab initio* calculations (Moores,

Golden and Sampson 1980; see also Vriens and Smeets 1980). Equation
(6.21) is in the small $\beta$ or $u$ limit equivalent to equations (6.19) and (6.20)
of Drawin and Emard (1977) and is also similar to the formula used by
Bates, Kingston and McWhirter (1962a) for hydrogenic ions, except that
these authors used $e^{-\beta}$ instead of the last factor and had a numerical
coefficient smaller by a factor $\sim 1.7$. Their use of the same formula also
for $n = 1$ may not have led to larger errors either, judging from (6.18)
which has an additional factor $\beta^{-1} = kT/E_\infty$. It is typically well below
1 for groundstate ionization, probably overcompensating for the above
factor of 1.7.

As a matter of fact, a number of experiments in transient plasmas
(Kunze 1972, Griem 1988a, Wang et al. 1988) have suggested ionization
rate coefficients larger than those given by (6.18) or, rather, by its nonhydrogenic version discussed above. This should only in part be due to
the focussing of incoming ions by the long-range Coulomb interaction
with target ions. Ionization cross sections calculated in the Coulomb-Born
approximation (Rudge and Schwartz 1965, 1966) suggest that this effect
can be allowed for by multiplying, e.g., (6.18) with a factor

$$F = \left(1 + \frac{kT}{E_\infty}\right)^{1/2}. \tag{6.22}$$

This factor corresponds to the ratio of the $G_i(\beta)$ factors for ions and
atoms in Sobel'man, Vainshtein and Yukov's (1981) parameterization of
calculated ionization coefficients. In their chapter 6, these authors also
give the angular momentum factors required for ionization from specified
initial states, whereas the coefficients discussed here imply averages over
such states. Sobel'man et al. (1981) do not use the logarithmic factor,
which is in accordance with theory (Bethe 1930, Rudge and Schwartz
1965, 1966) at high temperature, but approximate this behavior by a
constant.

For more detailed discussions of calculations and measurements of
electron collisional ionization, up to hydrogen- and helium-like uranium
(Marss et al. 1994), the reader is referred to the original literature, with
bibliographies provided by Itikawa et al. (1984) and by Itikawa (1991),
and more recent measurements of ionization cross sections summarized
by Gregory and Bannister (1994). Examples of unusual processes required
for accurate kinetic models are excitation-autoionization (Pindzola, Griffin
and Bottcher 1986, Griffin, Pindzola and Bottcher 1987) of magnesium-
like ions, and double ionization of barium (Tinschert et al. 1991) and of
Kr (Lablanquie and Morin 1991) ions. The former ionization process may
to some extent be represented by Lotz's empirical formula, although the
state of the resulting ion could be different. Innershell ionization quite
commonly leads into particular excited states; and the corresponding

cross sections (Henry 1981) are thus essential for accurate modeling. Rate coefficients for this process have been calculated, e.g., for the excitation of He-like (Sampson and Zhang 1988), Li-like ions (Itikawa, Kato and Sakimoto 1995), Ne-like (Sampson and Zhang 1987, Goldstein et al. 1987) and Ni-like (Hagelstein 1986, Sampson, Zhang and Mohanty 1988) levels.

Leaving any further discussion of complex collision processes to the following sections, we now return to the ionization of excited states, for which the small $\beta$ limit of (6.18), i.e., the ionization rate coefficient (6.21) should be applicable. In LTE, this rate would be balanced by three-body recombination and we have

$$c_n N_i' N_e^2 = S_n N_n' N_e \qquad (6.23)$$

in the notation used in (6.9), except for $m \to n$, and indicating by primes that resulting ion densities $N_i$ and densities in the principal quantum number $n$ state of the recombined ion are related by the Saha equation (chapter 7, equation 24), namely

$$\frac{N_i' N_e}{N_n'} = \frac{1}{2n^2} \left( \frac{kT}{\pi E_H} \right)^{3/2} a_0^{-3} \exp \left( -\frac{z^2 E_H}{n^2 kT} \right). \qquad (6.24)$$

This assumes that the level structure is close to hydrogenic and that the statistical weight of the recombining ion is cancelled by a corresponding factor for the recombined ion. The three-body recombination coefficient, which must be multiplied with $N_e^2$ and $N_i$ to obtain the rate, is finally

$$
\begin{aligned}
c_n &= S_n \frac{N_n'}{N_i' N_e} \\
&= \frac{(4\pi)^2 n^4}{z^6} \left( \frac{2E_H}{m} \right)^{1/2} a_0^5 \left( \frac{z^2 E_H}{kT} \right)^2 \\
&\quad \times 0.69 \left( \ell n \frac{n^2 kT}{z^2 E_H} - 0.577 \right) \exp \left( \frac{z^2 E_H}{n^2 kT} \right).
\end{aligned} \qquad (6.25)
$$

This expression differs from the estimate in equation (6-87) of Griem (1964) by a factor of $0.69(\cdots)/n^2$, which is quite small for large $n$, almost regardless of the value of the logarithm in $(\cdots)$. One reason for this is that the earlier estimate was for an effective (section 6.1) rate coefficient into the ground state, involving therefore a summation of the contributing elementary processes. Moreover, the principal quantum number $n'$ in this effective rate coefficient corresponds to the collision limit (Wilson 1962), not to some particular final state. A more meaningful comparison would be with specific rates used, e.g., by Bates, Kingston and McWhirter (1962a). Comparison of (6.21) with their expression (37) for ionization of one-electron ions then shows that the three-body recombination coefficient suggested here is actually larger by a factor of $1.7(\cdots) \exp(z^2 E_H / n^2 kT)$ than that adopted for $z > 1$ ions in early kinetic models. Corresponding factors between effective recombination coefficients may well be closer to 1, because collisional deexcitation is then the controlling rate process. For a further discussion of effective ionization rate coefficients, see Sasaki et al. (1994). Specific calculations of ionization out of and three-body recombination into various states of atomic oxygen were performed by Chung, Lin and Lee (1994). Strong-coupling calculations of effective ionization and recombination coefficients for very dense hydrogen plasmas were made by Schlanges and Bornath (1993) and Bornath and Schlanges (1993) for ground and excited states, respectively, and for dense alkali plasmas by Bornath et al. (1994).

Although electron collisions are mostly responsible for ionization, strong laser-fields and other electric fields can also contribute to the removal of bound electrons, see, e.g., Penetrante and Bardsley (1991), Geltman (1994) and Kato, Nishiguchi and Mima (1994). High velocity atoms moving across magnetic fields might be ionized in the $\mathbf{v} \times \mathbf{B}$ fields, especially if they are in excited states.

### 6.3    Collisional excitation and deexcitation

As already indicated by the discussion of simultaneous collisional ionization and excitation in the case of two- or more-electron target atoms and ions, there is considerable similarity between excitation and ionization. One important example for this is the high electron energy behavior of cross sections for electric dipole allowed transitions (Bethe 1930), namely $\sigma \sim E^{-1} \ell n E$. This behavior can, e.g., also be gleaned from the semiclassical expressions used in line broadening calculations (section 4.7). Other relationships (Read 1984, Hammond et al. 1985) can be used to infer (Fujimoto and McWhirter 1990) excitation rates for hydrogen from measured ionization rates (McGowan and Clark 1968).

The experimental and theoretical study of collisional excitation and deexcitation is now a major subfield of atomic physics and has therefore been the subject of several reviews. Measurements of excitation cross sections (Crandall 1983) are generally more difficult than ionization measurements, but have contributed considerably to a better understanding, e.g., of the effects of resonances. The body of theoretical work as reviewed by Henry (1981) and extensively discussed by Sobel'man, Vainshtein and Yukov (1981, 1995) and in Drake and Hedgecock (1996) is very large and cannot possibly be discussed in any detail here. For our purposes it is appropriate to begin with Van Regemorter's (1962) semiempirical formula for electric dipole transitions,

$$\sigma_{mn} = \frac{8\pi^2}{\sqrt{3}}a_0^2 f_{mn}\frac{E_H^2}{E_{mn}E}\,\bar{g}. \tag{6.26}$$

Here $f_{mn}$ is the absorption oscillator strength (section 2.5 and chapter 3) for the $n$ to $m$ transition and $E_{mn}$ is the difference $\Delta E$ between the two energy levels. For deexcitation, the emission oscillator strength (section 2.6) must be used and the absolute value of $E_{mn}$. For electron energies $E < \Delta E$, the excitation cross section vanishes, but not the deexcitation cross section.

The last factor in (6.26), often called the effective Gaunt factor or g-bar, is a function of $(E/\Delta E)$. It was originally estimated by comparison with other data to be $\bar{g} \approx 0.2$ at threshold, at least for ions as targets, for which already the Coulomb-Born approximation leads to finite cross sections at threshold (Seaton 1962). However, various calculations (Bely 1966 and Blaha 1969a) soon suggested larger threshold values. For neutral atoms as targets, the cross sections increase only gradually, corresponding to $\bar{g} = 0$ at threshold. Suggested $\bar{g} \equiv \gamma(u)$ factors as function of $u = E/\Delta E$ can be found in table 5.1 of Sobel'man, Vainshtein and Yukov (1981) for $z = 1$ (atoms) and $z > 1$ (ions), together with a formula for the corresponding excitation rate coefficients and a short table (5.2) for a function $p(\beta)$ arising from the Maxwell average. Note, however, that this table should not be used for $\Delta n = 0$ transitions, for which threshold values are close to 1 (Davis, Blaha and Kepple 1975, Younger and Wiese 1978).

For near threshold excitation, i.e., $\beta = \Delta E/kT \gg 1$, it is reasonable to use a constant effective Gaunt factor for ions as targets. In this approximation the excitation rate coefficient becomes

$$X_{mn} = \langle \sigma_{mn}v \rangle$$

$$\approx 16\pi \left(\frac{2\pi E_H}{3m}\right)^{1/2} a_0^2 f_{mn}\frac{E_H}{\Delta E}\left(\frac{E_H}{kT}\right)^{1/2}\bar{g}\exp\left(-\frac{E_{mn}}{kT}\right), \tag{6.27}$$

and comparison with equation (5.1.49) of Sobel'man et al. (1981) suggests that this simple expression may be used up to $kT \approx \Delta E$, i.e., for most situations where excitation from the ground state of an ion is important. The corresponding expression for deexcitation is even simpler, because the exponential factor disappears upon detailed balancing, and the statistical weight factors are accounted for by the use of emission oscillator strengths, as can be seen from (2.51). Bates et al. (1962a) used an expression corresponding to (6.27) with $\bar{g} = 0.4$, whereas Drawin and Emard's (1977) equation (33a) corresponds to a different cross section with an additional factor $E/\Delta E$. Their expressions (28)–(33) for the excitation of hydrogen contain this factor and an additional factor $(E - \Delta E)/E$ to account for the different behavior near threshold.

Subsequent measurements of cross sections (Crandall 1983) and of rate coefficients (Griem 1988b) and many *ab initio* calculations (Henry 1981) confirmed that the threshold value of the effective Gaunt factor is often substantially larger than 0.2, in particular for transitions not involving a change in principal quantum number for which $\bar{g} \approx 1$ is more typical. For the excitation of neutral atoms, (6.27) should not be used, however. Various formulas have been proposed by Drawin (1968), Mewe (1972) and Gryzinsky (1959, 1965) that can be used for estimates in the absence of calculated or measured atomic data. For excitation from low levels of hydrogen, Johnson's (1972) semiempirical formula continues to be useful, while for excitation from highly excited states calculations by Vriens and Smeets (1980) are recommended.

Except for the effective Gaunt factor, the cross section (6.26) is essentially the same as that obtained by making the Born and dipole interaction (multipole $\Lambda = 1$) approximations (Hey and Breger 1982), i.e., as that obtained by Bethe (1930). Assuming the dipole and Born approximations to be valid for all collisions, comparison of Bethe's result with (6.26) then gives for large electron energies

$$\bar{g} = \frac{\sqrt{3}}{\pi} \ell n \left( \frac{E}{\Delta E} \right).  \tag{6.28}$$

Actually, the dipole approximation is not necessarily valid for close collisions, and higher order Born terms may be important as well. The latter are frequently accounted for by a unitarization procedure (Seaton 1962) and are most important for excited ions or atoms as collision targets.

To avoid overestimates from an inappropriate use of the dipole approximation, Bethe (1930) had introduced a parameter $q_0 \approx 1/R_0$, $\hbar q_0$ being the maximum momentum transfer consistent with the dipole approximation and $R_0 \approx n^2 a_0/z$ the minimum distance for this approximation to be used. For large collision energies this leads to

$$\bar{g} \approx \frac{\sqrt{3}}{\pi} \ell n \left[ \frac{(E/E_H)^{1/2}}{(R_0/a_0)(\Delta E/E_H)} \right].  \tag{6.29}$$

The argument of the logarithm corresponds to the ratio of maximum and minimum impact parameters used in semiclassical calculations of cross sections (Seaton 1962) or electron collisional broadening (Griem et al. 1962, see also chapter 4). These analogies suggest to also consider $R_c \approx (z-1)e^2/mv^2$ as the Coulomb cutoff radius, $R_u \approx \hbar n^2/zmv$ as the minimum distance to avoid violations of unitarity and, of course, $R_\lambda \approx \hbar/mv$ for $n^2 < z$ when the de Broglie wavelength is the larger quantity. From these considerations and using (6.13), (6.26), (6.27) and (6.29), the Maxwell average of the effective Gaunt factor for $E \gg \Delta E$

collisions is finally

$$\bar{g} = \frac{\sqrt{3}}{\pi} \exp\left(\frac{\Delta E}{kT}\right) \int \exp\left(-\frac{E}{kT}\right) \ell n \left[\frac{(E/E_H)^{1/2}}{(R_0/a_0)(\Delta E/E_H)}\right] \frac{dE}{kT}, \quad (6.30)$$

assuming $R_0 \approx n^2 a_0/z$ to be the relevant quantity.

Since (6.29) should only be used if the argument of the logarithm is large, it seems appropriate to neglect its energy dependence. Replacing $E$ in the various expressions for the argument in terms of $R_0$, $R_c$, $R_u$ and $R_\lambda$, say, by $kT$, suitable $\bar{g}$ values are therefore

$$\bar{g} \approx g_{th} + \frac{\sqrt{3}}{\pi} \ell n [\cdots]_{\min}. \quad (6.31)$$

Here we must choose

$$[\cdots]_{\min} = \text{Min} \left(\frac{(kT/E_H)^{1/2}}{(n^2/z)(\Delta E/E_H)}, \frac{(kT/E_H)^{3/2}}{(z-1)(\Delta E/E_H)}, \frac{kT/\Delta E}{n^2/z}, \frac{kT}{\Delta E}\right), \quad (6.32)$$

depending on which of $R_0$, $R_c$, $R_u$ or $R_\lambda$ give the largest minimum distance for our approximation to be valid. We also add a threshold value $g_{th} \approx 1$ to account for close collisions in the semiclassical sense, i.e., low angular momentum partial waves of the incoming electrons.

Use of (6.31) and (6.32) in (6.27) results in excitation rate coefficients for the frequently most important $n \to n \pm 1$ transitions that agree for $kT/z^2 E_H \gtrsim 1$, i.e., for reasonably large values of the required smallest argument according to (6.32), to within $\sim 20\%$ with the quasiclassical, i.e., high $n$ rate coefficients according to tables 3.5 (hydrogen) and 3.6 (hydrogenic ions) of Sobel'man, Vainshtein and Yukov (1981). For hydrogenic ions, similar agreement is obtained even down to $kT/z^2 E_H = 0.16$. Otherwise, cross sections and rate coefficients according to equations 6.1.7 and 6.1.8, respectively, of Sobel'man et al. (1981) are preferable, together with their tables 6.4–6.8. As discussed by Golden and Sampson (1971), effective Gaunt factors for $\Delta n = 1$ transitions are not necessarily the best choice for $\Delta n \geq 2$ transitions. These authors also provide results for cross sections and rate coefficients between states specified by both principal and orbital angular momentum quantum numbers.

More accurate cross sections, especially near threshold, are obtained by close-coupling calculations, which were already mentioned in section 4.7. The improved accuracy, e.g., for 2s-2p excitation in Li-like oxygen, is supported by merged beam measurements (Bell et al. 1994). For a review of various calculations and measurements of the excitation, e.g., of nitrogen and nitrogen-like ions, the reader is referred to Kato (1994). A series of papers has been edited by Lang (1994), containing extensive reviews of data for excitation cross sections and rate coefficients. For further review papers, see Drake and Hedgecock (1996) and for bibliographies Itikawa et al. (1984) and Itikawa (1991).

The emphasis so far has been on collisional transitions corresponding to electric dipole transitions, although it has been long known (Seaton 1962) that $\Lambda \neq 1$ terms in the multipole expansion of the electron-target interaction Hamiltonian can be quite important. The monopole ($\Lambda = 0$) term in this expansion is responsible, e.g., for the 1s-2s excitation in hydrogen and hydrogenic ions. This cross section is comparable to that for 1s-2p, i.e., dipole ($\Lambda = 1$) excitation in the near-threshold region. Large predicted $\Lambda = 0$ cross sections, e.g., for 2p-3p excitations were the basis for proposals and early successful experiments on soft x-ray laser generation (Elton 1990) from plasmas containing neon-like ions. In subsequent experiments, nickel-like ions were excited via 3d-4d ($\Lambda = 0$ and 2) transitions. Many of the required cross sections have been calculated in a relativistic, distorted wave approximation by Hagelstein (1986), Hagelstein and Jung (1987), Hagelstein and Dalhed (1988), Bar-Shalom, Klapisch and Oreg (1988), Zhang and Sampson (1993), and others. Various measurements of relative line intensities of neon-like Cl VIII and Ar IX (Elton et al. 1989, 1990, Preissing et al. 1993) in well-diagnosed $\theta$-pinch plasmas are in reasonable agreement with the corresponding excitation rate coefficients.

For excitation from the ground state, small angular momenta of the incoming electron contribute most to the cross sections, making exchange of free and bound electrons an important process. This effect is included in most *ab initio* calculations (Seaton 1962, Henry 1981, Sobel'man et al. 1981) and accounts for the fact that, e.g., collisional excitation from the $(1s^2)^1S$ ground state of helium-like ions to the $(1s2p)^3P$ level is about as likely as that to the $(1s2p)^1P$ level (Itikawa, Kato and Sakimoto 1995). This explains why in relatively low density, e.g., solar corona, plasmas, resonance and intercombination lines of these ions have comparable intensities, even though their transition probabilities differ by large factors.

As already mentioned near the end of section 6.1, excitation or de-excitation between fine-structure levels, including those of the ground configuration, can be very important for multielectron atoms or ions for conditions where the usual assumption of statistical populations may not be appropriate. Corresponding ion-ion collisional rate coefficients were already discussed and referenced above. Electron collisional rates tend to be of the same order, and calculated rates for quite a number of systems can be found in the astrophysical literature (Seaton 1955a, Saraph, Seaton and Shemming 1969, Blaha 1969b, Petrini 1969, Pradhan 1978).

Resonances in the electron-ion scattering can contribute substantially to excitation and deexcitation rates (Fujimoto and Kato 1985), but their contributions at high densities may be reduced by collision-induced transitions from the resonance states (Fujimoto 1987).

## 6.4 Autoionization and dielectronic recombination

Two- and more-electron atomic systems, besides the usual excited states corresponding to the excitation of a single valence electron, also have doubly- and multiply-excited states. These additional excited states are situated above the normal ionization limit corresponding to the removal of a valence electron (see figure 6.1). In most cases, these states interact with states corresponding to the next higher ion and a free electron, which have the same total energy. This interaction between degenerate states causes the Auger (1925) effect, i.e., autoionization by the spontaneous ejection of an electron accompanied by a rearrangement of the remaining bound electron or electrons. A simple example is that of a $(2s2p)^{1,3}P$ state of He I decaying to the $1s^2S$ ground state of He II plus a free electron carrying the appropriate energy, angular momentum and spin.

The transition probability for autoionization can be obtained from Fermi's golden rule (2.28), as were the radiative transition probabilities. Since the final state of the system is represented by the product of final ion and free electron wave functions, the density of state factor can be included in the normalization of the free electron wave function, as was done in chapter 5 for photoionization. In this way the decay rate constant for autoionization becomes (Cowan 1981)

$$A^a_{mj} = \frac{2\pi}{\hbar} \sum_i |\langle i|H|j\rangle|^2, \tag{6.33}$$

if $j$ is the initial atom or ion in a certain doubly-excited state of total angular momentum $J_j$ and $m$ is the final state of the resulting ion with total angular momentum $J_m$. The states $i$ are angular-momentum-coupled product wavefunctions represented by $(J_m, \ell's')J_i$, such that $J_i = J_j$ because of overall momentum conservation. The sum in (6.33) is thus restricted to quantum numbers $\ell'$ and $s'$ of the free electron, whose coupling to $J_m$ results in $J_i = J_j$.

The interaction matrix element in (6.33) corresponds to the Coulomb interaction between the two electrons undergoing the transition. They are therefore essentially equal to matrix elements describing the interaction between two bound electron configurations, except that now one of the radial wave functions needed is an energy-normalized continuum function. The two-electron interaction is an even operator, in contrast to the electromagnetic field, atomic dipole interaction given by the limit of (2.38) for $\mathbf{k} \cdot \mathbf{r} \to 0$, which is then an odd operator. For autoionization, the parities of initial and final states are therefore equal.

As discussed by Cowan (1981), (6.33) can be evaluated numerically using a suitable modification of a "configuration interaction" calculation. As in the case of radiative transition probabilities, considerable angular momentum algebra (Cowan 1981, Sobel'man, Vainshtein and Yukov 1981, 1995) is required to obtain results in the most appropriate coupling scheme, an important consideration for the determination of effective rate coefficients for the inverse process, dielectronic recombination. Before taking up this topic, we note that the $A^a$ values also determine the widths of the Beutler(1935)-Fano(1961) resonances in photo-ionization cross sections (chapter 5).

Now consider a steady state balance between autoionization and dielectronic recombination or radiationless capture, namely

$$A_{mj}^a N_j^{z-1} = d_{jm} N_m^z N_e, \qquad (6.34)$$

if we introduce $d = \langle \sigma v \rangle$ as a Maxwell-averaged rate coefficient and indicate by the superscripts $z - 1$ and $z$ the change in ionic charge. According to the principle of detailed balance, the equality only holds if the ion densities obey a Saha equation analogous to (6.24). Using this equation one obtains

$$
\begin{aligned}
d_{jm} &= \langle \sigma_{jm} v \rangle \\
&= 4\pi^{3/2} a_0^3 \frac{g_j}{g_m} \left( \frac{E_H}{kT} \right)^{3/2} A_{mj}^a \exp \left( -\frac{E_{jm}}{kT} \right) \qquad (6.35)
\end{aligned}
$$

in terms of statistical weight factors for the two states and the difference of the two bound-state energies. This difference corresponds to the excitation energy of state $j$ of ion $z - 1$ relative to the excitation energy of state $m$ of ion $z$, minus the normal ionization energy of ion $z - 1$.

In much of the original work on dielectronic recombination, the resonant capture cross section occurring in (6.35), or, rather its average over the resonances, is obtained more directly by extrapolating the excitation cross section of ions in state $m$ to energies for the colliding electrons that are below threshold for the bound-electron excitation corresponding to state $j$ of the recombined ion. [See the review by Seaton and Storey (1976) for a detailed discussion of such calculations.] For capture into weakly bound $n, \ell$ Rydberg levels, one could begin with (6.26) for the excitation cross section of the recombining ion from state $m$ to, say, $m'$ and extend it to energies below threshold by using a constant Gaunt factor. The initial energy $E$ of the colliding electron is related to the normally required excitation energy $E_{m'm}$ by

$$E \approx E_{m'm} - z^2 E_H / n^2, \qquad (6.36)$$

· *z* here being the ionic charge of the recombining ion. Also, free electrons in the interval $\Delta E \approx 2z^2 E_H/n^3$ can be captured into all $n, \ell$ levels of a given principal quantum number in the sense of an averaged cross section. Using (6.13), (6.26) and (6.36) this gives for the $n, \ell$ contribution to the capture rate coefficient for $\ell \leq \ell_m$

$$
\begin{aligned}
d_{n\ell} &\approx \frac{2\ell + 1}{\sum(2\ell + 1)} \sigma_{m'm} v f(E) \Delta E \\
&= \frac{2\ell + 1}{\ell_m^2} 2^5 \pi a_0^2 \left(\frac{2\pi E_H}{3m}\right)^{1/2} \frac{z^2 E_H}{E_{m'm}} f_{m'm} \frac{\bar{g}}{n^3} \\
&\quad \times \left(\frac{E_H}{kT}\right)^{3/2} \exp\left(-\frac{E_{m'm} - z^2 E_H/n^2}{kT}\right),
\end{aligned}
\tag{6.37}
$$

and $d_{n\ell} = 0$ otherwise, since capture into $\ell > \ell_m \approx n_m$ levels is unlikely if $n_m$ is the principal quantum number of the initial state.

Because of angular momentum considerations (Cowan 1981, Sobel'man, Vainshtein and Yukov 1981, 1995), there is no simple relationship between $d_{n\ell}$ and the dielectric recombination rate coefficient $d_{jm}$ in (6.35), which applies to recombination into fully specified states $j$ of the recombined ion. However, an approximate mean value for the autoionization rate follows, nevertheless, from (6.35) and (6.37) to

$$
(A_{mj}^a)_{av} \approx \frac{8(E_H/\hbar)g_m}{\sqrt{3} n^3 g_{m'}} \left(\frac{n_1}{\ell_m}\right)^2 f_{m'm} \bar{g},
\tag{6.38}
$$

if we express $z^2 E_H/E_{m'm}$ in terms of the square of the lowest accessible principal quantum number $n_1$ for the recombined electron and estimate the statistical weight of the doubly-excited state by

$$
g_j \approx 2(2\ell + 1)g_{m'}
\tag{6.39}
$$

in terms of the statistical weight of the excited state $m'$ of the recombining ion. Comparison with (2.43) and (2.48) shows that (6.38) can also be expressed in terms of the radiative transition probability $A_{mm'}$,

$$
(A_{mj}^a)_{av} \approx \frac{8}{\alpha^3 \sqrt{3}} \left(\frac{n_1}{\ell_m}\right)^2 \left(\frac{E_H}{E_{m'm}}\right)^2 A_{mm'} \bar{g}/n^3,
\tag{6.40}
$$

where $\alpha \approx 1/137$ is again the fine-structure constant.

For $\Delta n \neq 0$ excitations, i.e., $A_{mm'} \sim z^4$, $E_{m'm} \sim z^2$, autoionization rates are thus nearly independent of $z$, with slightly different trends, e.g., for $3\ell 3\ell'$ states of two-electron ions suggested by three different calculations (Nilsen et al. 1994). The principal quantum number $n_s$ of the $n\ell$ electron defined by $(A_{mj}^a)_{av} = A_{mm'}$ is estimated by

$$
n_s \approx \frac{2\bar{g}^{1/3}}{3^{1/6}\alpha} \left(\frac{n_1 E_H}{\ell_m E_{m'm}}\right)^{2/3},
\tag{6.41}
$$

corresponding, e.g., to the more specific equation (5.2.12) of Sobel'man et al. (1981). Especially for $\Delta n = 0$ excitations $m \to m'$, very large values of $n$ are thus contributing to effective rate coefficients for dielectronic recombination, as first pointed out by Burgess (1964). This remains true even for fairly highly ionized atoms, for which $n_s$ then scales as $z^{-2/3}$, because of $E_{m'm} \sim z$ in that case. Consider as an example the recombining ion Fe XXI in the groundstate configuration $1s^2 2s^2 2p$, with the $\Delta n = 1$ excitation $2s \to 3p$ ($z^2 E_H / E_{m'm} \approx 5$) or the $\Delta n = 0$ excitation $2s \to 2p$ ($z^2 E_H / E_{m'm} \approx 74$). Using $\bar{g} \approx 0.2$, $\ell_m = 3$, and $\bar{g} \approx 1.0$, $\ell_m = 6$, respectively, (6.41) then gives $n_s(\Delta n = 1) = 5.9$ and $n_s(\Delta n = 0) = 94$, whereas $\ell$-resolved specific calculations suggest $n_s \approx 4$ and 70 (see figure 18-15 of Cowan 1981) for the most important $\ell$ values ($\ell \lesssim 3$ or 6, respectively).

To obtain effective rate coefficients $\alpha_d$ for dielectric recombination in low density plasmas, the capture rate coefficient, e.g., from (6.37) and summed over $\ell$, is multiplied with the branching ratio for radiative decay to occur after capture, rather than autoionization. Neglecting other than $m'$ to $m$ transitions, this branching ratio, $A/(A + A^a)$, gives a factor $[1 + (n_s/n)^3]^{-1}$, so that we obtain for the effective rate coefficient corresponding to $d_{jm}$ in (6.34) and (6.35)

$$
\alpha_d \approx 8\pi^{3/2} \alpha^3 z^4 \frac{E_H}{\hbar} a_0^3 \left(\frac{E_H}{kT}\right)^{3/2} \sum_{m'} f_{m'm} \exp\left(-\frac{E_{m'm}}{kT}\right)
$$

$$
\times \frac{\ell_m'^2}{(n_1')^4} \sum_{n > n_1', 1}^{\infty} [1 + (n/n_s')^3]^{-1} \exp\left(\frac{z^2 E_H}{n^2 kT}\right) \tag{6.42}
$$

with $n_1'$ defined by

$$
n_1' = (z^2 E_H / E_{m'm})^{1/2} \tag{6.43}
$$

and using (6.41) to simplify the expression (6.42). The primes on $\ell_m'$, $n_s'$ and $n_1'$ indicate that these quantities depend on the excited level $m'$, and our estimate is readily seen to be consistent with equations (5.2.11) and (5.2.12) of Sobel'man et al. (1981) for the contribution from one excitation, called there $\alpha_0 \to \alpha$ instead of our $m \to m'$. However, instead of using the $\ell_m$ cutoff, these authors have another sum over $\ell < n$, with any apparent divergence avoided by their $n_s$ values going to zero for large $\ell$. As already indicated above, $\ell_m \approx 1.5 n_m$ or $\ell_m \approx 3 n_m$ are values suggested by *ab initio* calculations (Cowan 1981) for $\Delta n = 1$ or 0 excitations, respectively.

The numerical prefactor in (6.42) is smaller than the prefactor in the expression (6.4) for the radiative recombination rate for completely stripped ions by a factor $3^{3/2}/16\pi \approx 0.1$, and $z^4 (E_H/\hbar)(E_H/kT)^{3/2}$ is a common factor. However, while the following sum in (6.4) is at most of order 1, the (double) sum in (6.42) tends to be very large at temperatures for which the first Boltzmann factor is not too small. This requirement and the relatively large values of $\ell_m$ and $n_s$ are responsible for the especially large ratios of $\Delta n = 0$ dielectronic recombination rates relative to radiative recombination rates. Remember also that the latter are much smaller than for recombination with fully stripped ions, as was already discussed in the paragraph following (6.4) and (6.5).

If large $n \approx n_s$ values are indeed most important, the sum over $n$ can be approximated by an integral and the last exponential factor be replaced by an approximate mean value, say, by its value for $n = n_s$. In this manner one estimates

$$
\sum_n [1 + (n/n_s)^3]^{-1} \exp\left(\frac{z^2 E_H}{n^2 kT}\right)
$$

$$
\approx \exp\left(\frac{z^2 E_H}{n_s^2 kT}\right) n_s \int_{x_1}^{x_2} \frac{dx}{1 + x^3}
$$

$$
\approx \left(\frac{\sqrt{3}}{4}\pi - x_1 - \frac{1}{2x_2^2}\right) n_1 \exp\left(\frac{z^2 E_H}{n_s^2 kT}\right), \tag{6.44}
$$

with $x_1$ and $x_2$ defined by

$$
x_1 = \text{Max}\left(\frac{n_1}{n_s}, \frac{1}{n_s}\right), \tag{6.45}
$$

$$
x_2 = \frac{n_{c\ell}}{n_s}, \tag{6.46}
$$

and $x_1$ and $1/2x_2^2$ both assumed to be small compared with $\sqrt{3}\pi/4 \approx 1.36$, which is the value of the integral from $x = 0$ to $x = \infty$. The collision limit $n_{c\ell}$ (Wilson 1962) is introduced to avoid counting capture into states for which collisional excitation and ionization are more likely than radiative decay. Values for $n_{c\ell}$ can either be inferred from the kinetic models discussed in section 6.1 or from the criterion for partial LTE derived in section 7.6.

From (6.41)–(6.44) and assuming all the underlying assumptions are reasonably valid, we finally estimate the effective dielectronic recombination rate coefficient as

$$
\alpha_d \approx 4\pi^{5/2}\alpha^2 \frac{E_H}{\hbar} a_0^3 \left(\frac{z^2 E_H}{kT}\right)^{3/2}
$$

$$
\times \sum_{m'} \left(\frac{3\overline{g}'}{z}\right)^{1/3} \frac{(\ell_m')^{4/3}}{(n_1')^2} f_{m'm}
$$

$$
\times \exp\left(-\frac{\overline{E}_{m'm}}{kT}\right)\left(1 - 0.73x_1' - \frac{0.37}{x_2'^2}\right), \tag{6.47}
$$

introducing effective excitation energies $\overline{E}_{m'm} = E_{m'm} - z^2 E_H / n_s^2$. Pursuing the above Fe XXI example further and using the data in table 18-3 of Cowan (1981) and $\ell'_m$ values as discussed below (6.43), one estimates from (6.47) for $kT = 1$ keV a total dielectronic recombination coefficient at low densities ($n_{c\ell} \gg n_s$) of $\sim 1.0 \times 10^{-11}$ cm$^3$/s, similar to a value of $\sim 1.6 \times 10^{-11}$ cm$^3$/s from the original Burgess (1965) formula and $\sim 0.6 \times 10^{-11}$ and $\sim 0.7 \times 10^{-11}$ cm$^3$/s from two versions of Cowan's relations which treat the $\ell$-average more carefully. The individual contributions 2s $\rightarrow$ 3p and 2s $\rightarrow$ 2p, which correspond to different $m' - s$ in the notation used here, differ by similar factors. A value of $\sim 1.2 \times 10^{-11}$ cm$^3$/s was deduced experimentally (Wang et al. 1988) at the insignificantly lower nominal temperature of 900 eV from effective recombination rates obtained from the modulation of line intensities by sawtooth oscillations in a tokamak plasma. (A calculated radiative recombination coefficient of $0.2 \times 10^{-11}$ cm$^3$/s was subtracted from the total recombination coefficient determined experimentally.) Unresolved is a factor of 2 between the Burgess values quoted in this paper and that in Cowan's (1981) table 18-3. Other experiment-theory comparisons can be found in Griem (1988b). They also suggest that relatively simple expressions like (6.47), which require, in addition to oscillator strengths and excitation energies, also an independent estimate of $\ell$-values making significant contributions to the capture cross sections, are quite realistic.

This additional parameter is not required for calculations based on the original Burgess (1965) formula or its numerous extensions (Cowan 1981, Sobel'man et al. 1981, 1995, Janev et al. 1985, Shevelko and Vainshtein 1993). For more systematic calculations of effective dielectronic rate coefficients, it is important to treat the various capture or excitation modes separately (Hahn 1993, 1994), with attention being paid to the $\ell$-averages. Corresponding formulae allowing for five different excitation modes for ions containing up to 12 bound electrons are estimated by Hahn to be reliable to within $\pm 50\%$ or better. For O- and F-like ions, Dasgupta and Whitney (1994) have presented $Z$-scaled data allowing interpolation and some extrapolation along isoelectronic sequences. Behar et al. (1995) have calculated dielectronic recombination rates for Ni-like ions, including effects of high-lying configurations. For the modeling of spectra, not only the total recombination coefficients discussed so far are needed but also those for recombination into specific excited states. Examples are dielectronic recombination coefficients into $n = 2$ and 3 states of C I (Dubau and Kato 1994) or into doubly-excited states responsible for satellites to He-like lines of S, Ca and Fe (Itikawa, Kato and Sakimoto 1995).

In some cases, radiative decay of the doubly-excited state produced in the original capture process first leads to another doubly-excited state which may still be capable of autoionization (Blaha 1972, see also Jacobs et al. 1977b), thus causing a reduction in the effective recombination coefficient. More important numerically are usually collisional or electric field effects on the original doubly-excited states. Collisions leading to further excitation and ionization, before radiative stabilization can take place, were already mentioned and result in a reduction of the recombination rate coefficient if the collision limit $n_{c\ell}$ is not much larger than $n_s$, with the relative correction estimated here to be $\sim -0.37(n_s/n_c)^2$, see (6.47). However, collisions changing the $\ell$-values (Burgess and Summers 1969, Seaton and Storey 1976, Weisheit 1975, Jacobs, Davis and Kepple 1976) or, equivalently, electric microfields (Jacobs and Davis 1976) have the opposite effect: they transfer the captured electrons to larger $\ell$-values from which autoionization is less likely than in the absence of such $\ell$-redistribution, while radiative stabilization rates are not significantly affected. Corresponding effects were found to be very important in crossed-beam measurements of dielectronic recombination cross sections (Dunn 1992).

The doubly-excited states formed during the dielectronic recombination also can be produced by charge transfer, e.g., to metastable He-like ions, which then become doubly-excited Li-like ions (Bliman, Cornille and Katsonis 1994). Such states can produce satellites similar to the stabilizing radiative decay in dielectronic recombination. Other processes possibly affecting satellite line intensities are shake-off or shake-up reactions involving outer-shell bound electrons after inner-shell photoionization (Carlson and Krause 1965, Krause, Carlson and Dismukes 1968).

Finally, the separate treatment of radiative and dielectronic recombination may lead to some errors because of interference between the corresponding amplitudes. A more unified treatment (Kukushkin and Lisitsa 1992, Jacobs, Cooper and Hahn 1994) is therefore desirable and also possible.

## 6.5 Heavy particle collisions

Collisional interactions involving only atoms and ions, rather than electrons colliding with atoms or ions, are often omitted from kinetic models because of the much smaller relative velocities. This is generally justified, provided the two sets of cross sections are similar in magnitude. That this is not necessarily true can be gleaned already from the approximate expression (6.26) for electric dipole excitation cross sections, observing that its dominant energy dependence with $1/E$ derives from a factor $2/m_e v^2$, with $v$ being the relative velocity of the (singly) charged projectile. However, $m_e$ is the electron mass and not, say, the reduced mass of the colliding particles. Assuming comparable effective Gaunt factors, one thus finds collision rates proportional to $v^{-1}$, i.e., a dominant role for heavy particle collisions. This was recognized by Pengelly and Seaton (1964) in their work on H and He$^+$ recombination spectra on the basis of semiclassical calculations of $n, \ell \rightarrow n, \ell \pm 1$ collisional cross sections. As in the corresponding spectral line broadening calculations (Griem, Kolb and Shen 1959, see also chapter 4), these cross sections turn out to be infinite, unless one allows for Debye shielding (section 4.4) and, in the present case, also for radiative decay during large impact parameter collisions.

All of this is true only in the limit $E_{mn} \rightarrow 0$, and the reader may therefore be concerned about the factor $(E_{mn})^{-1}$ in (6.26). Actually, when combined with the oscillator strength in (2.48) and using $\omega = E_{mn}/\hbar$, a finite result is obtained, and the divergence is only through the Gaunt factor, i.e., logarithmic, see, e.g., (6.28). For highly charged ions, Coulomb repulsion instead of attraction diminishes the ratio of $E_{mn} \approx 0$ ion-ion and electron-ion cross sections some. For finite $E_{mn}$, the different threshold behaviors of ion and electron collisional cross sections very quickly reverse the ratio to the more common situation. This general trend can also be obtained by comparing semiclassical, impact-parameter-method calculations for ion-collision-induced transitions between two arbitrarily spaced levels (Vainshtein, Sobel'man and Presnyakov 1962, Sobel'man, Vainshtein and Yukov 1981, Walling and Weisheit 1988) with electron-atom or -ion inelastic cross sections.

By a natural extension of these observations, one infers that ionization due to heavy particle collisions is generally also not important, except in very weakly ionized gases (McDaniel 1964) or, e.g., for proton-helium collisions (Olson 1987) at very high proton energies. For atom-atom collisions, corresponding cross sections (Phelps 1990, 1991, 1992) often are orders of magnitude below gas-kinetic cross sections. Notable exceptions are the cross sections for excitation, deexcitation and ionization of hydrogen by atom-atom collisions (Kunc 1987) or by ion-atom collisions (Janev et al. 1994), and of argon by hydrogen atoms (Phelps 1992). Particularly large cross sections, as large as gas kinetic values, have been invoked for excitation energy transfer from argon to hydrogen or nitrogen (Gordon and Kruger 1993), even though the resonance condition, which is fulfilled in case of resonance broadening (see section 4.8), is violated by 1.3 eV. Ionization of energetic hydrogen atoms by collisions with, e.g., fully stripped ions (Janev, Ivanovski and Solov'ev 1994) is also important. In these cases, cross sections again come close to or even exceed typical gas kinetic cross sections (see also Salzborn 1990).

Returning to excitation and deexcitation cross sections, the reader is reminded of the importance of collisional transitions between fine-structure levels, and other very closely spaced levels, for kinetic models as discussed in section 6.1. An early example is a calculation of the proton-induced 2s-2p transition rate in hydrogen (Seaton 1955b), which was extended to excitation in various complex ions by Bahcall and Wolf (1968), Seaton (1964b), Bely and Faucher (1970), Masnov-Seeuws and McCarroll (1972), Landman (1973 a,b), and Kastner (1977).

A unique role in kinetic processes in plasmas is played by charge transfer, also called charge exchange, collisions (Bates and McCarroll 1962, McDowell and Coleman 1970, Mapleton 1972). The basic reaction is

$$A^{z_A} + B^{z_B} \rightleftharpoons A^{z_A+1} + B^{z_B-1} \qquad (6.48)$$

if $z_A$ and $z_B$, etc., are the initial ionic charges and an electron is transferred from ion or atom A to B in the left-to-right version of the reaction. A very useful way to discuss the corresponding cross section is to consider the quasimolecule $(AB)^{(z_A+z_B)}$, whose wave function would present the system during the collision (Sobel'man, Vainshtein and Yukov 1981, 1995, Janev, Presnyakov and Shevelko 1985). This picture is particularly appropriate, e.g., for the $H + C^{6+} \rightarrow H^+ + C^{5+}(n)$ reaction. In that case, the electronic and Coulomb energies involved are related as

$$-E_H = \frac{z_B - 1}{R/a_0} 2E_H - \frac{z_B^2}{n^2} E_H \qquad (6.49)$$

with $z_B = 6$, and $n$ being the principal quantum number of the captured electron. The electronic term of the system $H^+ + C^{5+}(n)$ therefore intersects the ground term of the original system at an internuclear distance

$$R_n = \frac{2(z_B - 1)}{(z_B/n)^2 - 1} a_0, \qquad (6.50)$$

and at relatively small velocities charge transfer is affected near this distance by nonadiabatic transitions (Landau-Zener effect, see, e.g., Mc-Daniel 1964) with a probability approximately given by $1/z_B$ (Janev et al. 1985). Corresponding cross sections are therefore particularly large for $n \lesssim z_B$, and various models suggest that the total capture cross sections from hydrogen into all $n$-levels can be as large as $\pi a_0^2 z_B^2$ (Vinogradov and Sobel'man 1973, Sobel'man, Vainshtein and Yukov 1981, Fritsch and Lin 1984). Analogous quasimolecule formation and level crossings seem responsible for the large nonresonant atom-atom excitation transfer cross sections mentioned earlier in this section (Chambaud, Levy and Pernot 1985, Vance and Gallup 1980).

The generalization of (6.50) for arbitrary initial charges and ionization energy $E_A$ of the "donor" ion is

$$R_n = \frac{2(z_B - z_A - 1)}{[(z_B/n)^2 - E_A/E_H]} a_0, \qquad (6.51)$$

and the Landau-Zener effect would now suggest large cross sections for transfer into $n < (E_H/E_A)^{1/2} z_B$. A proviso in this case is, however, that the Coulomb energy, i.e., $2z_A z_B E_H a_0/R_n$ be smaller than $kT$, so that there is no effective barrier for the ions to approach each other.

The corresponding rate coefficients for populating excited states are often large enough to cause enhancements in the corresponding line radiation. This was recognized by Dixon and Elton (1977) for laser-produced carbon plasmas (see also Dixon et al. 1978), and for completely stripped impurity ions in hydrogen plasmas by Isler (1977) and has since become the basis of H, $C^{6+}, \ldots$, charge exchange spectroscopy of tokamak plasmas (Fonck et al. 1983, Fonck, Darrow and Jaehnig 1984, Rice et al. 1986, von Hellermann, Mandl and Summers 1990), which will be the subject of section 12.1. Another interesting reaction, namely charge transfer to metastable ions, was discussed already at the end of the preceding section.

For charge transfer cross section measurements, the reader is referred to two short reviews (Salzborn 1990, Melchert 1993) and also to some of the references in section 12.1. The first of these reviews describes crossed-beam measurements for positive ions representing the simplest systems, involving only one or two bound electrons and relatively small ionic charges. In the second review the emphasis is on negative $H^-$ ions colliding with each other and with positive ions of charge state up to $z_B = 8$. Very large cross sections were measured, which is not surprising in view of $E_A/E_H = 0.055$ and the considerations of the electronic level shifts discussed above. Agreement with some of the calculated cross sections was found to be very good. For hydrogen collisions with multiply-charged ions (see, e.g., Hardie and Olson 1983), Midha and Gupta (1994) have proposed an empirical relation which should be useful below 10 keV. A review of the data base for collisions of hydrogen atoms and molecules and of helium atoms with metallic impurity ions has been presented by Fritsch et al. (1991).

Molecular reactions (Brown 1959, 1994) can be important as well, e.g., in the edge regions of high temperature plasmas or in low density, low temperature plasmas. An important process of this kind is the dissociative excitation of Lyman and Balmer lines by electron $-H_2$ collisions (Fujimoto, Sawada and Takahata 1989a, Moran 1995).

# 7

# Thermodynamic equilibrium relations

In view of the complexity of the various kinetic models discussed near the beginning of the previous chapter and of the vast array of cross sections and other atomic data required for their implementation, there is a great need for a more basic theory. Such a theory will have a restricted region of validity but can be used with confidence for the interpretation of plasma spectroscopic experiments performed under appropriate conditions. This basic theory is provided by equilibrium statistical mechanics or thermodynamics. It, together with the various conditions for its validity, is the subject of the present chapter. Before introducing this new topic, a second role for the more basic theory should be emphasized, namely its service in testing kinetic models under conditions for which thermodynamic equilibrium should hold.

The practical goal of any of the kinetic models and of the local thermodynamic equilibrium (LTE) relations introduced here is the calculation of populations of the various states of atoms or ions and of the free electron density for specified temperature(s), pressure or mass density, and assumed chemical composition. As already discussed in the preceding chapter, one deals with local, and usually also transient, equilibria, because strict thermodynamic equilibrium would either require an unbounded, spatially and temporally homogeneous plasma or a plasma enclosed in an ideal blackbody hohlraum (cavity).

By LTE in laboratory plasma spectroscopy one usually understands a situation allowing the use of thermal equilibrium relations for level populations, particle velocity distributions, etc., although the radiation field is much weaker than the blackbody intensity (2.54) at the electron temperature, $T_e$. Such LTE can only prevail at high electron densities such that collisional rates exceed radiative rates by, say, at least an order of magnitude. Corresponding estimates of the required electron densities can be found in section 7.6, including estimates for the validity of (partial) thermal equilibrium (PLTE) of excited state populations relative to the groundstate population of the following ion.

An effect of considerable importance for the population of low-lying excited states is the self-absorption of corresponding resonance line radiation. Although a self-consistent solution of the level population equations then also requires a solution of the radiative transfer problem to be discussed in the following chapter, this effect is usually modeled by reducing the radiative decay rates of these levels in estimates for the minimum electron density for LTE to hold. It is not uncommon that the existence of complete LTE for all levels, including groundstate populations, depends on this reduction of effective radiative decay rates, which is often accounted for by multiplying the radiative rates with the escape factor introduced in section 8.4. This is quite important in many experiments, both for temperature measurements and for the determination of basic atomic data from quantitative measurements of line and continuum intensities.

To ensure LTE in the presence of rapid time variations and steep spatial gradients, further conditions have to be placed on the electron density. In the former case, the collisional-radiative relaxation times for level populations must be significantly shorter than the time scale for, say, electron temperature or density variations. In the latter case of strong spatial gradients, one may compare these relaxation times with the time required, e.g., for diffusion across some assumed gradient of macroscopic plasma parameters.

Underlying all of this is the assumption that the free-electron distributions are close to Maxwellian or Fermi distributions. This assumption is normally consistent with having near-LTE conditions for boundstate populations, because electron-electron energy and momentum transfer cross sections are much larger than the inelastic electron-atom or electron-ion cross sections required in boundstate-population rate equations.

## 7.1  Thermodynamic equilibrium and statistical mechanics

To demonstrate the basic simplicity of the methods of equilibrium quantum statistics and thermodynamics (Tolman 1979, McQarrie 1976, Reif 1965), we first consider the distribution of $N$ electrons over the various one-electron states, $n$, with energies $E_n$, of $N$ atoms, all of the same species. The macroscopic description of the system is provided by the set of $N_n$ values, each giving the number of atoms in state $n$. There are

$$W = \frac{N!}{\Pi_n N_n!} \qquad (7.1)$$

distinct microscopic distributions associated with this set, $\{N_n\}$, if we remember that exchanges of electrons among atoms in the same state must not be counted because of their indistinguishability.

For large $N_n$ and therefore still larger

$$N = \sum_n N_n \tag{7.2}$$

the function $W$ has a very sharp maximum, and the corresponding $\{N_n\}$ is the desired equilibrium distribution. It is mathematically much easier to consider instead $\ell n W$, which will also have a maximum for the equilibrium distribution, the logarithm being a slowly varying monotonic function. The logarithms can be approximated by Stirling's formula, e.g.,

$$
\begin{aligned}
\ell n N! &= \sum_1^N \ell n x \approx \int_1^N \ell n x \, dx \\
&= [x(\ell n x - 1)]_1^N \approx N(\ell n N - 1).
\end{aligned} \tag{7.3}
$$

With this approximation $\ell n W$ becomes

$$
\begin{aligned}
\ell n W &\approx N(\ell n N - 1) - \sum_n N_n(\ell n N_n - 1) \\
&= -N \sum_n \frac{N_n}{N} \ell n \left( \frac{N_n}{N} \right),
\end{aligned} \tag{7.4}
$$

if we use (7.2) twice in the last step.

If we now take $N_n/N$ to correspond to the equilibrium distribution, then its variation $\delta(N_n/N)$ must result in $\delta(\ell n W) = 0$, i.e.,

$$
\begin{aligned}
\delta(\ell n W) &= -N \sum_n \left[ \ell n \left( \frac{N_n}{N} \right) + 1 \right] \delta \left( \frac{N_n}{N} \right) \\
&= -N \sum_n \left[ \ell n \left( \frac{N_n}{N} \right) \right] \delta \left( \frac{N_n}{N} \right) = 0, 
\end{aligned} \tag{7.5}
$$

again using (7.2) to simplify the result for a given total number of atoms or ions. This condition is written as

$$N = N \sum_n \left( \frac{N_n}{N} \right), \tag{7.6}$$

and we also assume that the total energy

$$E = N \sum_n E_n \left( \frac{N_m}{N} \right) \tag{7.7}$$

is not changed by varying the distribution. This assumption is certainly reasonable for a large isolated system, i.e., a system represented by Gibbs' microcanonical ensemble (see, e.g., McQuarrie 1976) of statistical mechanics.

The problem of finding the thermal equilibrium population ratios $N_n/N$, i.e., the simultaneous solutions of (7.5), (7.6) and (7.7), is dealt with by the method of Langrange multipliers. One multiplies the variations of (7.6) and (7.7) with $-\ell n\alpha$ and $\beta$, respectively, and adds the resulting expressions to (7.5), obtaining

$$\sum_n \left[\ell n\left(\frac{N_n}{N}\right) - \ell n\alpha + \beta E_n\right] \delta\left(\frac{N_n}{N}\right) = 0. \qquad (7.8)$$

So far the multipliers are arbitrary constants, and all $\delta(N_n/N)$ can now be considered independent. In equilibrium, [...] must therefore be zero, giving

$$\frac{N_n}{N} = \alpha\exp(-\beta E_n)$$

$$= \frac{\exp(-\beta E_n)}{Z_a} \qquad (7.9)$$

with the internal partition function

$$Z_a = \sum_n \exp(-\beta E_n) = \frac{1}{\alpha} \qquad (7.10)$$

determined such that (7.6) is fulfilled.

In order to find the value of the second Lagrange multiplier, one relates it to the total energy from (7.7),

$$E = \frac{N}{Z_a}\sum_n E_n\exp(-\beta E_n) \qquad (7.11)$$

and considers its thermodynamic analog, namely, the internal energy. According to the definitions of heat $Q$ and entropy $S$, and also allowing for possible changes in the volume of our system, which is assumed to be in thermal equilibrium with a surrounding large heat bath of temperature $T$, one has from the first law of thermodynamics

$$đ Q = TdS = dE + pdv. \qquad (7.12)$$

In other words, any increase in the internal energy $E$ of our system and work done by the system is made up by heat transfer from the thermal bath. In any case, the thermodynamic definition of $T$ or, rather, of $1/T$ is

$$\frac{1}{T} = \left(\frac{\partial S}{\partial E}\right)_v, \qquad (7.13)$$

i.e., the partial derivative of the entropy with respect to the internal energy at constant volume.

According to Boltzmann, the connection between the entropy, which is an additive thermodynamic quantity, and statistical mechanics is provided by his famous equation

$$S = k\ell n W, \tag{7.14}$$

$k$ being the Boltzmann constant. Using (7.4), (7.6), (7.7), (7.8) and (7.11), this can be written as

$$
\begin{aligned}
S &= -kN \sum_n \frac{N_n}{N} \ell n \left( \frac{N_n}{N} \right) \\
&= -kN \sum_n \frac{N_n}{N} (\ell n \alpha - \beta E_n) \\
&= kN \ell n Z_a + k_B \beta E.
\end{aligned} \tag{7.15}
$$

The partial derivative,

$$\frac{\partial S}{\partial E} = kN \frac{1}{Z_a} \frac{\partial Z_a}{\partial \beta} \frac{d\beta}{dE} + kE \frac{d\beta}{dE} + k\beta \tag{7.16}$$

can be simplified considerably using (7.10) and (7.11), namely,

$$
\begin{aligned}
\frac{N}{Z_a} \frac{\partial Z_a}{\partial \beta} &= -\frac{N}{Z_a} \sum_n E_n \exp(-\beta E_n) \\
&= -E.
\end{aligned} \tag{7.17}
$$

The first two terms in (7.16) therefore cancel each other, and comparison with (7.13) finally gives

$$\beta = \frac{1}{kT}. \tag{7.18}$$

We conclude that the LTE distribution over the states of an atom or ion obeys

$$\frac{N_n}{N} = \frac{g_n}{Z_a(T)} \exp(-E_n/kT), \tag{7.19}$$

if we further introduce degeneracy or statistical weight factors $g_n$ ($= 2n^2$ for one-electron systems with negligible fine structure or $= 2J_n + 1$ more generally, with $J_n$ being the total angular momentum quantum number). With the same changes in (7.10), the atomic or internal partition function becomes

$$Z_a(T) = \sum_n g_n \exp(-E_n/kT). \tag{7.20}$$

So far we implicitly assumed the atoms to be independent of each other, i.e., the volume per atom to be larger than the volume occupied by bound electrons. Such assumption is not only inconsistent with the notion of LTE being facilitated by collisions, but is clearly wrong for atoms in highly excited states, often called Rydberg levels. However, the discussion of these finite atomic volume effects is best postponed until we extend the

Boltzmann factor relations (7.19) and (7.20) to positive energy states, i.e., to free electrons.

## 7.2    Ionization equilibrium equations

To extend Boltzmann factor relations like the LTE population ratios in two bound states $n$ and $m$, which may or may not be the ground state of our atom or ion, i.e., according to (7.19)

$$\frac{N_n}{N_m} = \frac{g_n}{g_m} \exp\left(-\frac{E_n - E_m}{kT}\right) \qquad (7.21)$$

to continuum states, a number of methods can be used (Herzfeld 1916, Saha 1920, Planck 1924, Menzel 1933, Theimer and Kepple 1970). The most basic difference among the methods is in the use of either the chemical or the physical picture (Rogers 1994). In the first picture, e.g., for hydrogen, one deals with atoms, free protons and free electrons and at lower temperatures also with $H^-$, $H_2$, etc. In the physical picture, the only constituents are protons and electrons, which form bound pairs, three-body states, etc.

If we take as our physical system a single electron-proton pair, we can interpret (7.21) as the relative LTE probability for this pair to be in the two bound states $n$ and $m$. The generalization of (7.21) to continuum states of energy $E_k$ and wave number $k$ (not to be confused with the Boltzmann constant) then involves the replacement of $N_n$ by $dN_k$, the number density of free electron-ion pairs with the electrons in wavenumber interval $k$, $k + dk$, and of $g_n$ by the number $dg_k$ of free electron states in this interval. This quantity corresponds to (2.32) for photons, the two spin directions now included instead of only one polarization. With these replacements and allowing for possible degeneracy $g_i$ of the resulting ion, (7.21) turns into

$$\frac{dN_k}{N_m} = \frac{V g_i}{\pi^2 g_m} \exp\left(-\frac{E_k - E_m}{kT}\right) k^2 dk. \qquad (7.22)$$

Using $E_k = \hbar^2 k^2 / 2m_e$ and integrating over $k$ we get the free electron density $N_e = \int dN_k$, or

$$\begin{aligned}
\frac{N_e}{N_m} &= \frac{V g_i}{\pi^2 g_m} \exp\left(\frac{E_m}{kT}\right) \int_0^\infty k^2 \exp\left(-\frac{\hbar^2 k^2}{2m_e kT}\right) dk \\
&= V \frac{2 g_i}{g_m} \left(\frac{m_e kT}{2\pi \hbar^2}\right)^{3/2} \exp\left(\frac{E_m}{kT}\right).
\end{aligned} \qquad (7.23)$$

It remains to determine the appropriate normalization volume for our proton (ion)-free electron pair. The correct choice is $V = 1/N_i$, the average volume per resulting ion, i.e., the volume available per free-electron proton (ion) pair. (See, e.g., Planck 1924, who evaluated and minimized the free energy.) If we also combine the various constants into $E_H$ and $a_0$, we obtain the Saha (1920, 1921) equation

$$\frac{N_e N_i}{N_m} = \frac{2g_i}{g_m a_0^3} \left(\frac{kT}{4\pi E_H}\right)^{3/2} \exp\left(-\frac{E_{mi}}{kT}\right), \qquad (7.24)$$

which Saha actually wrote in the form of the corresponding mass-action law and then applied to relate stellar spectra and temperatures. In (7.24), the lower boundstate energy $E_m$ in the exponential was replaced by (minus) the ionization energy $E_{mi}$ of state $m$ relative to the ion $i$. If $i$ is an excited state of the resulting ion, its excitation energy must be added to the (lowest) ionization energy of state $m$ in order to obtain $E_{mi}$. Combining (7.19) and (7.24) we finally obtain the Saha equation linking total "atom" and "ion" densities, $N$ and $N_i$, respectively,

$$\frac{N_e N_i}{N} = \frac{2Z_i(T)}{Z_a(T) a_0^3} \left(\frac{kT}{4\pi E_H}\right)^{3/2} \exp(-E_\infty/kT), \qquad (7.25)$$

where $Z_i(T)$ is analogous to $Z_a(T)$, [see (7.20)], and $E_\infty$ is the usual ionization energy of the "atom." This form of the equation was, for the special case of hydrogen, apparently first obtained by Planck (1924).

The Saha equation linking total densities of adjacent ionization stages is particularly useful for LTE plasma composition calculations, but its validity regime is very restricted, not only by our use of Boltzmann rather than Fermi statistics (Chandrasekhar 1930, 1931 and Milne 1930) for the free electrons, but also by a number of other high density effects to be discussed next. At low densities, restrictions arise from the various departures from LTE, which are the subject of the final section of this chapter.

## 7.3 High density corrections

Boltzmann factor relations and Saha equations described so far cannot be expected to yield realistic boundstate populations and charge state distributions unless the electron density is high enough for collisional rates to be larger than radiative rates. Nevertheless, we have so far treated our composite plasmas as if they were noninteracting, i.e., ideal gases. This inconsistency is the reason for the well-known divergence of the internal partition functions according to (7.20) caused by the convergence of the unperturbed $E_n$ towards the ionization energy $E_{mi}$ of the ground state $m$, and by the rapidly increasing statistical weights $g_n$ for highly excited states. The elimination of this divergence will be discussed in section 7.4.

The general theory of dense plasmas with significant interactions between the constituents is a fruitful area for applications of many-body

theory, classical and quantum-mechanical. We will here use a more intuitive approach, but refer the interested reader to several monographs
(Ecker 1972, Ebeling et al. 1976, Kraeft et al. 1986, Fortov and Iakubov
1990 and Ebeling et al. 1991) which deal with this subject much more
thoroughly. The prototype of such theories is the Debye-Hückel (1923)
treatment of interactions between ions in electrolytes. If one considers
the time-averaged and spatially-smoothed potential $V_z(r)$ in the vicinity
of an ion of charge $z - 1$, the corresponding equilibrium charge density
$\rho_z(r)$ can, for nondegenerate plasmas, be expressed in terms of Boltzmann
factors analogous to (7.19). In this manner we obtain

$$
\begin{aligned}
\rho_z(r) &= e\left[-N_e \exp\left(\frac{eV_z(r)}{kT}\right) + \sum_{z',a} z'N_{z'a} \exp\left(-\frac{ez'V_z(r)}{kT}\right)\right] \\
&\approx -\frac{e^2 V_z(r)}{kT}\left(N_e + \sum_{z',a} z'^2 N_{z'a}\right) \\
&\doteq -\frac{\epsilon_0 V_z(r)}{(\rho'_D)^2},
\end{aligned}
\tag{7.26}
$$

after the exponentials are replaced by the first two terms in their power
series and the macroscopic neutrality condition is used to cancel the zero-
order terms. Also, $\rho'_D$ is the Debye radius as in (3.38), except that the
sum is now both over charge states $z'$ and species $a$. If the degeneracy of
free electrons is important, appropriate Fermi functions must be inserted
(Graboske et al. 1969, Cooper and DeWitt 1973, Rogers 1991).

A second relation for the effective charge densities and potentials is
provided by Poisson's equation

$$
\frac{d^2 V_z(r)}{dr^2} + \frac{2}{r}\frac{dV_z(r)}{dr} = -\frac{1}{\epsilon_0}\rho_z(r).
\tag{7.27}
$$

Its appropriate solution (Debye and Hückel 1923) is with (7.26)

$$
V_z(r) = \frac{(z-1)e}{4\pi\epsilon_0 r}\exp\left(-\frac{r}{\rho'_D}\right).
\tag{7.28}
$$

The electrostatic energy associated with one ion or electron ($z = 0$) and
its surrounding Debye cloud is from (7.26) and (7.28)

$$
\begin{aligned}
E_z &= 4\pi \int_0^\infty r^2 V_z(r)\rho_z(r)dr \\
&\approx -\frac{(z-1)^2 e^2}{4\pi\epsilon_0(\rho'_D)^2}\int_0^\infty \exp\left(-\frac{2r}{\rho'_D}\right)dr \\
&= -\frac{(z-1)^2 e^2}{8\pi\epsilon_0\rho'_D}.
\end{aligned}
\tag{7.29}
$$

If an atom or ion $z, a$ is ionized into ion $(z + 1), a$ and an electron, with the ionic charge $(z - 1)$ replaced by $-1$, the total electrostatic energy accordingly decreases by

$$
\begin{aligned}
\Delta E_{zi} &= [z^2 + 1 - (z - 1)^2] \frac{e^2}{8\pi\epsilon_0 \rho'_D} \\
&= \frac{2ze^2}{8\pi\epsilon_0 \rho'_D} \\
&= 2z E_H \left(\frac{a_0}{\rho'_D}\right).
\end{aligned}
\tag{7.30}
$$

This $\Delta E_{zi}$ is the reduction of the ionization energy of an ion $z$ in the present approximation, because bound states with binding energies $\sim z^2 E_H/n^2$ less than $\Delta E_{zi}$ would correspond to electron-ion pairs with more total energy than the next ion plus the new free electron at zero kinetic energy. Thermodynamic considerations (Theimer 1957, Griem 1962) also yield (7.30). However, this result is not applicable if the free electrons are degenerate. If $kT$ is small compared to the Fermi energy $E_F$ given by (4.66) and if $E_F$ is well below $m_e c^2$, Chandrasekhar's (1930, 1931) and Milne's (1930) modification of the Saha equation (7.25) amounts to replacing the factor $2a_0^{-3}(kT/4\pi E_H)^{3/2}$ by the electron density, cancelling the $N_e$ factor on the left-hand side, and to increasing the ionization energy by $E_F[1 - (\pi^2/12)(kT/E_F)^2 + \cdots]$. See Sommerfeld (1928) and McDougall and Stoner (1939) for the asymptotic expansions of the corresponding Fermi integrals and note that other changes in the effective continuum edge are usually more important (see, e.g., Graboske et al. 1969 and Seidel et al. 1995). In the early work on this problem, Coulomb interactions giving rise to the reduction in ionization energy and truncation of the partition functions had been neglected, causing unrealistically low degrees of ionization.

Before discussing the various approximations made in ionization equilibrium calculations including Coulomb interactions, we note that for singly-charged negative ions, e.g., $H^-$, the reduction of the ionization energy according to (7.30) is zero, a reasonable result because the Debye-Hückel charge cloud surrounding such an ion is the same as that of the resulting free electron, and because there is no such charge cloud associated with neutral atoms. The latter point is only true to the extent that shielding of the proton by the bound electron is essentially complete. This is usually a reasonable assumption if the atom is in the ground state or a low-lying excited state, but certainly not for highly excited states (Jackson and Klein 1969, Theimer and Kepple 1970), as will be discussed below.

Another approximation is the replacement of discrete point charges by smoothed charge distributions, which is predicated on having at least a small number of charged particles in a Debye sphere (Berlin and Montroll 1952, Duclos and Cambel 1961). Except for a numerical factor and the use of the electron Debye length defined by (4.59) rather than of $\rho'_D$ from (3.38) here, this condition is equivalent to having a small plasma parameter (Krall and Trivelpiece 1986). A more restrictive approximation is normally the linearization of the exponentials in (7.26), especially that in the ion contribution. Corresponding errors for $r \approx \rho'_D$ and the ion $z + 1$ are with (7.28) seen to be of order $zz'(E_H/k_B T)(a_0/\rho'_D)$. The ensuing relative error in the reduction of the ionization energy according to (7.30) is smaller by a factor $2/z'$.

The relative error in the reduction of the ionization energy $\Delta E_{zi}$ thus may be estimated by $\Delta E_{zi}/k_B T$. Not surprisingly, the linearized Debye theory therefore is only reasonably accurate if one can replace the exponential in the Saha equation (7.25) by

$$\exp(-E_\infty/kT) \rightarrow \exp(-E_\infty/kT)\left[1 + \frac{\Delta E_{zi}}{kT} + \cdots\right] \qquad (7.31)$$

and neglect the higher order terms. This is not to say that one could not estimate higher order terms using empirical and self-consistency arguments (Gündel 1970), which suggest a next term of $\sim (0.6 \text{ - } 0.7) \times (\Delta E_{zi}/kT)^{3/2}$ in the square bracket.

Since ion-ion (or ion-electron) separations $r \approx \rho'_D$ are most important in (7.29) and (7.30), the earlier assumption of complete shielding by bound electrons is only warranted for boundstate radii $r_n \approx n^2 a_0/z$ that are much smaller than $\rho'_D$. For excited states near the reduced ionization limit, with binding energy $\Delta E_n \approx z^2 E_H/n^2$, we have from (7.30)

$$r_n \approx \frac{1}{2}\rho'_D. \qquad (7.32)$$

For states below the reduced ionization energy, shielding by these bound electrons is therefore indeed nearly complete (Jackson and Klein 1969, Theimer and Kepple 1970, Rogers 1990), consistent with our notion of equal reductions of the ionization energies for almost all remaining bound states. An equivalent statement is to say that differences between boundstate level shifts, either from differences in the shielding or from higher order effects, tend to be very small in the validity regime of the linearized Debye theory. This point was first made by Theimer and Kepple (1970) and is entirely consistent with relatively small line shifts even in dense plasmas (see section 4.10).

The reader should also realize that the effective series limits $n_{\max}$ caused by the overlapping of Stark-broadened lines are usually below the reduced ionization energies. The maximum principal quantum number $n_c$ for bound states below the reduced ionization energy is according to (7.32) and, again, using the hydrogenic boundstate radius $r_n$

$$n_c \approx \left( \frac{z \rho'_D}{2a_0} \right)^{1/2} . \tag{7.33}$$

Comparison with the Inglis-Teller (1939) estimate for neutral atoms, i.e., with (5.24a), or with (5.24b) for ions at relatively low temperatures, gives

$$\frac{n_c}{n_{max}} \approx \frac{(3/4\pi)^{2/15}}{2^{1/4} a^{2/5}} \left( \frac{\rho_D}{a_0} \right)^{1/10} \tag{7.34a}$$

or

$$\frac{n_c}{n_{max}} \approx \left( \frac{\rho'_D}{2a_0} \right)^{1/2} \left[ N_p a_0^3 \frac{2\pi z_p kT}{z(z-1)E_H} \right]^{1/10} , \tag{7.34b}$$

respectively, with the parameter $a$ as defined by (4.61). In the first relation, i.e., for $z = 1$, we assumed $z' = 1$, i.e., singly-charged ions as perturbers, so that $\rho'_D$ is simply $\rho_D / \sqrt{2}$. Since the first factor in (7.34a) is usually close to 1, and the second factor is often substantially larger than that, this estimate confirms the statement concerning the effective series limits for neutral atoms. Note that such atoms cease to exist when $\rho_D$, or its equivalent for degenerate plasmas, approaches $a_0$, corresponding to the Mott (1961) transition in solids (see, e.g., Kraeft et al. 1986, Seidel et al. 1995).

In the relation (7.34b) for ions, an average value $z_p$ was used for the charges of perturbing ions, so that with $z_p N_p = N_e$ and $r_e$ as defined by $4\pi r_e^3 N_e / 3 = 1$, the second factor in (7.34b) can also be written as

$$[\cdots]^{1/10} = \left( \frac{a_0}{r_e} \right)^{3/10} \left[ \frac{3kT}{2z(z-1)E_H} \right]^{1/10} . \tag{7.35}$$

The last factor here is practically always very close to 1, and the first factor is usually below 1. Multiply-charged radiators in stellar interiors and those used for diagnostics in laser fusion experiments are exceptions to this rule. Since for such plasmas the first factor in (7.34b) can be well below 1, this suggests that in these plasmas lines may disappear without an extension of the corresponding recombination continua to the line positions. Before accepting this conclusion, one should reconsider the effects of electron impacts, which were neglected in the estimates of $n_{max}$, and verify that (5.24b) was indeed appropriate. If not, (7.34a) divided by $(z^{1/10}/z_p^{1/5}) (1 + z_p)^{1/4}/2^{1/4}$ should be used, with $r_1$ in (4.61) replaced with the ion-sphere radius defined by (7.36). Various estimates of $n_c$ and $n_{max}$ based on several different models for hydrogen plasmas have been evaluated by Kunc and Soon (1992). Their results are all consistent with our conclusion concerning the $n_c/n_{max}$ ratio for neutral atoms, which they find to be as large as $\sim 10$ at $N_e \approx 10^{10}$ cm$^{-3}$.

For very dense plasmas, the Debye-Hückel theory is not always valid, however. A possible reason at relatively low temperatures would be the degeneracy of free electrons, which reduces the shielding relative to the classical result of Debye and Hückel (1923). This reduction (Graboske et al. 1969, Cooper and DeWitt 1973, Rogers 1991) is well known in solid state physics (see, e.g., Kittel 1963) but not very important numerically in plasmas containing multiply-charged ions, because then the ion contribution to the Debye radius (3.38) dominates anyway. More critical are often violations of the conditions for the validity of the linearization of the exponentials in (7.26) and of the smearing of the charges in the screening cloud surrounding a given ion or electron. The most general method available as a first-principle calculational technique for such situations is the density functional theory (Hohenberg and Kohn 1965, Kohn and Sham 1965, Mermin 1965, Perrot 1982, Dharma-wardana and Perrot 1982), which in turn may be viewed as an extension of the finite temperature Thomas-Fermi (-Dirac) model (see, e.g., Cowan 1981 or More 1983, 1991) of an atom confined in a high density medium. Even more realistic are probably quantum mechanical molecular dynamics simulations using, e.g., tight-binding or density-functional methods to obtain the interatomic forces (Kwon et al. 1994, Collins et al. 1995). Intermediate to these methods of calculation is the INFERNO model of Liberman (1971, 1979, 1982), which may be considered as a fully quantum mechanical and relativistic version of the time-honored ion-sphere model for atoms or ions in dense plasmas. Various general properties of such theoretical models have recently been discussed by Murillo and Weisheit (1995).

The ion-sphere model, which is closely related to the Wigner-Seitz cell model used in condensed matter theory, may be traced back to an early model of Herzfeld (1916) for an atom enclosed by a sphere of radius $R_0$, in order to simulate the effects of neighboring atoms. Various versions of this model, which correspond to schematic charge distributions around some nucleus as shown in figure 7.1, have been used to estimate or calculate reductions of ionization energies in dense plasmas (see, e.g., Unsöld 1948, Carson, Mayers and Stibbs 1968, Zimmerman and More 1980, More 1982, Burgess and Lee 1982), also called pressure ionization (More 1985, 1986 and 1989).

Normally the external region is taken to be uniform, i.e., the free electron charge density is compensated by a fictitious, smeared out, positive charge representing the surrounding ions. At the center of the cell is a point nucleus $Z$, and the cell radius $R_0$ is defined by

$$\frac{4\pi}{3}R_0^3 N_i = 1, \qquad (7.36)$$

i.e., by the average volume per ion, if we assume a one-species plasma for

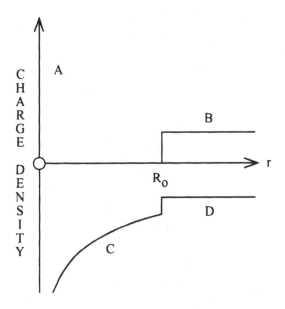

Fig. 7.1. Schematic charge distribution (after Liberman 1982) used in ion-sphere models: A, a point nucleus at the center of a spherical cavity; B, a constant positive charge density outside the cavity which represents the surrounding ions; C, a spherically symmetric electronic charge density inside the cavity; D, a volume-averaged electronic charge density outside the cavity of radius $R_0$. Both regions are electrically neutral.

simplicity. The cell contains both bound and free electrons, also the latter generally being of nonuniform density. The ion-sphere is assumed to be electrically neutral, i.e., the electric field and therefore $dV/dr$ vanish on the boundary. Usually one also assumes the potential $V(r)$ to vanish for $r \geq R_0$. The electron wave functions can be determined from self-consistent solutions of one-particle Schrödinger or Dirac equations (Liberman 1971, 1979 and 1982), with exchange corrections, and of Poisson's equations, with charge densities in the cell calculated from the absolute values squared of the wave functions, and assuming LTE populations according to Fermi statistics. As discussed by More (1985), there are several ways to define the average ionization state $Q \triangleq Z^* = Z - n_b$, where $n_b$ is the number of bound ($E < 0$) electrons per ion. Consistent with the ion-sphere model version discussed here is to relate $Q$ to the electron density at the outer edge of the spherical cell, i.e.,

$$Q = \frac{4\pi}{3} R_0^3 N_e(R_0). \qquad (7.37)$$

However, this choice is not unique (More 1985).

In calculations of this kind there is no need for a separate generalized Saha equation or determination of the reduction in ionization energies, etc., as emphasized by Liberman (1982). Before using the ion-sphere model for a simple estimate of the effective reduction of ionization energies, it is interesting to note that the definitions of $Q$ or $Z^*$ are also important issues in solid state physics (Anderson and McMillan 1967). For other versions of the ion-sphere model allowing for some structure of the surrounding medium, see, e.g., Dharma-wardana and Perrot (1982), Davis and Blaha (1982), Perrot (1982), Cauble, Blaha and Davis (1984), Mancini and Fontán (1985), Perrot and Dharma-wardana (1995) and Stein and Salzmann (1992).

For a very approximate estimate of effective ionization energies at very high densities but not so high temperatures, one can assume the free electron density to be constant also in the interior of the ion sphere. This assumption is particularly reasonable if these electrons are Fermi-degenerate, i.e., if $kT$ is less than the Fermi energy $E_F$ given by (4.66). The perturbation potential produced by such a constant density screening cloud can be obtained from Gauss' law for the perturbing radial field

$$F(r) = -\frac{e}{3\epsilon_0} N_e r^3 / r^2, \tag{7.38}$$

namely

$$V(r) = \frac{e}{6\epsilon_0} N_e r^2. \tag{7.39}$$

In first-order perturbation theory and assuming hydrogenic wave functions corresponding to an effective charge of $Q$, an outer bound state of principal quantum number $n$ is therefore upshifted by

$$\Delta E \approx \frac{n^4}{Q^2} E_H \left( \frac{a_0}{r_e} \right)^3, \tag{7.40}$$

with $r_e$ being the mean separation of free electrons. Such a hydrogenic level will move into the continuum if $\Delta E$ becomes larger than its unperturbed binding energy, i.e., $Q^2 E_H / n^2$. Equating these expressions gives for the principal quantum number of the highest bound state

$$n_c \approx Q^{2/3} \left( \frac{r_e}{a_0} \right)^{1/2} = Q^{1/2} \left( \frac{R_0}{a_0} \right)^{1/2}, \tag{7.41}$$

and for the effective reduction in the ionization energy or "continuum lowering"

$$\Delta E \approx Q^{2/3} E_H \frac{a_0}{r_e} \approx Q E_H \frac{a_0}{R_0}. \tag{7.42}$$

In the last expression, quasineutrality was invoked in order to express $\Delta E$ in terms of the ion-sphere radius defined by (7.36).

Except for numerical factors close to 1 (see, e.g., More 1986 or Kunc and Soon 1992) the estimate (7.42) is the same as results obtained using a variety of slightly different physical arguments. Comparison with (7.30) suggests to combine these relations into

$$\Delta E_\infty = QE_H \text{Min} \left( \frac{2a_0}{\rho_D'}, \frac{3a_0}{2R_0} \right), \qquad (7.43)$$

if we identify $Q = Z^*$ with the charge of the resulting ion $z$ as used at the beginning of this section and insert a 3/2 factor to allow for the change in the cavity radius (More 1986) when a particular ion is ionized at constant electron density. (This change is required to preserve the overall neutrality of the cell.) Other formulas for $\Delta E_\infty$ (Ecker and Weizel 1958, Duclos and Cambel 1961, Ecker and Kröll 1963 and Stewart and Pyatt 1966) may be viewed as interpolations between the two versions of (7.43), but are difficult to reconcile with the requirement of thermodynamic consistency (Griem 1962, Zimmerman and More 1980).

An appealing feature of the ion-sphere model is that according to (7.41) the Bohr radius of the level with principal quantum number $n_c$ is equal to $R_0$. It is again of interest to compare this $n$-value with the $n_{\text{max}}$ value obtained from considerations of line merging due to broadening from ion-ion interactions, i.e., in the present case with (5.24b). This gives

$$\frac{n_c}{n_{\text{max}}} \approx \left( \frac{Q}{Q+1} \right)^{3/5} \left( \frac{3kT}{2Q^2E_H} \right)^{1/10} \left( \frac{R_0}{a_0} \right)^{1/5}. \qquad (7.34c)$$

All three factors are generally close to one. Extremely dense plasmas are exceptions and can have again $n_c < n_{\text{max}}$, as discussed following (7.35). Then ion-ion dipole interactions, which are not contained in the ion-sphere model, would not cause level and line merging before the level in question moves into the continuum, lending physical reality to the continuum resonances to be discussed in the following section. Also, of the last two versions of $n_c/n_{\text{max}}$, namely (7.34b) or (7.34c), the larger of the two should be used in any given situation.

In the domain of the ion-sphere and related models, it is generally not useful to find a Saha-like equation to determine $Q$. Given, e.g., the mass density, one would instead begin with $R_0$ and therefore a value of $n^2a_0/Q$ for the highest bound state. At relatively low temperatures, all lower bound states would be occupied, resulting in

$$n_b = Z - Q = \sum_1^n 2n^2$$

$$\approx \frac{2}{3}(n^3 - 1) \approx \frac{2}{3}n^3 \approx \frac{2}{3} \left( \frac{R_0}{a_0}Q \right)^{3/2} \qquad (7.44)$$

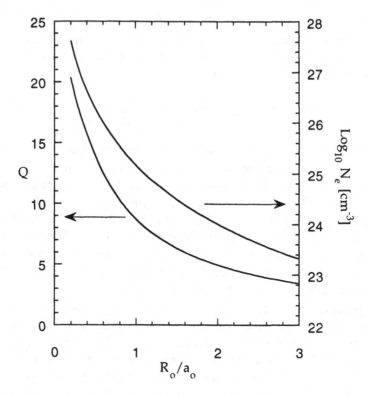

Fig. 7.2. Ion-sphere model prediction for effective ion charge $Q$ and decimal logarithm of the free-electron density $N_e$ of a low-temperature iron plasma as a function of the ion-sphere radius $R_0$, in units of $a_0$ (see text).

for the number of bound electrons. Solutions of this equation for iron ($Z = 26$) as a function of $R_0/a_0$ are shown in figure 7.2 together with the corresponding free electron density from (7.37). Such approximate solutions can then be used as the initial condition for an iterative calculation including finite temperature effects and other refinements as mentioned above. The results can be expressed in terms of correction factors to the Saha equation and excited state populations. This was done by Shimamura and Fujimoto (1990) for hydrogen plasmas.

For very transient dense, strongly coupled plasmas, one also requires effective rate coefficients as discussed near the end of section 6.2. For degenerate electrons, the ratio of such ionization and recombination coefficients (Schlanges and Bornath 1993) would reduce to Chandrasekhar's (1930, 1931) and Milne's (1930) generalization of the Saha equation mentioned at the end of section 7.2, modified for Coulomb interactions as discussed below (7.30).

## 7.4   Partition functions

For LTE plasma composition calculations based on Saha equations like (7.25) and Boltzmann factors as in (7.19), atomic or internal partition functions as given by (7.20) must be evaluated, while thermal equilibrium properties are governed by the total or canonical partition function $Z$ via the Helmholtz free energy $F$ (see, e.g., McQuarrie 1976, who uses $Q$ instead of $Z$)

$$F = -kT \ell n Z. \tag{7.45}$$

It has been known for a long time (Herzfeld 1916) that the internal partition function for, e.g., hydrogen atoms, diverges if the sum is not restricted to a finite number of terms. Another problem arises in composition calculations at some given pressure, if the ideal gas equation of state must be corrected for Coulomb and other interactions. The handling of the partition function divergence and of the pressure corrections requires considerable care to retain thermodynamic consistency.

The internal partition function, say, for hydrogen, is

$$Z_i = 2 \sum_n n^2 \exp(E_H/n^2 kT), \tag{7.46}$$

if we factor $\exp(-E_H/kT)$ from (7.20), as is customary in the theoretical literature. We may essentially follow Planck (1924) by extending the sum only to $n_1$ such that $E_H/n_1^2 kT \approx 1$ and by estimating the remainder quasiclassically. (Planck actually assumed $E_H/n_1^2 kT \ll 1$.) Since bound states of the electron-proton pair, in the physical picture, are numerically important only for, say $kT \lesssim E_H/9$, i.e., $n_1 \gtrsim 3$, such a quasiclassical estimate for the $n \gtrsim 4$ contributions in (7.46) is usually sufficient.

For one electron-proton pair in some volume $V$, Planck wrote the partition function associated with the relative motion as

$$Z_{sc} = 2 \left(\frac{\mu}{2\pi\hbar}\right)^3 \int v^2 dv d\Omega_v \int r^2 dr d\Omega_r \cdot \exp\left(-\frac{\epsilon}{kT}\right), \tag{7.47}$$

except that we inserted the spin factor 2, used $v$ as the velocity variable instead of $q$, and replaced $h$ by $2\pi\hbar$ in the phase space factor. The $\Omega_v$ and $\Omega_r$ are solid angles in velocity and coordinate space, $\mu \approx m_e$ is the reduced mass, and $\epsilon$ is the energy, kinetic and potential, i.e.,

$$\epsilon = \frac{1}{2}\mu v^2 - \frac{e^2}{4\pi\epsilon_0 r}. \tag{7.48}$$

Since Planck was interested in the contribution of states with negative energies in the range

$$-kT \lesssim \epsilon < 0, \tag{7.49}$$

he replaced $v$ by $\epsilon$ as a variable of integration, obtaining

$$Z_{sc} = \frac{(2\mu)^{3/2}}{(2\pi\hbar)^3} \int \exp\left(-\frac{\epsilon}{kT}\right) \left(\epsilon r^2 + \frac{e^2}{4\pi\epsilon_0}r\right)^{1/2}$$
$$\times r\, dr\, d\epsilon \; d\Omega_r d\Omega_v. \tag{7.50}$$

Because of (7.49), the exponential factor can be omitted in an estimate of the contribution $Z_2$ to the partition function from states in the range (7.49), as can be the first (negative) term under the square root. In this way Planck estimated

$$Z_2 \lesssim \frac{(2\mu)^{3/2}}{(2\pi\hbar)^3} \left(\frac{e^2}{4\pi\epsilon_0}\right)^{1/2} kT \int r^{3/2} dr\, d\Omega_r d\Omega_v$$
$$\approx \frac{24(2\pi\mu)^{3/2}}{5\sqrt{\pi}(2\pi\hbar)^3} \left(\frac{e^2}{4\pi\epsilon_0 R}\right)^{1/2} kTV, \tag{7.51}$$

with $4\pi R^3/3 = V$. This estimate of contributions from highly excited bound states may be compared with the $\epsilon > 0$ contribution to the partition function, called $Z_1$ by Planck. Again using (7.50), but now assuming the first term under the square root to dominate, this gives the usual one-particle translational partition function, namely

$$Z_1 = \frac{(2\mu)^{3/2}}{(2\pi\hbar)^3} \int \exp\left(\frac{-\epsilon}{kT}\right) \epsilon^{1/2} d\epsilon\, r^2 dr\, d\Omega_r d\Omega_v$$
$$= 2(2\pi\mu kT)^{3/2} \frac{V}{(2\pi\hbar)^3}, \tag{7.52}$$

using $\int_0^\infty \epsilon^{1/2} \exp(-\epsilon/kT) d\epsilon = \sqrt{\pi}(kT)^{3/2}/2$ and again $4\pi R^3/3 = V$. Except for the statistical weight and exponential factors, this is the same expression as that used in (7.23), with $\mu \approx m_e$.

From (7.51) and (7.52) follows for the ratio of $Z_2$ to $Z_1$

$$\frac{Z_2}{Z_1} \lesssim \frac{12}{5\sqrt{\pi}} \left(\frac{e^2}{4\pi\epsilon_0 RkT}\right)^{1/2}. \tag{7.53}$$

This ratio is well below 1 if the Coulomb interaction energy for typical (free) electron-proton separations is small compared with $kT$. A quantum-mechanical estimate for $Z_2$ is given below.

Since Planck's choice of $n_1$, which corresponds to $n_0$ as used in (7.56) below, differs from our preliminary choice below (7.46), his estimate of the ambiguity caused by the inclusion or omission of the $n \approx n_1$ term

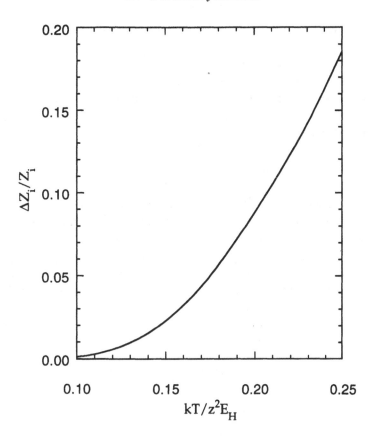

Fig. 7.3. Relative error in the internal partition function of hydrogen atoms or hydrogenic ions associated with the states whose binding energy is near $kT$.

in the boundstate contribution (7.46) is not appropriate here. Instead, we estimate this error by a comparison with the sum of groundstate and lower-excited state contributions as

$$\frac{\Delta Z_a}{Z_a} \approx \frac{n_1^2 \exp(E_H/n_1^2 kT)}{\sum_n^{n_1} n^2 \exp(E_H/n^2 kT)}$$

$$\approx 2.7\frac{E_H}{kT} \left( \sum_{n=1}^{n_1} n^2 \exp(E_H/n^2 kT) \right)^{-1}$$

$$= 2.7\frac{E_H}{kT} \exp\left(-\frac{E_H}{kT}\right) \left( \sum_{n=1}^{n_1} n^2 \exp\left[-\left(1-\frac{1}{n^2}\right)\frac{E_H}{kT}\right] \right)^{-1},$$

$$(7.54)$$

if we return to the use of excitation energies. As can be seen from figure 7.3, this error is usually quite small for partially ionized hydrogen LTE plasmas.

Planck also generalized his results to the case with $N$ protons and one electron in the volume $V$, and finally to $N$ protons and $N$ electrons. He obtained, for $Z_2$ to be small, an expression similar to (7.53), except that $R$ had to be replaced by the (proton) ion-sphere radius $R_0$ according to (7.36). He emphasized that

$$\frac{e^2}{4\pi\epsilon_0 R_0} = 2E_H\left(\frac{a_0}{R_0}\right) \ll kT \tag{7.55}$$

is also a necessary condition for an electron to be either bound to a particular proton or to be free. If this condition is violated, one would have to consider molecules, etc., in the language of the chemical picture. Also, for multiply-charged ions, $E_H$ should be multiplied by $Q^2$, the effective charge acting on highly excited bound electrons as discussed, e.g., in the previous section or, more experimentally, as deduced from the line spectrum, say, of O VI, giving $Q = 6$.

Before accepting this conclusion regarding $Z_2$, it is necessary to consider the quantum mechanical version of (7.51), namely

$$\begin{aligned} Z_2' &= 2\sum n^2 \exp\left(\frac{E_H}{n^2 kT}\right) \\ &\approx 2\sum n^2 \\ &\approx \frac{2}{3}n_0^3, \end{aligned} \tag{7.56}$$

with the highest principal quantum number $n_0$ estimated by equating the corresponding Bohr radius $n_0^2 a_0$ to the ion-sphere radius $R_0$ according to (7.36). This results in

$$Z_2' \approx \frac{2}{3}\left(\frac{R_0}{a_0}\right)^{3/2}, \tag{7.57}$$

or in a ratio of quantum mechanical to semiclassical contribution from highly excited states of

$$\frac{Z_2'}{Z_2} \approx \frac{5\pi}{6\sqrt{2}}\frac{E_H}{kT}\frac{a_0}{R_0}, \tag{7.58}$$

if we again use (7.55). For most LTE hydrogen plasma conditions, one finds $Z_2'/Z_2 \lesssim 0.1$, so that Planck's upper limit turns out to be very generous. This further supports his main point that $Z_2$ is usually small compared with $Z_1$, the free-electron contribution.

Boundstate partition functions calculated by summing the terms in (7.20) to states whose binding energies are close to $kT$ have become known as Planck (1924)–Larkin (1960) partition functions (Rogers 1986). Often empirical energies are available for the lower states, but care should be used in checking the completeness of the data by verifying the sum of the degeneracies or statistical weight factors. For higher bound states near the Planck-Larkin limit, it is usually sufficient to use hydrogenic energies.

A more rigorous justification of the Planck-Larkin procedure leads, e.g., for a one-electron atom with (negative) energies $E_{n\ell}$ to

$$Z_{PL} = 2\sum_{n,\ell}(2\ell + 1)[\exp(-E_{n\ell}/kT) - 1 + E_{n\ell}/kT], \qquad (7.59)$$

if we include the spin factor into Rogers' (1986) expression in terms of principal and orbital angular momentum quantum numbers. The sum over $n$ and $\ell$ is no longer restricted, but the additional terms in the square bracket are practically equivalent to the previous cutoff near $|E_{n\ell}| \approx kT$. However, as emphasized by Rogers, $Z_{PL}$ is not a true partition function in the sense of giving the free energy in (7.45), etc., but rather a quantity arising in a systematic cluster (or virial) expansion of corrections to the ideal gas equation of state (Ebeling, Kraeft and Kremp 1976, Rogers 1986). In particular, the Planck-Larkin partition function should not be used directly in the Boltzmann and Saha relationships of the previous section in order to calculate level populations, e.g., for the calculation of synthetic spectra. However, it will result in an effective electron density that allows for pressure contributions from highly excited bound states, at least to some extent.

The basic physical reason for much of this is that in a fully quantum mechanical theory the electron wave functions for low energy continuum states are really superpositions of the usual plane waves or, rather, Coulomb waves, with wave functions describing resonance scattering states. On the other hand, the wave functions describing highly excited bound states acquire admixtures of free electron wave functions as the density increases. For a more formal consideration of these connections it is instructive to begin with an evaluation of the electron density of states in some spherical volume of radius $R \gg R_0$, with the electron interacting with the ion via a central potential that falls off faster than $1/r$ at large distances, say, due to Debye screening.

More (1985) gives three derivations of the density of states (in his Appendix C), of which the simplest begins with the asymptotic expression (Mott and Massey 1965) for the radial continuum wave function

$$w_\ell(p,r) \sim \sin\left[\frac{1}{\hbar}pr - \frac{1}{2}\ell\pi + \delta_\ell(p)\right]. \qquad (7.60)$$

Here $p/\hbar$ is the wave number $k$, related to the energy $E_k$ as stated below (7.22), $\ell$ the orbital quantum number, and $\delta_\ell(p)$ the phase shift. From the boundary condition $w_\ell(p, R) = 0$, the number $N'_\ell$ of states with energies $E_k \leq p^2/2m_e$ is determined by

$$\pi N'_\ell = \frac{1}{\hbar}pR - \frac{1}{2}\ell\pi + \delta_\ell(p), \tag{7.61}$$

because the argument of the sine function must be a multiple of $\pi$ at $r = R$. Multiplication of $N'_\ell$ with $2(2\ell + 1)$ and summing over $\ell$ gives the total number of states with such energies. Differentiation with respect to $E_k$ then results in the free electron density of states

$$
\begin{aligned}
\frac{dN}{dE} &= \frac{2}{\pi}\sum_\ell (2\ell + 1)\left[R + \frac{d}{dp}\delta_\ell(p)\right]\frac{dp}{dE} \\
&\approx \frac{2}{\pi}\ell_{max}^2 R\frac{1}{\hbar}\left(\frac{m_e}{2E}\right)^{1/2} + \frac{2}{\pi}\sum_\ell (2\ell + 1)\frac{d\delta_\ell}{dE} \\
&= (1/2\pi^2)(2m_e/\hbar^2)^{3/2}VE^{1/2} + \frac{2}{\pi}\sum_\ell (2\ell + 1)\frac{d\delta_\ell}{dE}. \tag{7.62}
\end{aligned}
$$

Here we used $\ell_{max} = (2/3)^{1/2}m_e vR/\hbar = (2/3)^{1/2}pR/\hbar$ as an estimate for the maximum $\ell$-value in the assumed volume $V = 4\pi R^3/3$, in order to obtain the correct vacuum value for the density of states as used, e.g., in deriving the Saha equation. To avoid this procedure, one would have to use exact, rather than asymptotic, wave functions for large $\ell$. However, the second term, which is of interest here, is not sensitive to deviations from the asymptotic wave functions, because significant phase shifts occur only for small $\ell$-values.

The partition function corresponding to the density of states (7.62),

$$
\begin{aligned}
Z'_1 &= \int \frac{dN}{dE}\exp\left(-\frac{E}{kT}\right)dE \\
&= (1/2\pi^2)(2m_e/\hbar^2)^{3/2}V\int_0^\infty E^{1/2}\exp\left(-\frac{E}{kT}\right)dE \\
&\quad + \frac{2}{\pi}\sum_\ell (2\ell + 1)\int_0^\infty \frac{d\delta_\ell}{dp}\exp\left(-\frac{p^2}{2m_e kT}\right)dp \\
&= Z_1 + \Delta Z_1 \tag{7.63}
\end{aligned}
$$

is an improvement over (7.52) in that $\Delta Z_1$ allows for the distortion of the usually assumed plane wave states by the scattering in the effective central potential. In analogy to Rogers' (1986) discussion of the second cluster coefficient, we may use integration by parts to write the scattering correction as

$$\Delta Z_1 = -\frac{2}{\pi} \int (2\ell + 1)\delta_\ell(0) + \frac{2}{\pi m_e kT} \sum (2\ell + 1)$$
$$\times \int_0^\infty p\delta_\ell(p) \exp\left(-\frac{p^2}{2m_e kT}\right) dp$$
$$= -2\sum(2\ell + 1)N_\ell + \Delta Z_1', \tag{7.64}$$

if the celebrated Levinson (1949) theorem is used to express the $p = 0$ phase shift in terms of the number $N_\ell$ of bound states having angular momentum $\ell$, namely

$$\delta_\ell(0) = \pi N_\ell. \tag{7.65}$$

The first term in $\Delta Z_1$, if combined with the usual boundstate partition function (7.20), results in the $-1$ term in the Planck-Larkin partition function (7.59). A further integration by parts gives the $E_{n\ell}/kT$ term (Rogers 1986), on account of higher-order Levinson theorems (Rogers 1977, Bollé 1981), and instead of $\Delta Z_1'$ a remainder that now depends on the actual effective potential (Rogers 1979).

In most cases LTE population calculations can be done using the partition function according to (7.20) truncated at the excited states whose zero density binding energy is just below the reduced ionization energy, with the reduction estimated according to (7.30), (7.42) or (7.43). Also, for the lowered continuum, the deviations from $Z_1$ in (7.52) just discussed are normally not important. This schematic procedure is usually satisfactory for calculations of line and continuum spectra, because any detail is typically obscured by line broadening, as indicated in connection with (7.34a, 7.34b and 7.34c). In the exceptional cases in which line broadening is not dominant, it is necessary to consider shifts of the highest bound states and to consider reductions in their statistical weights in order to account for the wave function components that are no longer localized, thus contributing to the free electron density. Such modifications have been included in the ion-sphere model by Zimmerman and More (1980) and for low density Debye plasmas, e.g., by Gündel (1970, 1971), by Gurovich and Engel'sht (1977), and by Däppen, Anderson and Mihalas (1987). A more physical way of arriving at boundstate populations which are increasingly reduced from the usual Boltzmann populations is to calculate pair distribution functions and effective interaction potentials self-consistently, and then to extract effective level populations from the inferred screening by bound electrons (Rogers 1990).

## 7.5   Equations of state for dense plasmas

Turning finally to pressure corrections (Eliezer et al. 1986) to the ideal
gas law, it is first of all important to keep in mind the desirability of
thermodynamic consistency, which is ensured if one uses the same free
energy expressions, e.g., to determine the ionization equilibrium and the
pressure correction (see, e.g., Griem 1962). This approach will be used
here, although there are at least two other ways (More 1985) to obtain
this pressure, e.g., by direct calculation of the momentum or stress tensor
or by relating the pressure to kinetic and potential energy densities using
the virial theorem. If thermodynamic equilibrium cannot be assumed, the
more physical direct calculation would be necessary.

One requires a free energy expression, and therefore according to (7.45)
a partition function, which includes the effects of particle interactions and
screening. The free energy of a partially ionized ideal hydrogen gas as
derived by Planck (1924) and discussed, e.g., by Hill (1960), which results
in the Saha equation (7.25), must therefore be modified by considering
the energy $U$ to be also a function of relative coordinates of the various
constituents of the system. Following McQuarrie (1976) we write the
exponentials arising from (7.45) with and without the various charges
"turned-on" and take ratios, namely

$$\exp[-(F - F_0)/kT] = \frac{Z}{Z_0}$$

$$= \frac{\int \ldots \int \exp(-U_n/kT) d\mathbf{r}_1 \ldots d\mathbf{r}_n}{\int \ldots \int \exp(-U_n^0/kT) d\mathbf{r}_1 \ldots d\mathbf{r}_n}, \tag{7.66}$$

where $F$, $F_0$ or $Z$, $Z_0$ and $U_n$, $U_n^0$ are free energies, configurational partition
functions and energies with and without charged particle interactions,
respectively. Such separation implies that the internal partition function,
e.g., (7.46) for atomic hydrogen, is not changed by this "charging-up"
process (Ebeling, Kraeft and Kremp 1976). This assumption is only
approximately correct, as already discussed in the preceding section, and
is not made in equation-of-state calculations based on the physical picture
and using grand canonical ensemble methods (see, e.g., Rogers 1991, 1994).

The potential energy between the $n$ charged and neutral particles in the
system may be written as

$$U_n = \frac{1}{2} \sum_{i,j} \frac{z_i' z_j' e^2}{4\pi\epsilon_0 r_{ij}} + \frac{1}{2} \sum_{i,j} u(r_{ij}) = \frac{1}{2} \sum_{i,j} \frac{z_i' z_j' e^2}{4\pi\epsilon_0 r_{ij}} + U_n^0, \tag{7.67}$$

with $z_i' e$ being the charge of the i-th particle and $r_{ij}$ the separation of
particles $i$ and $j$. In a primitive model $u(r_{ij})$, and therefore also $U_n^0$, is
taken to present rigid spheres of radius $a_s$, to model, e.g., exchange forces

if a free electron comes close to a filled subshell of some atom or ion. The first term in (7.67) can be expressed in terms of the electrostatic potential $V_j$ acting upon the j-th ion,

$$
\begin{aligned}
\Delta U_n &= U_n - U_n^0 \\
&= \sum_j z_j' e V_j(\mathbf{r}_j).
\end{aligned}
\tag{7.68}
$$

The ensemble average of $V_j$ can according to (7.66) and (7.68) be written as

$$
\begin{aligned}
< V_j > &= \frac{\int \ldots \int V_j(\mathbf{r}_j) \exp(-U_n/kT) d\mathbf{r}_1 \ldots d\mathbf{r}_n}{\int \ldots \int \exp(-U_n/kT) d\mathbf{r}_1 \ldots d\mathbf{r}_n} \\
&= \frac{1}{e} \frac{\partial F}{\partial z_j'},
\end{aligned}
\tag{7.69}
$$

if temperature and volume are kept constant. This important result is easily verified by differentiating both sides of (7.66) and by using the original equation to eliminate $Z_0$. Writing $dz_j' = z_j' d\lambda$, the free energy becomes

$$
F = F_0 + e \sum_j z_j' \int_0^1 \langle V_j(\lambda) \rangle d\lambda
\tag{7.70}
$$

according to this "charging-up" procedure.

To lowest order in the density expansion of the Coulomb interaction term, we can use the Debye-Hückel result as in (7.29), except that the region $r < a_s$ must now be excluded from the integral and that (7.28) must be multiplied by $\exp(a_s/\rho_D')$, because there are no screening charges in $r < a_s$. Assuming $a_s/\rho_D' \lesssim 1$ and replacing $z - 1$ by $z_j$, this gives

$$
F \approx F_0 - \frac{e^2}{4\pi\epsilon_0} \sum_j z_j'^2 \int_0^1 \frac{\lambda d\lambda}{\rho_D'(\lambda) + a_s},
\tag{7.71}
$$

because one of the $z_j'$ must be replaced by $\lambda z_j$, as must be the charges in $\rho_D'$ from (3.38). The last step amounts to $\rho_D'(\lambda) = \rho_D/\lambda$, so that integration leads to

$$
\begin{aligned}
F &\approx F_0 - \frac{e^2}{12\pi\epsilon_0\rho_D'} \left( 1 - \frac{3}{4} \frac{a_s}{\rho_D'} + \cdots \right) \sum_j z_j'^2 \\
&= F_0 - \frac{kT}{12\pi\rho_D'^3} \left( 1 - \frac{3}{4} \frac{a_s}{\rho_D'} + \cdots \right) V,
\end{aligned}
\tag{7.72}
$$

if finally the sum is expressed in terms of densities and the volume $V$ of the system. The pressure correction follows then from the thermodynamic connection (see, e.g., McQuarrie 1976)

$$\Delta p = -\frac{\partial}{\partial V}(F - F_0)$$

$$= -\frac{kT}{24\pi\rho_D'^3}\left(1 - \frac{3}{2}\frac{a_s}{\rho_D'} + \cdots\right), \qquad (7.73)$$

if care is taken to include the volume dependence of $\rho_D' \sim V^{1/2}$ at fixed particle numbers and temperature.

The leading term in this correction to the ideal gas pressure is in agreement with systematic calculations of pressure corrections based on an activity (density) expansion as reviewed by Rogers (1994). The expansion parameter in these calculations for one-component plasmas is

$$\Lambda = \frac{z^2 e^2}{4\pi\epsilon_0 \rho_D'' kT}, \qquad (7.74)$$

with $z$ being the nuclear charge of the completely stripped ions and $\rho_D''$ as in (3.38) without the electron contribution. As can be seen from figure 3 in Roger's review, the Debye correction is accurate only for $\Lambda \lesssim 0.2$, while a calculation including also a term of order $\Lambda^2$ agrees well with numerical simulations (Slattery, Doolen and DeWitt 1982) up to $\Lambda \approx 2$, corresponding to an ion-ion coupling parameter $\Gamma_{ii} \approx 1$ according to (4.65). The relative correction to the ideal gas pressure of high $z$ ions is equal to $\Lambda/6$ for $a_s = 0$, i.e., should not exceed $\sim 3\%$ before becoming an overestimate of the pressure correction. The $a_s$ term works in the same direction. It corresponds to the core repulsion effect associated, e.g., with neutral argon atoms (Rogers 1981) in partially ionized argon plasmas.

While strong coupling effects can be quite important for calculations of ion contributions to the pressure as just discussed, this is generally not the case for the free electron contribution. This is similar to the situation in microfield calculations (see section 4.3); but besides the Fermi degeneracy effects (Cooper and DeWitt 1973), other quantum mechanical effects may have to be considered as well (Rogers 1991, More 1985, Kwon et al. 1994, Perrot and Dharma-wardana 1995, Blenski and Ishikawa 1995).

## 7.6   Validity conditions for local thermodynamic equilibrium

As discussed in the introduction to this chapter, any laboratory plasma can, at best, only be expected to be near a state of local thermodynamic equilibrium (LTE) corresponding to a temperature varying in space and time. For the interpretation or prediction of spectra, we are mostly concerned with the accuracy of the Saha and Boltzmann relations for level populations in situations where not all constituents of the system are in thermodynamic equilibrium at some common temperature. Besides the usual underpopulation of photons with respect to the blackbody, atom and ion velocity distributions are frequently not the Maxwellians corresponding to the electron temperature. Although these deviations from equilibrium naturally affect the total pressure, they may still not invalidate the Saha-Boltzmann equations for level populations of atoms and ions. This is a consequence of the usual predominance of electron collisional processes as discussed in sections 6.2, 6.3 and 6.4 over the heavy particle collisional effects discussed in 6.5.

Hence, as already mentioned near the end of the introduction to the present chapter, the basic assumption for the validity of LTE is to have Maxwellian or, perhaps, Fermi distributions for the free electrons. This assumption is made in most of the kinetic models reviewed in 6.1, and these models can then be used to find the regions in electron density, temperature and chemical plasma composition parameter space in which LTE populations are within $\pm 10\%$ of the actual populations. Fujimoto and McWhirter (1990) have taken this approach for partially-ionized hydrogen plasmas and for plasmas having only one-electron and completely stripped ions of nuclear charge $z$. ($H_2$ molecules and helium-like, etc., $z \geq 2$ ions were not included.) They use these results to examine and improve approximate LTE validity criteria (Wilson 1962, Byron, Stabler and Bortz 1962, Griem 1963, 1964, McWhirter 1965, Drawin 1969), some of which have been widely used in plasma spectroscopy (Griem 1964). Fujimoto and McWhirter (1990) give a thorough review of the deviations from LTE in various situations, some of which will be discussed shortly. For a systematic discussion of various steady state solutions for excited state populations, the interested reader is referred to a review by van der Mullen (1990) and also to an earlier monograph by Biberman, Vorobev and Yakubov (1987). Much less important in the present context are the atom and ion distribution functions or the corresponding temperatures. There have been attempts, e.g., to generalize the Saha equation for arbitrary ratios of electron and ion temperatures. However, these generalizations are usually unphysical and unnecessary, because it is normally correct (van der Mullen et al. 1994) to simply use the electron temperature.

For complete LTE for the populations of all levels, including the ground state, a necessary condition is that electron-collisional rates for a given transition exceed the corresponding radiative rate by about an order of magnitude. For optically thin plasmas, this condition is usually most stringent for the resonance line, with excitation energy $E_2$, and gives for one-electron ions (Griem 1963, 1964)

$$N_e \gtrsim \frac{5}{8\sqrt{\pi}} \left(\frac{\alpha}{a_0}\right)^3 z^7 (E_2/z^2 E_H)^3 (kT/z^2 E_H)^{1/2}$$
$$\approx 3.9 \times 10^{17} z^7 (kT_e/z^2 E_H)^{1/2} [\text{cm}^{-3}], \tag{7.75}$$

on the basis of exact $A$-values (see chapter 3) and approximate deexcitation cross sections. For hydrogen, i.e., $z = 1$, (7.75) gives a substantial underestimate of the critical electron density, which is then only marginally below $N_e = 10^{18}\,\text{cm}^3$ and less sensitive to the temperature than indicated by the square-root factor. This is according to presently accepted collisional rate coefficients for hydrogen atoms (Johnson 1972, Fujimoto and McWhirter 1990) and also to a calculation (Numano 1995) of the electron density at which CR models agree with the Saha equation (7.24). Such an increased value for the critical density in hydrogen plasmas is consistent with laser scattering measurements (Snyder, Lassahn and Reynolds 1993) in a free-burning argon arc, if one uses the scaling (Griem 1963) of the critical electron density with $(4E_2/3E_H)^3$, $E_2$ in this case being the 3p-4s excitation energy of argon ($\sim 11.5\,\text{eV}$). However, it does not explain the observed difference between electron and gas (atom) temperatures. This is probably related to insufficient rates of energy transfer to neutral atoms and to radial diffusion. For one-electron ions, (7.75) remains a valid estimate.

Many plasmas of practical interest do not come close to being in complete LTE, even if the resonance lines are optically thick, in which case the critical density may be reduced by as much as an order of magnitude. However, as already indicated in section 6.1, there may still be partial thermodynamic equilibrium (PLTE) in the sense that populations of sufficiently highly excited levels are related to the next ion's ground-state population by Saha-Boltzmann relations as (6.10), or to the total population in all fine-structure levels of the groundstate configuration. For hydrogen and hydrogenic ions, but also for any other atom or ion with simple Rydberg level structure, various criteria were advanced for the minimum principal quantum number $n_{c\ell}$ for the lowest level, often called the thermal or collision limit, for which PLTE remains valid, say, to within 10%. The criteria were arrived at using various assumptions concerning the particular collision process which failed to dominate the unbalanced radiative decays.

One consideration was based on the observation (Byron, Stabler and Bortz 1962) that in rapidly-recombining plasmas, electron collisions would easily dominate the downward flow of boundstate populations through principal quantum number space until level energies became comparable to $kT$, resulting in a "bottleneck" level

$$n_B \approx (z^2 E_H/3kT)^{1/2}. \tag{7.76}$$

Below this level, deviations from PLTE were presumed likely, although detailed CR model calculations (see, e.g., figure 4 of Fujimoto and McWhirter 1990) of the critical quantum number $n_{cr}$ for $\sim 10\%$ deviation from LTE do not confirm this conjecture. The consideration that $n_{cr}$ could be estimated (Griem 1963) by requiring collisional depopulation rates to be ten times the radiative decay rate, which led to

$$n_{cr} \approx \left[ \frac{10}{2\sqrt{\pi}} \frac{z^7}{N_e} \left( \frac{\alpha}{a_0} \right)^3 \right]^{2/17} \left( \frac{kT}{z^2 E_H} \right)^{1/17}, \qquad (7.77)$$

turned out to be of more general validity. As already discussed, the corresponding cross sections are particularly crude for $z = 1$, recent estimates suggesting $n'_{cr} \approx 99\, N_e^{-1/9}$, with $N_e$ in cm$^{-3}$, for hydrogen (van der Mullen 1990) instead of $n_{cr} \approx 141 \times N_e^{-2/17}$ at $T = 10^4 K$ from (7.77). The ratio $n'_{cr}/n_{cr} \approx 0.70 \times N_e^{1/153}$ is surprisingly close to 1 for many plasmas of practical interest. Also, the implicit assumption that practically all electrons have enough energy to induce $n_{cr} \to n_{cr} + 1$ transitions is usually not a serious restriction, although multiplication of (7.77) with the originally neglected factor $\exp\{(2/17)z^2 E_H[1/n^2 - 1/(n+1)^2]/kT\}$ would improve the agreement with van der Mullen's results, which are compared in figure 7.4 with earlier estimates for the densities corresponding to $n_{cr} = 2, 3$ and 4.

Still, even for recombining plasmas, $n_{cr}$ from (7.77) or the very similar values of Wilson (1962), McWhirter (1965), Drawin (1969), or Biberman, Vorobev and Yakubov (1979), do not agree very well with CR model calculations. At temperatures low enough for an over-ionized plasma to recombine rapidly, CR models (Fujimoto and McWhirter 1990) typically give $n_{cr}$ values which are larger by $\sim 20\%$ than those estimated by (7.77). On the other hand, at temperatures exceeding $\sim 20z^2$ eV, the CR model suggests a LTE population distribution for all excited levels even at very low densities, due to a near equality of population in- and outflow for all excited levels (Fujimoto 1980). This is referred to as the capture radiative cascade (CRC) regime by van der Mullen (1990). Except for very high $z$-ions, this situation is not of practical interest, because in such high temperature plasmas very few ions would recombine before reaching ionization balance. Of greater interest is the observation that at low temperatures and densities high enough for complete LTE, the CR model for a recombining plasma gives rather high $n_{cr}$ values, because then electrons do not have enough energy for the $n \to n + 1$ excitations. This results in underpopulation of low-lying excited states relative to the Saha-Boltzmann populations, provided that groundstate populations do not become completely dominant, or that recombining ion densities become vanishingly small.

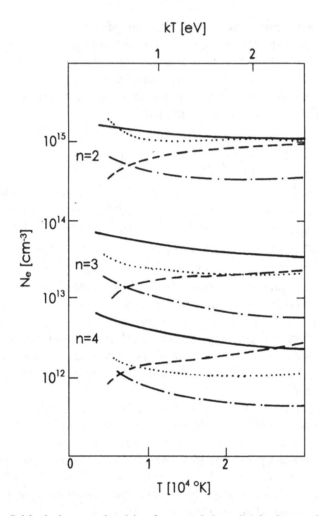

Fig. 7.4. Critical electron densities for $n = 2, 3$ and $4$ hydrogen levels beyond which electron collisions control level populations according to van der Mullen (1990). Some earlier estimates, all as function of temperature, are also indicated: Griem (1963) (dashes); Drawin (1969) (dots); and Biberman et al. (1979) (dash - dot).

For plasmas close to steady state ionization balance, not to mention for ionizing plasmas, PLTE criteria similar to (7.77) were recognized quite early to be insufficient (Fujimoto 1973, Engelhardt 1973). This occurs basically because they do not account for collisional excitation from the ground state, which is represented by the term $r_1 N_1$, rather than $r_0 N_i N_e$ in the CR model expression for excited state populations in quasisteady state conditions (McWhirter and Hearn 1963) as discussed in section 6.1. Direct groundstate excitation is dominant in the low density, corona limit, leading to (6.6) for relative excited-state populations. As the electron density increases, multistep collisional excitation and also collisional deexcitation become increasingly important, possibly yielding a new excitation equilibrium with very characteristic population distributions, e.g., excited level populations proportional to $n^{-4}$ rather than the $n^2$ dependence of LTE populations. The $n^{-4}$ behavior is a consequence of the multistep ladder-like excitation scheme (Fujimoto (1979b) and corresponds to the excitation saturation balance (ESB) solution of the CR model (van der Mullen 1990). Although being the result of one of the improper balances, the ESB distribution can be given a very physical interpretation (van der Mullen, van der Sijde and Schram 1983) in that the corresponding distribution of populations with respect to the binding energy are Maxwellians at the electron temperature, if one treats this energy as a continuous variable.

This interesting behavior of excited-state populations is, as pointed out by van der Mullen (1990), only possible for $n$-values exceeding $n_{cr}$ from (7.77) or its refinements. The extent of the ESB regime then depends on the relative importance of recombination contributions, i.e., on the actual ratio of (completely stripped) ion and groundstate densities and on electron densities and temperatures. No simple and general criterion has been devised for the validity of PLTE for ionizing plasmas; but for plasmas that are in ionization balance the principal quantum number $n$ for $\sim \pm 10\%$ PLTE is found to be below our $n_{cr}$ at temperatures lower than $\sim 2.5z^2\,\mathrm{eV}$ and above $n_{cr}$ at higher temperatures (see figure 6 of Fujimoto and McWhirter 1990).

Besides the cross sections for the multitude of collisional processes, the validity of our conclusions concerning the deviations from LTE or PLTE populations obtained by comparisons with CR model calculations also relies on whether or not the assumption of quasisteady state (QSS) populations in the excited states is compromised by any changes in plasma conditions. This question was discussed especially by McWhirter and Hearn (1963) and by Drawin (1970, 1974) for near-LTE conditions and will be resumed below. First, the reader should be aware that for large deviations from LTE, relaxation times are substantially larger (Cacciatore, Capitelli and Drawin 1976, see also section 6.1). Furthermore, most of the estimates and calculations refer to relatively simple atomic systems, e.g., do not involve metastable or doubly-excited states. As a matter of fact, so far we have mostly considered the question as to whether or not collisions by electrons with thermal energy or velocity distributions were sufficient to overcome the deviations from thermodynamic equilibrium populations caused by unbalanced radiative decays. As already discussed in section 6.5, some heavy particle induced reactions may also occur at rather high rates, e.g., near-resonance charge and excitation energy transfer reactions.

If corresponding populations are not in LTE at the same temperature as the electrons, these reactions can drive level populations away from Saha-Boltzmann equilibrium. A drastic example is provided by experiments using hydrogen atomic beams to transfer electrons to completely stripped impurity ions (resonance charge transfer) , e.g., in tokamak plasmas. This leads to rather selective population of specific levels in the corresponding one-electron ions and is the basis of charge exchange recombination spectroscopy (CHERS), to be discussed in chapter 12.

A more subtle example is the observation by Gordon and Kruger (1993) of a change in the PLTE of argon plasmas due to small admixtures of hydrogen or nitrogen. Because of the larger average transition probability of the hydrogen resonance line, $4.7 \times 10^8 \, \text{sec}^{-1}$ versus the weighted average of $5.2 \times 10^8$ and $1.2 \times 10^8 \, \text{sec}^{-1}$ for the argon, $j = \frac{3}{2}$ and $\frac{1}{2}$, resonance lines (Federman et al. 1992), and because the latter are much more likely to be self-absorbed in this situation, deviations of Ar 4s first excited level populations from PLTE are less than those for the $n = 2$ first excited hydrogen level. The excitation transfer would thus lead to a decrease in Ar 4s,4p and, e.g., 5p populations, as observed. Other processes, namely associative charge transfer between $H_2$ and $Ar^+$ into $Ar H^+$, followed by dissociative recombination giving Ar and H, can also contribute (Meulenbroeks et al. 1994). In cases where electron and atom or ion collisions are of comparable importance in establishing relative populations, but electron and atom, etc., kinetic temperatures differ, one cannot use the Saha equation because of the temperature ambiguity and the corresponding lack of thermodynamic equilibrium. This difficulty may be responsible for significant deviations from LTE in an atmospheric-pressure argon/hydrogen plasma arc discharge (Snyder et al. 1995).

However, in general, the velocity distributions, etc., of heavy collision partners are not very important for level and charge state populations so that, e.g., electron, atom and possibly various ion temperatures may differ even in plasmas that are near Saha-Boltzmann equilibrium. Such situations are possible because of the relatively small energy transfer cross sections, e.g., between electrons and ions, not to mention atoms, and because the plasma heating may act primarily on the electrons. All of these processes and also charge exchange, etc., must be allowed for in the modeling, e.g., of transient and inhomogeneous plasmas in order to calculate the total pressure and internal energy, but are usually not of direct importance for the radiative properties.

Continuing the discussion of time-dependent and spatiallyinhomogeneous plasmas, and possible deviations from LTE due to relaxation effects, we first note that spatial relaxation effects can usually be estimated by comparing the product of particle diffusion velocities and the appropriate relaxation times with the scale length, e.g., for the radial distribution of the electron temperature. This gives a measure of the associated temperature error and of the population uncertainties. It is therefore sufficient to discuss the various relaxation times. For the establishment of complete LTE during plasma production, excitation from the ground state to the upper level of the strongest resonance line tends to be the rate-limiting process. However, since in near-complete LTE, more likely than not, this state is further excited or ionized, it is not sufficient to only excite a Boltzmann fraction, e.g., $4\exp[(3/4)E_H/kT]$ in the case of hydrogen, but rather a fraction of the order $N_i/(N_a + N_i)$, when $N_i$ and $N_a$ are the equilibrium ion or atom densities. In any case, since for complete LTE deexcitation rates are larger than radiative transition probabilities, and since the corresponding excitation rates are related to the deexcitation rates by detailed balancing, we should for conditions near complete LTE expect a relaxation time of

$$\tau_i \lesssim \frac{0.1 g_1}{A_{12} g_2} \left( \frac{N_i}{N_a + N_i} \right) \exp \left( \frac{E_2}{kT} \right), \tag{7.78}$$

"1" and "2" standing for ground and resonance levels, respectively. This gives, e.g., $\tau \lesssim 0.3$ $\mu$sec for a $kT = 1\,\text{eV}$, $N_e \geq 10^{17}\,\text{cm}^{-3}$ hydrogen plasma, for which $N_i \lesssim 0.2\,N_a$ from the Saha equation, or $\tau \lesssim 0.05\,\mu\text{sec}$ for a $kT = 4\,\text{eV}$, $N_e \gtrsim 1.5 \times 10^{19}\,\text{cm}^{-3}$ mostly singly- and doubly-ionized helium plasma. In this case the $N_i/(N_a + N_i)$ factor is $\sim 0.5$, the "atom" now being He$^+$. Similar ionization relaxation times can be inferred from the effective ionization coefficients $S_{CR}$ of the collisional-radiative models described in section 6.1, i.e., from

$$\tau_i \approx (S_{CR} N_e)^{-1} \frac{N_i}{N_a + N_i}. \tag{7.79}$$

For our two examples, we thus obtain $\tau_i \approx 1\mu$sec and 0.1 $\mu$sec, respectively.

For rapidly-recombining plasmas and near-LTE conditions, an analogous estimate using the collisional-radiative recombination coefficients, without a factor corresponding to $N_i/(N_a + N_i)$, is appropriate, since in such situations usually most of the completely stripped ions would recombine. A corresponding relation for the recombination relaxation time, i.e.,

$$\tau_r \approx (\alpha_{CR} N_e)^{-1} \tag{7.80}$$

gives $\tau_r \lesssim 0.3\,\mu$sec or $\tau_r \lesssim 1$ nsec for our two examples, according to the coefficients shown in figure 6.3.

These small relaxation times may not seem to pose much of a problem. However, the reader should realize that their product with typical diffusion velocities may well be comparable to temperature scale lengths, indicating some violation of the concept of *local* thermodynamic equilibrium. Moreover, although (7.78) indicates that the near-LTE ionization relaxation times scale with $z^{-4}$ (McWhirter and Hearn 1963) along the one-electron isoelectronic sequence (suggesting, e.g., $\tau_i \lesssim 1$ nsec for $z = 6$ at relatively low temperature LTE conditions), this is certainly too long for ionization in short-pulse laser-produced plasmas. Then sufficiently rapid ionization requires temperatures and densities above minimum values for LTE, say by factors of 2 and 10, respectively, in which case $\tau_i \approx 10$ psec would be achieved. Recombination times after some expansion, say, to $N_e = 10^{22}$ cm$^{-3}$ and $kT = 50$ eV of completely stripped carbon to the one-electron ion according to (7.80) are as short as $\sim 1$ psec, but remain above $\sim 10$ psec if the temperature stays above $\sim 100$ eV. As emphasized by Cacciatore, Capitelli and Drawin (1976), and already mentioned above, actual relaxation times associated with, e.g., the establishment of LTE ionization balance for $z \geq 2$ ions are larger by factors of 2-10 than those estimated here.

In cases where a quantitative description of the relaxation of excited level populations is required, CR model calculations (see section 6.1) must be performed, e.g., with assumed abrupt changes of plasma parameters. This was done for hydrogen plasmas by Sawada and Fujimoto (1994) who discussed their results in terms of relaxation times for the population of levels that are near LTE, i.e., have principal quantum numbers somewhat below $n_{cr}$ as estimated by (7.77).

# 8

# Radiative energy transfer

In the preceding chapters, the basic radiative and collisional processes governing local radiative properties of a plasma were introduced; and their quantitative evaluation or acquisition was discussed to help the reader in the selection of data needed for analysis or prediction of a spectrum. We also learned about various kinetic or thermodynamic models designed for comprehensive descriptions of level populations.

There are two reasons for this seemingly all-encompassing body of theory and basic data to be insufficient, nevertheless, for both analysis and predictions. First, one generally cannot measure local values of the plasma emission, but must infer them from some distance and averaged over the various contributing volume elements. Second, and even more fundamentally, radiative processes also influence level populations so that the emission or absorption in one location depends on the radiation flux coming from the rest of the plasma.

Therefore a self-consistent treatment of radiation transport and level populations is necessary, requiring in general a nonlocal and nonlinear theory. Such theory has been developed over many years, mainly by astronomers and astrophysicists. Much progress has been made after the two basic treatises (Chandrasekhar 1950 and Sobolev 1963) were written, mostly by computational methods (see, e.g., Athay 1972, Kalkofen 1984, Crivellari, Hubeny and Hummer 1991), but also through more or less analytic models (see, e.g., Thomas and Athay 1961, Jefferies 1968, Ivanov 1973).

A third aspect of radiative energy transfer calculations arises from its close connection with radiative heating or cooling of plasmas, and especially in astrophysics also with pressure balance. In such situations the self-consistency requirements are even further reaching than those for populations and radiative intensities. Classical examples for this broader class of problems are models of stellar atmospheres (Aller 1953,

Unsöld 1955, Mihalas 1978). Analogous situations arise in the radiation-hydrodynamics of laser-produced plasmas (Rubenchik and Witkowski 1991). Any detailed discussion of such models in this broader class would be well beyond the scope of the present book. However, both the reader and the writer can profit greatly from studying the preceding astrophysical research. Astrophysics, in turn, benefits from laboratory measurements, e.g., of absorption coefficients of plasmas containing the elements and ions of interest at electron densities and temperatures that are relevant for astrophysical applications (see, e.g., Rogers and Iglesias 1992, and Seaton et al. 1992). References to some of the laboratory experiments can be found in the introduction to chapter 9.

## 8.1  Effective absorption coefficients

The total radiation field for our purposes is described best by the directional intensity $I(\omega)$ first introduced in (2.45). This intensity is generally not only a function of the angular frequency, or its equivalent, but also of location, direction and time. In many situations, the explicit time-dependence can be neglected, i.e., the radiation field can be assumed as in a quasisteady state consistent with the instantaneous values of the radiative properties of the plasma. Of these radiative properties, the effective absorption coefficient $\kappa'$ is most basic, namely, the difference of absorption and induced emission coefficients. These coefficients correspond to Einstein's (1917) $B$ coefficients, except that these were originally defined with respect to the spectral energy density $u_\omega$ of an isotropic radiation field. If $du_\omega/d\Omega$ is known, also as function of location and direction, and if any dispersion is neglected, we can use the obvious relation

$$I(\omega,\Omega) = c \, du_\omega(\Omega)/d\Omega \tag{8.1}$$

to obtain for the energy density

$$
\begin{aligned}
u_\omega &= \frac{1}{c} \int I(\omega,\Omega) d\Omega \\
&= \frac{4\pi}{c} J(\omega).
\end{aligned}
\tag{8.2}
$$

Here $J(\omega)$ is the mean intensity or, for a layered planar plasma, the zeroth moment of the intensity with respect to $\mu = \cos\theta$, $\theta$ being the polar angle contained in $\Omega$. Higher order moments will be discussed in the following section in connection with the radiative transfer equation.

If only absorption and induced emission are to be considered, the change of intensity on traversing a thin layer of thickness $ds$ is

$$dI(\omega) = -\kappa' I(\omega) ds, \tag{8.3}$$

with $\kappa'$ expressed in terms of cross sections for absorption and induced emission and lower and upper level densities. For the line spectrum, one has

$$\kappa'_\ell(\omega) = \sum \sigma^a_{mn}(\omega) N_n - \sum \sigma^i_{nm}(\omega) N_m \qquad (8.4)$$

with (2.47) and (2.50) for the cross sections, e.g.,

$$\sigma^a_{mn}(\omega) = 2\pi^2 r_0 c f_{mn} L_{mn}(\omega). \qquad (8.5)$$

The $f_{mn}$ are the dipole absorption oscillator strengths discussed in chapter 3, while $L_{mn}(\omega)$ is the normalized line shape function introduced in chapter 4. For a given frequency, the line shape functions effectively control the ranges of the (double) summations, and contributions from overlapping lines are usually simply superimposed. This procedure is not always valid, as already mentioned in section 4.7 in connection with the quasistatic microfield mixing of states with different principal quantum numbers (Kilcrease, Mancini and Hooper 1993). A similar interference effect also occurs in dynamical theories, as easily demonstrated in the one-electron approximation (Baranger 1962, Griem 1974, Stehlé, Voslamber and Feautrier 1989). Both effects cause a reduction in the absorption coefficient between lines, relative to the usual result of linear superposition.

Non-allowed electric dipole transitions normally contribute little to line absorption. Notable exceptions to this rule would be forbidden Stark components at high densities (section 4.7), and possibly some forbidden transitions in high Z elements. Much more important is usually continuum absorption. In solar plasmas, photoionization of $H^-$ and inverse bremsstrahlung on H atoms gives a significant contribution to $\kappa'$, as discussed near the end of section 5.3; but at higher temperatures photoionization of atoms and ions and inverse bremsstrahlung on ions are more important. Corresponding effective absorption coefficients $\kappa'_c(\omega)$ for one-electron systems can usually be obtained by multiplying (5.12) with $1 - \exp(-\hbar\omega/kT)$, provided the electron velocity distribution is thermal, or for two- and more-electron systems from calculations as discussed in chapter 5 (Seaton et al. 1992, Rogers and Iglesias 1992).

As with the linear superposition of line profiles, there is also an interference effect between electron-collision (Stark) broadened lines and (inverse) bremsstrahlung (Burgess 1968, Griem 1974). The total absorption coefficient in the vicinity of lines subject to this effect may become asymmetric, and the spectrally integrated radiation transport could be slightly enhanced because it essentially involves the inverse of $\kappa$ (section 8.3). Note that this interference effect is different from that causing the asymmetries of Fano-type resonances in photoionization involving interference with inner-shell excitation due to configuration interaction, as discussed near the end of section 5.1.

A third process leading to the attenuation (extinction) of the intensity described by (8.3) is scattering, i.e., a two-photon rather than one-photon transition in the language of the basic quantum theory of radiation (chapter 2). The largest scattering cross sections are those for resonance scattering, or fluorescence, as discussed in section 2.9. However, it would be unphysical simply to add this cross section to the line absorption cross section. This follows because the line absorption cross section already accounts for the first step in the resonant two-photon process, so that addition of the resonance scattering cross section would amount to double-counting. This leaves Thomson scattering on free electrons, see (1.30) or the first term in (2.80), with relativistic corrections (Sampson 1959, Huebner 1986) and inclusion of collective effects (Boercker 1987), if necessary, and Rayleigh scattering.

An approximate formula for the Rayleigh scattering cross section can be obtained from (2.80) by expansion in powers of $(\omega/\omega_{n''n})^2$. The zero order terms cancel on account of the $f$-sum rule (section 3.4), so that the Rayleigh scattering coefficient becomes

$$
\begin{aligned}
\kappa_R &= \sum_n \sigma_{nn}(\omega) N_n \\
&= \frac{8\pi}{3} r_0^2 \sum_n \left( \sum_{n'} f_{n'n} \frac{\omega^2}{\omega_{n'n}^2} \right)^2 N_n,
\end{aligned}
\tag{8.6}
$$

to be used only on the far low-frequency side of the resonance lines. Assuming natural line broadening (section 2.7) to be dominant far from a resonance line, the line absorption contribution to $\kappa'$ according to (8.4) and (8.5) would be

$$
\begin{aligned}
\kappa_\ell(\omega) &= \pi r_0 c \sum_{n,n'} f_{n'n} \frac{A_{nn'}}{(\omega_{n'n} - \omega)^2} N_n \\
&= 2\pi r_0^2 \sum_{n,n'} \frac{g_n}{g_{n'}} f_{n'n}^2 \left( \frac{\omega_{n'n}}{\omega_{n'n} - \omega} \right)^2 N_n,
\end{aligned}
\tag{8.7}
$$

using (3.1) with $\omega = \omega_{n'n}$ in the second version. Comparison of (8.6) and (8.7) would suggest equal contributions from line absorption and Rayleigh scattering for frequency detunings of

$$
|\Delta\omega| \approx \left( \frac{3g_n}{4g_{n'}} \right)^{1/2} |\omega_{n'n}|,
\tag{8.8}
$$

i.e., typically only very far from the line where, strictly speaking, also (8.7) should be modified to allow for the explicit frequency dependence of $\kappa_\ell(\omega)$. Additional line broadening processes must also be considered as discussed in chapter 4, especially near the end of section 4.7. Higher order calculations, in powers of $(\omega/\omega_{n'n})^2$, are also required to extend the

Rayleigh scattering cross section closer to the resonance lines (Seaton et al. 1994). Cross sections for atomic hydrogen and helium have been fitted to such power series expansions by Kurucz (1970), but should be replaced by the Thomson scattering cross section, corresponding to $\sum(\cdots)^2 = n_e$ in (8.6), with $n_e$ being the number of equivalent electrons in the outer shell, if this latter cross section would be exceeded (Rogers and Iglesias 1992). Rayleigh scattering on heavier atoms has been calculated by Kissel, Pratt and Roy (1980) for photon energies from 100 eV to 10 MeV.

To summarize this discussion of the various loss terms for the intensity as described by the coefficient $\kappa'$ in (8.3), we may write the total effective absorption or attenuation coefficient as

$$\kappa'_{\text{total}} = \kappa'_\ell + \kappa'_c + \kappa_{Th} + \kappa_R. \tag{8.9}$$

The first two terms describe the net effect of line and continuum absorption, respectively, and the corresponding induced emission processes; while the last two terms account for scattering on free and bound electrons, omitting resonant terms in the latter case. Fortunately the ambiguities concerning near-resonant scattering by bound electrons are normally insignificant, e.g., compared to uncertainties in the absorption line profiles. This latter point should be clear from figure 8.1, which shows the various contributions to the absorption coefficient of a dense hydrogen plasma. Note finally that here $\kappa'_{\text{total}}$, etc., are specified per unit mass, as is customary in the astrophysical literature.

Using different methods, Clausset et al. (1994) have calculated absorption coefficients of lower density hydrogen plasmas typical for stellar envelopes (see their figures 8 and 9). Fast and accurate computer models of H- and He-like line shapes for radiative transfer calculations have been developed by Gilles and Peyrusse (1995).

## 8.2 Effective emission coefficients and redistribution functions

One consequence of the relatively simple treatment of absorption, induced emission, and scattering losses is the considerable intricacy in formulating reasonably general expressions for local additions to the directional intensity. We will designate the corresponding volume emission coefficient as $\epsilon(\omega)$, sometimes replaced by the mass emission coefficient $j_\omega$ in the astrophysical literature. This $\epsilon(\omega)$ is the spectral power per volume and solid angle elements, and it may also depend on location, direction and time.

In LTE or PLTE plasmas (section 7.1) scattering is normally not important, and the emission coefficient is simply related to the spontaneous transition probability and the population in the upper level, or to corresponding quantities for free-bound or free-free emissions. Such con-

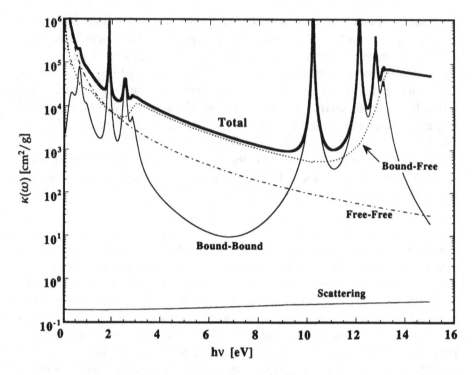

Fig. 8.1.  Contributions to the total absorption coefficient expressed in units of $cm^2/g$ of an $N_e = 10^{18}\,cm^3$, $kT = 2.5\,eV$ hydrogen plasma according to Rogers and Iglesias (1992). Note that there are essentially only three or two or barely one line, respectively, in the Lyman, Balmer and Paschen series, and that the bound-free edges are extended and smoothed consistently with the line broadening and merging (see chapters 4, 5 and 7).

tinuum emission coefficients were already discussed in section 5.4. An analogous expression for the line emission coefficient is obtained by combining the self-evident expression for the radiated power from $m \rightarrow n$ transitions per unit volume, solid angle and angular frequency intervals, namely

$$\epsilon_\ell(\omega) = \frac{\hbar\omega}{4\pi} L(\omega) A_{nm} N_m, \tag{8.10}$$

with (3.1) for the Einstein coefficient for spontaneous transitions and (7.24) for the Saha-Boltzmann relation for populations.

The resulting expression

$$\epsilon_\ell(\omega) = \omega L(\omega) \left(\frac{\hbar\omega}{E_H}\right)^2 \sqrt{\pi}\alpha^3 \left(\frac{E_H}{kT}\right)^{3/2} E_H \frac{g_n}{g_z} f_{mn}$$
$$\times a_0^3 N_e N_z \exp[(E_\infty - \Delta E_z - E_m)/kT] \tag{8.11}$$

is readily compared with the continuum emission coefficient, especially if we assume that the line shape $L(\omega)$ is dominated by thermal Doppler broadening. In this rather common situation, we have $\omega L(\omega) \lesssim c/v_i$, if $v_i$ is the ion thermal velocity. Moreover, for strong lines $g_n f_{mn}/g_z$ is typically of order 1, as are the Gaunt or Biberman factors in 5.16 or 5.21. Near a strong line, comparison with (8.11) therefore shows that line emission is more important than continuum emission by a factor

$$\frac{\epsilon_\ell(\omega_0)}{\epsilon_c} \lesssim \frac{c}{2v_i}\left(\frac{\hbar\omega}{z^2 E_H}\right)^2 \frac{\exp[(E_\infty - E_n)/kT]}{\sum n^{-3}\exp(z^2 E_H/kT) + (kT/2z^2 E_H)}, \quad (8.12)$$

if we use the hydrogenic formula for the continuum, neglect the corrections $\Delta E_z$ and $\Delta E_L$, and use $\hbar\omega \approx E_m - E_n$. Even if bound-free transitions down to $n = 1$ were to be included in the sum in the denominator, the last factor would be close to 1. In a more physical case, e.g., for Lyman lines, i.e., $n \geq 2$, it is well above 1, while $(\hbar\omega/z^2 E_H)^2$ is $\geq 9/16$, and we would typically indeed find $\epsilon_\ell(\omega_0)/\epsilon_c \approx c/v_i$ near the peak of a primarily Doppler-broadened resonance line. If Stark broadening dominates Doppler broadening, one would have to divide this estimate by the ratio of Doppler and Stark widths. See Gallagher (1996) for further discussions of the role of line shapes in radiative transfer.

Although not even PLTE holds down to, say, $n = 2$ in most plasmas, our estimate nevertheless explains why it is so important to use a good physical model for line emission coefficients. Before discussing the corresponding developments in the theory of non-LTE radiative transfer, we note that in strongly magnetized plasmas the isotropy of line emission according to (8.11) may become invalid as a result of Zeeman effects (Cowan 1981, Sobel'man 1992). Also, if electron velocity distributions are not spherically symmetrical in velocity space or, more important in the present context, if the radiation field is anisotropic, the line emission coefficient will have to be evaluated as a function of direction and polarization.

To describe non-LTE effects, it is useful to write the line emission coefficient semiphenomenologically as

$$\epsilon(\omega, \Omega) = p\int \kappa_\ell(\omega')R(\omega,\Omega;\omega',\Omega')I(\omega',\Omega')d\omega'd\Omega' + (1-p)\epsilon_\ell(\omega), \quad (8.13)$$

with $p$ denoting the probability for a photon to be absorbed and immediately reemitted in the line before other depopulation processes occur. The other key quantity is $R(\omega,\Omega;\omega',\Omega')$, called the redistribution function, i.e., the probability for (near) resonant scattering of photon $\omega',\Omega'$ into photon $\omega,\Omega$. The line absorption coefficient $\kappa_\ell(\omega)$ corresponds to the first term in (8.4), and $\epsilon_\ell(\omega)$ is as in (8.10). An intuitive estimate for the reemission probability $p$ is

$$p = A_{nm} \left[ A_{nm} + 4\pi \int \sigma_{nm}(\omega) J(\omega) \frac{d\omega}{\hbar\omega} + C_{nm} \right.$$

$$\left. + \sum_{n' \neq n} \left( A_{n'm} + 4\pi \int \sigma_{n'm}(\omega) J(\omega) \frac{d\omega}{\hbar\omega} + C_{n'm} \right) \right]^{-1}.$$

$$(8.14)$$

Here the $C_{nm}$, which correspond, e.g., to $X_{nm} N_e$ in chapter 6, are the collisional rates per atom in state $m$ for transitions originating in state $m$. The other quantities are induced emission and absorption transition probabilities, expressed through the cross sections already encountered in (8.4) and through the angle ($\Omega$)-averaged intensities $J(\omega)$. These terms require some discussion. First, the induced $m \to n$ transition appears in the denominator, because this process is already allowed for as negative absorption in (8.4). Second, it and all other absorption and emission rates depend on the radiation field, here characterized by $J(\omega)$, near all emission and absorption line transitions involving state $m$ as upper or lower level. This gives rise to the phenomenon of interlocking (Jefferies 1968, Athay 1972, Mihalas 1978, Mihalas and Mihalas 1984) of photon fluxes in different regions of the spectrum. If collisions are relatively rare compared with radiative rates, photons are then not only redistributed within a line but also among different lines and continua. These complicated processes are modeled only rather crudely by expressions like (8.13) and (8.14) and by the line source function discussed in the next section, to which a continuum contribution must be added in case of weaker lines.

The redistribution function in (8.13) has been the subject of much theoretical research, e.g., by Burnett et al. (1980), Burnett and Cooper (1980), and by Cooper, Ballagh and Hubeny (1989). It can be studied experimentally using laser resonance fluorescence techniques (Carlsten, Szöke and Raymer 1977, Osterhold, Himmel and Schlüter 1989, Adler, Dengra and Kelleher 1993). The reader will notice that this resonance scattering, which was deliberately left out in the evaluation of effective absorption and scattering coefficients in the preceding section, is now taken into account after all. It is very closely related to the resonant two-photon process discussed in section 2.9, except that we there finally assumed that the spectrum of incoming photons, i.e., $I(\omega'\Omega')$ in (8.13) was constant across the absorption line shape $L(\omega)$. Had we not made this assumption, (2.81), (2.82) and (2.83) would suggest (Weisskopf 1933)

$$R \sim \frac{A}{A + \sum A'} \delta(\omega - \omega') + \frac{\sum A'}{A + \sum A'} \frac{(A + \sum' A)/\pi}{(\omega - \omega_0)^2 + (A + \sum' A)^2}, \quad (8.15)$$

if only natural line broadening due to spontaneous radiative decays of the line in question, $A$, and other lines, $A'$, is important. This form of $R$ corresponds rather closely to resonant Raman scattering (Penney 1969, Koningstein 1972).

The first term describes coherent scattering, with an angular dependence as in (1.32) or (2.80). The second, incoherent scattering term corresponds to redistribution, and both terms are readily generalized to include collisional effects (Huber 1969) even beyond the range of the impact approximation (Voslamber and Yelnik 1978, Voslamber 1993) for relatively large frequency detunings $\Delta\omega$ and $\Delta\omega'$ from the unperturbed resonance frequency $\omega_0$. This generalization essentially amounts to adding the collisional line width, both in the denominators of the two weight factors for coherent and incoherent scattering and in the numerator for the latter, and naturally also to the width of the profile of redistributed photons. In general, the line profile of the redistributed light differs from that of the absorption coefficient, $L(\omega)$. Also, if the detuning is large, the quasistatic approximation can be used, and the scattering is mostly coherent for a given perturber configuration.

To some extent this is analogous to the treatment of redistribution in the case of pure Doppler broadening (Unno 1952a, Hummer 1962, Jefferies 1968). However, already a relatively small natural, and by extension also collisional, width will lead to a drastic change of the otherwise effectively complete redistribution of $\omega'$ photons on the line wings, in that coherent scattering occurs on the wings of the Voigt profile (Unno 1952b, see figure 5.2 of Jefferies 1968). If scattering in the atom's frame leads to complete redistribution, pure Doppler broadening (Hummer 1962, Jefferies 1968) interestingly enough does not result in complete distribution in the plasma frame. An atom absorbing, e.g., on the red side of the line, will also tend to reemit on this side of the line.

An implicit assumption here is the absence of velocity-changing collisions before reemission. The criterion for this to be justified would correspond to elastic atom-ion, etc., collision frequencies smaller than radiative emission probabilities. Examples of plasmas not fulfilling such conditions are those used for x-ray laser experiments (Koch et al. 1992), which do not exhibit the rebroadening of gain-narrowed lines expected in saturation (Casperson and Yariv 1972). Redistribution due to velocity-changing collisions appears to explain (Koch et al. 1994, Pert 1994) the absence of rebroadening; but the elastic ion-ion collision frequency is not large enough to cause collisional narrowing as discussed in section 4.6. This would require these collision rates to exceed the Doppler width, rather than only the sum of radiative and electron-collisional transition rates.

## 8.3    Radiative transfer equation and source function

For steady-state radiation fields and neglecting refractive effects (Bekefi 1966), balance of gain and loss terms gives for the incremental change of the directional spectral intensity

$$dI(\omega,\Omega) = [\epsilon(\omega,\Omega) - \kappa'I(\omega,\Omega)]dx \qquad (8.16)$$

with $\kappa'$ and $\epsilon$ given, e.g., by (8.9) and (8.13), respectively. It is customary to replace the distance $x$ along the line of sight by the optical depth via

$$d\tau = \kappa'dx \qquad (8.17)$$

so that the differential radiative transfer equation becomes

$$\frac{dI}{d\tau} = \frac{\epsilon}{\kappa'} - I$$
$$\equiv S - I, \qquad (8.18)$$

the second form giving the definition of the all-important source function $S$. The formal solution of the transfer equation is

$$I(\tau) = \int_0^\tau S(\tau')e^{-(\tau-\tau')}d\tau' + I(0)e^{-\tau}, \qquad (8.19)$$

if we take $\tau'$ as line integral of (8.17) from the far side of the plasma to some point $x'$ on this line of sight, with $\tau$ and $x$ corresponding to the exit point of the detected photons as shown on figure 8.2. The last term in (8.19) is due to any incident radiation on the far side. Also, the reader should be aware that astronomers usually count the optical depths into the stellar atmosphere, with $\tau = 0$ corresponding to the surface. They tend to consider planar models and have $\tau$, etc., as functions of $\mu = \cos\theta$, $\theta$ being the angle between the propagation direction and the normal of these planes.

Normally there is no incident radiation for laboratory plasmas, and if the source function variation can be approximated by $S(\tau) = S_0 + (\tau - \tau')f$ on the near side of an optically thick, $\tau \gg 1$, plasma we obtain from (8.19)

$$I(\tau) = S(\tau) + f, \qquad (8.20)$$

which is, by our assumption, the source function at $\tau' = \tau - 1$. This is the famous Eddington-Barbier relation, which has often been invoked to determine the temperature structure by measuring the emerging line emission from optically thick plasmas and using calculated optical depths as functions of frequency detunings and directions. However, such temperature determination is justified only in complete LTE, i.e., if $S$ can be replaced by the Planck function $I_T$ in (2.54) using Kirchhoff's law.

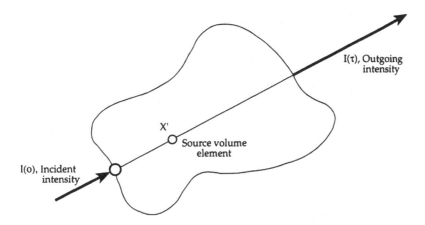

Fig. 8.2. Schematic for the basic quantities occurring in the formal solution of the radiative transfer equation.

Before considering non-LTE source functions, another simplification of the transfer equation which has found broad uses in both LTE and non-LTE problems and in radiation hydrodynamics (Mihalas and Mihalas 1984) should be mentioned. Especially in the astronomical literature, one often introduces moments of the directional intensity $I(\omega, \mu)$, namely besides

$$J = \frac{1}{2} \int_{-1}^{+1} I(\mu)d\mu \qquad (8.21)$$

also the corresponding first- and second-order moments $H$ and $K$. Moreover, by multiplying the right-hand side of the radiative transfer equation (8.18) with $\mu = \cos\theta$, one deals in case of planar atmospheres only with the optical depth in the normal, $\mu = 1$, direction. Taking zero- and first-order moments with respect to powers of $\mu$, this modified transfer equation then leads to

$$\frac{dH}{d\tau} = S - J, \qquad (8.22)$$

$$\frac{dK}{d\tau} = -H \qquad (8.23a)$$

or, more appropriately here,

$$\frac{d^2K}{d\tau^2} = J - S. \qquad (8.23b)$$

A very useful approximation introduced by Eddington (1926) is to assume

$$I(\mu) = a_0 + a_1\mu, \qquad (8.24)$$

i.e., to only include dipole-like deviations from spherical symmetry of the radiation field. In that case, the functions $J$ and $K$ become

$$J = a_0, \qquad (8.25)$$

$$K = \frac{1}{3}a_0, \qquad (8.26)$$

and we obtain a diffusion equation

$$\frac{1}{3}\frac{d^2J}{d\tau^2} = J - S \qquad (8.27)$$

for the desired mean intensity $J(\omega, \tau)$, provided the source function is isotropic. This dispenses with the need to solve the original transfer equation as a function of direction and directly provides a relation for $J(\omega)$, which is needed for the effective emission coefficient (8.13) for isotropic redistribution and also for the reemission probability (8.14). Even more important is the possibility to obtain the radiation flux $H$ by combining (8.23a), (8.25) and (8.26) to

$$
\begin{aligned}
H(\omega) &= -\frac{1}{3}\frac{d}{d\tau}J(\omega) \\
&\approx -\frac{1}{3}\frac{d}{d\tau}I_T(\omega),
\end{aligned} \qquad (8.28)
$$

again using the Eddington approximation. In the second version, we assume also LTE and a slow variation of the temperature with $\tau$ so that we obtain $J = S = I_T$ from (8.27). The signs in (8.28) are opposite to the astrophysical convention, because we have $\tau$ increasing along the emerging light beam.

In many applications, the spectrally and angle integrated flux of radiation is required, i.e., from (8.17) and (8.28)

$$
\begin{aligned}
F &= 4\pi \int H(\omega)d\omega \\
&\approx -\frac{4\pi}{3} \int \frac{dI_T(\omega)}{dT}\frac{d\omega}{\kappa'(\omega)}\frac{dT}{dx} \\
&\equiv -\frac{4\pi}{3\kappa'_R} \int \frac{dI_T(\omega)}{dT}d\omega\frac{dT}{dx},
\end{aligned} \qquad (8.29)
$$

with the factor $4\pi$ arising from the azimuthal integration and removal of the $1/2$ factor in the definition of the $\mu$-moments (Mihalas 1978). The last version provides a definition for the Rosseland (1924) mean of the effective absorption coefficient,

$$\kappa'_R = \int \frac{dI_T(\omega)}{dT} d\omega \Big/ \int \frac{dI_T(\omega)}{dT} \frac{d\omega}{\kappa'(\omega)}. \tag{8.30}$$

This is the quantity of primary astrophysical interest (Weiss, Keady and Magee 1990, Rogers and Iglesias 1992, Seaton et al. 1992, Seaton et al. 1994, Magee et al. 1995, Rickert et al. 1995) and is usually called the mean opacity and expressed in ratio to the mass density of the absorbing layer. Generally calculations of $\kappa'_R$ and $\kappa'(\omega)$ are now found to be in reasonable agreement with each other (Seaton et al. 1994). This is not to say that there are no differences, e.g., from omission or inclusion of some inner-shell photo-absorption by excited ions and differences in the treatment of excited state populations (Iglesias and Rogers 1995). For many-electron atoms and ions, the supertransition-array method (Bar-Shalom et al. 1989, 1995) greatly eases the computation of the complicated line spectra.

Returning to non-LTE transfer problems, we finally consider the source function as given by the ratio of (8.13) and (8.9), except that we omit the normally negligible Thomson and Rayleigh scattering in (8.9). In this way the source function becomes

$$
\begin{aligned}
S(\omega) &= \frac{p \int \kappa_\ell(\omega') R(\omega, \Omega; \omega', \Omega') I(\omega') d\omega' d\Omega'}{\kappa'_\ell(\omega) + \kappa'_c(\omega)} \\
&\quad + \frac{(1-p)\epsilon_\ell(\omega) + \epsilon_c(\omega)}{\kappa'_\ell(\omega) + \kappa'_c(\omega)} \\
&\approx \frac{p \int \kappa_\ell(\omega') J(\omega') d\omega' L(\omega) + (1-p)\epsilon_\ell(\omega) + \epsilon_c(\omega)}{\kappa'_\ell(\omega) + \kappa'_c(\omega)}, \tag{8.31}
\end{aligned}
$$

if we assume complete and isotropic redistribution corresponding to an emission line shape $L(\omega)$. This assumption is much more commonly applicable than Milne's (1928) assumption of coherent scattering, in which case the scattering contribution to the emission coefficient is simply $p\kappa_\ell(\omega)J(\omega)$. With this simplification, using (8.14), and representing the atom by a two-level system without a continuum, Milne's expression for the line source function can be written as (Jefferies 1968)

$$S_M(\omega) = \frac{J(\omega) + \epsilon' I_T(\omega)}{1 + \epsilon'} \tag{8.32}$$

in terms of the dimensionless parameter

$$\epsilon' = \frac{C_{12} - C_{21}g_1/g_2}{A_{12}}$$

$$= \frac{C_{12}}{A_{12}}\left[1 - \exp\left(-\frac{E_{21}}{kT}\right)\right]. \tag{8.33}$$

This parameter, not to be confused with the emission coefficient $\epsilon(\omega)$, is the ratio of net collisional deexcitation, i.e., deexcitation minus collisional excitation probability per atom in the upper state "2", and the spontaneous transition probability $A_{12}$. In order to obtain the corresponding contribution to the emission coefficient in terms of the Planck function $I_T(\omega)$ and also the second version of (8.33), a thermal distribution must hold, i.e., for the electrons which are responsible for excitation and deexcitation.

For complete redistribution, the two-level source function of Thomas (1957) and of Jefferies and Thomas (1958)

$$S_T(\omega) = \frac{\int J(\omega)L(\omega)d\omega + \epsilon' I_T(\omega)}{1 + \epsilon'}, \tag{8.34}$$

in which photoionization and recombination terms are omitted, is the equivalent of Milne's expression, with more generalizations required for realistic descriptions of radiative transfer in radiation-dominated, non-LTE plasmas. Of these generalizations, inclusion of continuum effects, i.e., emission, absorption, and photo-ionization and photo-recombination, besides the already-mentioned interlocking effects of different lines through the population equations, is probably most important. As emphasized by Athay (1972), there is no unique way of parameterizing the corresponding source functions, so that great care is needed in interpreting observed spectra by comparison with modeling results. However, two general conclusions bear repeating: First, although a measured intensity $I(\omega)$ from an optically thick plasma essentially gives the source function at $\tau(\omega) \approx 1$, this cannot be interpreted as a temperature, e.g., of the free electrons unless the parameter $\epsilon'$ is large, so that scattering is not important compared to emission that is uncorrelated to the excitation. Second, if scattering and other non-LTE effects are important, the coupled set of radiative transfer equations for a sufficiently large number of frequencies must be solved concurrently with the population equations, e.g., by the numerical method of complete linearization (Auer and Mihalas 1969, Athay 1972, Auer and Heasly 1976, Mihalas 1978, Mihalas and Mihalas 1984). For the treatment of transfer by pure scattering, the reader is also referred to a second monograph by Sobolev (1972).

In much of the astrophysical work, it is more convenient to write, e.g., the two-level atom line source function as

$$S_T(\omega) = (1 - \epsilon'') \int J(\omega)L(\omega)d\Omega + \epsilon''I_T(\omega). \qquad (8.35)$$

Comparison with (8.34) and use of (8.33) yields

$$\begin{aligned}
\epsilon'' &= \frac{C_{12} - C_{21}g_1/g_2}{A_{12} + C_{12} - C_{21}g_1/g_2} \\
&\approx \frac{C_{12}}{A_{12} + C_{12}},
\end{aligned} \qquad (8.36)$$

if we neglect the induced emission factor $[\cdots]$ in (8.33). The last version suggests to call $\epsilon''$ the quenching or photon destruction probability per scattering process (Hummer 1962, Avrett and Hummer 1965). It is readily generalized to multilevel atoms (Apruzese 1993) and also for the effects of continuous absorption, etc. (Athay 1972).

## 8.4 Transient problems and escape factors

In optically thin laboratory plasmas, the transit time of light is practically always negligibly small compared with characteristic time scales for variations in plasma conditions. However, if line radiation undergoes multiple scattering, or absorption and reemission, before escaping an optically thick plasma, the photon transport may be more akin to a diffusion process, requiring a more careful consideration based on the time-dependent equation of transfer instead of (8.18), namely (Sobolev 1963, Cooper and Zoller 1984),

$$\frac{\partial I}{\partial \tau} = S - I - \frac{1}{c\kappa'}\frac{\partial}{\partial t}I. \qquad (8.37)$$

The formal solution is again as (8.19), except that $S$ and $\tau - \tau'$ must be evaluated at the retarded time. Together with population and depopulation rates for the determination of the source function and the effective absorption coefficient as discussed in the preceding sections, the time-dependent transfer equation may then be solved, e.g., to model the decay of a laser-produced plasma (Apruzese 1993). A related, but nearly time-independent problem had been encountered in the so-called imprisonment of resonance radiation in decaying plasmas (Zemansky 1930, 1932, Mitchell and Zemansky 1934), with a theoretical interpretation provided well before the development of non-LTE radiative transport theory (Biberman 1947, 1948, Holstein 1947, 1951).

Assuming that heat conduction or other energy losses have cooled the electrons such that both collisional excitation and deexcitation are negligible, the decay of the population density $N(\mathbf{r}, t)$ in the upper state of a resonance line is governed by

$$\frac{\partial}{\partial t}N(\mathbf{r}, t) = -AN(\mathbf{r}, t) + \frac{4\pi}{\hbar\omega}\int \kappa(\omega)\langle I(\omega, \mathbf{r}, t, \Omega)\rangle_{av}d\omega. \tag{8.38}$$

Here $A$ is the spontaneous transition probability and $\kappa(\omega)$ the line absorption coefficient as given by (8.4) and (8.5), but neglecting induced emission. The directional average over $\Omega$ must be calculated using the formal solution (8.19) of the radiative transfer equation,

$$I(\omega, \mathbf{r}, t, \Omega) = \frac{\hbar\omega}{4\pi}AL(\omega)\int_0^{\rho_m} N(\mathbf{r}', t)\exp[-\kappa(\omega)\rho]d\rho \tag{8.39}$$

in the present case. Here $\mathbf{r}' = \mathbf{r} - \mathbf{u}\rho$ describes the location of the source point with $\mathbf{u}$ being a unit vector in the $\Omega$ direction, and we neglect any transit time effects and space- or time-variations in the line-shape function $L(\omega)$ and in $\kappa(\omega)$. In the terminology of the preceding section, these coupled equations for $N$ and $I$ imply complete and isotropic redistribution, while the assumption concerning $\kappa(\omega) \sim N_1 L(\omega)$ implies a uniform and constant groundstate density $N_1$, besides the assumption of equal absorption and emission line shapes.

The directional average in (8.38) can with $d\mathbf{r}' = \rho^2 d\rho d\Omega$ be written as a volume integral, and substitution of (8.39) then gives the integro-differential equation

$$\frac{\partial}{\partial t}N(\mathbf{r}, t) = -AN(\mathbf{r}, t) + A\int G(\mathbf{r}, \mathbf{r}')N(\mathbf{r}', t)d\mathbf{r}' \tag{8.40}$$

for the excited-state density, with the kernel

$$\begin{aligned} G(\mathbf{r}, \mathbf{r}') &= \frac{1}{4\pi\rho^2}\int L(\omega)\kappa(\omega)\exp[-\kappa(\omega)\rho]d\omega \\ &= -\frac{1}{4\pi\rho^2}\frac{d}{d\rho}\int L(\omega)\exp[-\kappa(\omega)\rho]d\omega. \end{aligned} \tag{8.41}$$

From the second form of $G(\mathbf{r}, \mathbf{r}')$, the integro-differential equation is seen to correspond to Holstein's (1947) result, which was obtained from probability arguments.

To solve the equations, one expands

$$N(\mathbf{r}, t) = \sum_m c_m n_m(\mathbf{r})\exp(-\beta_m At), \tag{8.42}$$

with the eigenfunctions $n_m(\mathbf{r})$ obeying

$$(1 - \beta_m)n_m(\mathbf{r}) = \int G(\mathbf{r}, \mathbf{r}')n_m(\mathbf{r}')d\mathbf{r}' \tag{8.43}$$

according to (8.40), and the $c_m$ coefficients to be determined from the initial excited-state density $N(\mathbf{r}, 0)$. The eigenfunctions and eigenvalues or escape factors $\beta_m$ depend on both geometry and line shapes. They can be found using standard procedures (Courant and Hilbert 1953). Often one is interested only in the asymptotic decay law, i.e., in the smallest eigenvalue $\beta_1 = \beta$. On account of the orthogonality of the eigenfunctions and because of (8.43) it may be written as

$$\beta = 1 - \frac{\int\int G(\mathbf{r}, \mathbf{r}')n(\mathbf{r})n(\mathbf{r}')d\mathbf{r}d\mathbf{r}'}{\int n^2(\mathbf{r})d\mathbf{r}}. \tag{8.44}$$

This relation can be used to obtain approximate $\beta$ values using the Ritz variational procedure (Holstein 1947, 1951). The method employs trial functions containing adjustable parameters, which are varied to minimize $\beta$. For Doppler broadened lines, typical $\beta$ values are $[\tau_0(\ell n \tau_0)^{1/2}]^{-1}$, if $\tau_0 = \kappa(\omega_0)X$ is the optical depth at line center, and if $X$ is a characteristic dimension, e.g., the radius $R$ of a plasma column or the thickness $D$ of a planar plasma sheath. These results only apply if $\beta$ is much smaller than 1, i.e., if effective decay rates are much smaller than the transition probability. Phelps (1958) has evaluated corresponding escape factors also for plasmas that are less optically thick. Escape factors for Lorentzian line shapes (Holstein 1951) are about $0.37\tau_0^{-1/2}$ with $\tau_0 = \kappa(\omega_0)D$, or $0.35\tau_0^{-1/2}$ with $\tau_0 = \kappa(\omega_0)R$, for planar or cylindrical geometry, respectively, again for $\beta \ll 1$.

As discussed by Irons (1979a), there are various quantities related to the escape factors and also more general definitions than that of Holstein (1947). Very useful in applications is the transmission factor $T$, which is closely related to the photon escape probability $P_e$, namely the probability that a photon in the line travels a distance corresponding to $\tau_0$. This probability is

$$\begin{aligned} P_e &\approx T(\tau_0) \\ &= \int_{-\infty}^{+\infty} \exp[-\tau(\omega)]L(\omega)d\omega, \end{aligned} \tag{8.45}$$

with $T(\tau) \sim [\tau_0(\pi \ell n \tau_0)^{1/2}]^{-1}$ and $(\pi\tau_0)^{-1/2}$ for Gaussian and Lorentzian line shapes, respectively. Returning to the escape factor, its original definition was clearly for excited state populations having a spatial distribution corresponding to the most slowly decaying mode (Irons 1979b) in the expansion (8.42). Using corresponding asymptotic results for other situations is therefore not quite correct.

In the astronomical literature, much use is made of the escape probability $P(e) = P_e$ as estimated by Osterbrock (1962) and Hearn (1963) by replacing (8.45) with

$$P(e) \approx 2 \int_{\omega'}^{\infty} L(\omega)d\omega \tag{8.46}$$

and determining $\omega'$, e.g., from

$$\kappa(\omega')R = 1. \tag{8.47}$$

Expressed in terms of the optical depth $\tau_0$ at line center, this condition is in terms of the line-shape function values at $\omega'$ and $\omega_0$

$$L(\omega') = L(\omega_0)/\tau_0. \tag{8.48}$$

These estimates and asymptotic results as quoted below (8.45), or by Athay (1972) also for Voigt profiles, agree within $\sim 30\%$ with exact calculations (Irons 1979c) for $\tau_0 \gtrsim 3$. For detailed calculations of escape probabilities for Doppler profiles, the work by Apruzese et al. (1980) should be consulted and that of Apruzese (1985) for Voigt profiles, or papers by Weisheit (1979) and by Mancini, Joyce and Hooper (1987) for Stark profiles. As shown by Clark et al. (1995) the escape-factor, probabilistic model can also be used to describe transport of continuum radiation. In case of significant Doppler shifts between emitting and absorbing layers, a method developed by Sobolev (1957) has been applied to expanding cylindrical plasmas by Shestakov and Eder (1989), and has been used by Eder and Scott (1991) for line transfer calculations in radially expanding plasmas. Such effects have been observed also in the laboratory (Wark et al. 1994) in plasmas produced by high power lasers.

For the case of radiative decay of resonance lines of helium-like ions in short-pulse laser-produced plasmas, Apruzese (1993) compared his numerical solution of (8.37), and of the required time-dependent and self-consistent populations of upper and lower levels, with an analytic model based on escape probabilities $P_e$ or $P(e)$ and photon destruction probabilities $P_d$ or $P(d)$ in Athay's (1972) notation. Actually, Apruzese used the closely related quenching or photon destruction probability $\epsilon''$ per scattering as defined in (8.36) in connection with the two-level atom source function. These $\epsilon''$ were calculated by multiplying the collisional transition probabilities $C_{i2}$ from the upper level to levels $i$ with $(1 - P_{ri})$, $P_{ri}$ being the relative probability of returning from level $i$ to the upper level of the line. In this way and by summing over $i$, the two-level model was adapted to the actual multilevel problem using collision rates, etc., of Duston and Davis (1981). For the $A\ell$ XII resonance line at $N_e = 10^{21}$ cm$^{-3}$, the resulting net probability of collisional quenching was found to be $\epsilon'' = 0.029$.

The probability that a resonance-line photon escapes after an excitation into level 2 is $(1 - \epsilon'')P_e$, while the probability that it remains in the plasma is $S = (1 - \epsilon'')(1 - P_e)$. Both probabilities are obtained by the usual multiplication of probabilities for the two processes involved, i.e., $(1 - \epsilon'')$ for reemission after a scattering with $P_e$ for escape, or with $(1 - P_e)$ for the photon not to escape. The probability for escape after $N$

scatterings is then

$$P_N = (1 - \epsilon'')P_e(1 + \sum_{n=1}^{N} S^n)$$

$$= (1 - \epsilon'')P_e \frac{1 - S^N}{1 - S}, \tag{8.49}$$

which is the sum of the probabilities of escape after the first, second and more subsequent scatterings, which can only occur if the photon had remained in the plasma after the preceding scattering, hence the factors $S^n$. The final version uses the formula for the sum of a geometric progression, and for $N \to \infty$, one obtains

$$P_u = \frac{1 - \epsilon''}{1 - S} P_e \tag{8.50}$$

for the ultimate probability of escape per original photon. The quantity $P_N/P_u = f_e$ is the fraction, of all eventually escaping photons,

$$f_e = 1 - S^N$$

$$= 1 - [(1 - \epsilon'')(1 - P_e)]^N, \tag{8.51}$$

which has escaped after $N$ scatterings, i.e., after a time $t \approx N/A_{12}$, if we neglect transit time effects.

Estimated escaped photon fractions according to the analytic model, i.e., (8.50), are compared in figure 8.3 with the results of the detailed atomic model, radiative transfer calculations of Apruzese (1993). At the two lower aluminum ion densities, $3 \times 10^{19}$ and $9.1 \times 10^{19} \text{cm}^{-3}$, a 100 $\mu$m thick plasma slab was assumed, and at $9.1 \times 10^{20} \text{ cm}^{-3}$ a 10 $\mu$m slab. The initial excitation was assumed to be concentrated in a relatively narrow central layer, and Voigt profiles were used for the absorption line profile. Note that the A$\ell$ XII resonance line has a radiative lifetime $A_{12}^{-1} = 0.036$ ps and that the emission is nearly complete at $t = 5$ ps. The analytic model gives for the times in which half of the ultimately escaping photons have left the plasma 0.3, 1.6 and 1.8 ps, respectively, for the highest intermediate and lowest densities. These times are in reasonable agreement with the curves in figure 8.3. It is also important to remember that at the assumed densities, i.e., $N_e \lesssim 10^{22} \text{ cm}^{-3}$, Doppler broadening corresponding to $kT = 200$ eV dominates over the Stark broadening.

## 8.5 Reconstruction of source distributions from intensities

A long-standing problem in applications of plasma spectroscopy is the derivation of local plasma properties from emission or absorption measurements, which are necessarily integrals along the line of sight. The

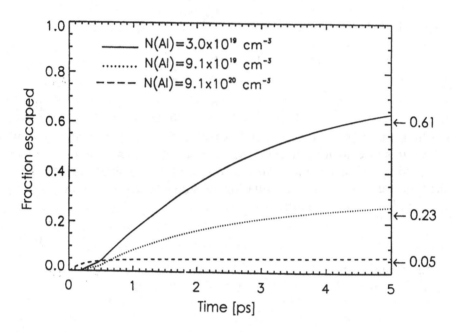

Fig. 8.3.  Fraction of A$\ell$ XII resonance line photons escaping ultra-short laser-heated plasmas (see text) according to Apruzese (1993). Also indicated are the (ultimately) escaped photon fractions according to his analytic model (see arrows on the right margin).

natural way to achieve some localization is to make the measurements along different lines of sight, basically as in tomography, and then to either perform a numerical inversion of these data or to compare them with model calculations for assumed source distributions. Given noisy data or somewhat arbitrary assumptions in the modeling method, there is usually considerable uncertainty in the inferred source distributions (Preobrazhenskii and Pikalov 1982), but also the opportunity to use optimized inversion or image reconstruction methods (Lewitt 1983, Censor 1983). There exist some purely mathematical arguments for the possibility of reconstructing even three-dimensional images from suitable line integrals (Radon 1917, Bracewell 1956, Cormack 1963), the question of the uniqueness of the inversion process for experimental data notwithstanding.

The inversion problem is greatly eased if the plasma has some simple symmetry, say, cylindrical for some laboratory plasmas or spherical for the sun and stars, and if the plasma is optically thin. The latter condition is not at all appropriate in the study of stellar atmospheres, hence the great interest in radiative transfer and non-LTE source functions, which were the principal subjects of the preceding sections. In laboratory studies, transfer problems are frequently important at least for strong lines (Preobrazhenskii 1971), a situation that can actually be put to some use for temperature measurements (Bartels 1950 a,b, Zwicker 1968, Karabourniotis 1986), provided the source function is close to the Planck function.

In the following, the plasma will be taken as optically thin, an assumption that is best verified by comparing single- and double-pass emission, the latter often involving a correction for the return mirror's reflectivity. The classical example is the Abel inversion (Hörman 1935) of the chordal emission $I(y)$ of an infinitely long cylindrical plasma column of radius $R$, and $y$ being the chordal height, or impact parameter, of the line of sight. The emission coefficient $\epsilon(r, \omega)$ is assumed to be only a function of the radial coordinate $r$, and we suppress any frequency dependence. The emerging intensity is therefore

$$
\begin{aligned}
I(y) &= \int_{-x_m}^{x_m} \epsilon(r)dx \\
&= 2\int_y^R \frac{\epsilon(r)rdr}{(y^2 - r^2)^{1/2}},
\end{aligned}
\tag{8.52}
$$

if we use $x = (r^2 - y^2)^{1/2}$ and assume the plasma is not illuminated by an external source. The inversion is then given by the Abel transform

$$
\begin{aligned}
\epsilon(r) &= \frac{1}{\pi} \int_r^R \frac{-I'(y)dy}{(y^2 - r^2)^{1/2}} \\
&= \frac{1}{\pi} \int_r^R \frac{-I'(x)dx}{(x^2 - r^2)^{1/2}},
\end{aligned}
\tag{8.53}
$$

the primes indicating derivatives with respect to the chord heights $y$ or $x$. To verify this inversion formula, one substitutes (8.53) into (8.52), changes the order of the integrations and performs the $r$ integral first, thus recovering $I(y)$, provided $I(R)$ vanishes.

The occurrence of the derivatives of $I(y)$, or of $I(x)$, is indicative of the sensitivity of the inversion to any noise in the measured intensities, not to mention to any deviations from cylindrical symmetry. It is particularly difficult to obtain accurate results for $\epsilon(r)$ near $r = 0$, because then the entire range of $y$ or $x$ values is required, leading possibly to an accumulation of errors. As in the more general image reconstruction methods for tomography, one can either use least-squares (Décoste 1985) or optimized (Holland, Powell and Fonck 1991) methods involving a maximum entropy principle (Frieden 1983, Meinel 1988) and physical constraints, e.g., $\epsilon(r) \geq 0$, and any information from other measurements.

Such improved inversion methods have been proposed or used to obtain two-dimensional images of x-ray emission of tokamak plasmas, which do not have any simple symmetry in the poloidal plane (Décoste 1985,

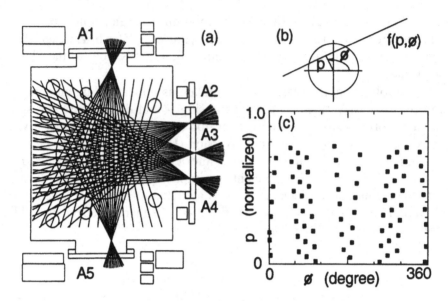

Fig. 8.4. (a) X-ray imaging system for tokamak plasmas (Janicki, Décoste and Simm 1989) with five arrays of sixteen surface-barrier detectors each. (b) Coordinate system for sight lines: $p$ is the chord impact parameter, $f(p, \phi)$ is the chord's brightness. (c) Location of the detectors in $(p, \phi)$ space, with $p$ normalized to unity at a minor radius of $a = 25$ cm.

Nagayama 1987, Holland, Powell and Fonck 1991). Because of the toroidal symmetry, using one wide-angle x-ray pinhole camera aimed tangentially to the plasma torus with a two-dimensional detector array provides, in principle, sufficient information for the image reconstruction. Instead of the parallel chords as function of chord height $y$ in case of cylindrical symmetry, one now has chords characterized, e.g., by the two coordinates of the corresponding pixels of the detector array. Alternatively, and especially if access is restricted, one can use several cameras aimed perpendicular to the toroidal direction from various locations, say, in one poloidal plane (Décoste 1985, Janicki, Décoste and Simm 1989, Janicki, Décoste and Nöel 1992), see figure 8.4. These authors actually employed a Fourier-Bessel harmonic reconstruction method to take advantage of basic tokamak symmetries.

An equivalent method to obtain local emission coefficients, still assuming toroidal symmetry, has been used in the Tormak IV bicusp experiment (Myers and Levine 1978) for measurements of spectral line intensities and profiles in the visible range. These authors performed four sets of one-dimensional line-of-sight measurements, each as for the usual Abel inversion, but displaced from each other and rotated relative to each other in steps of $45^o$.

As already mentioned, the reconstructed images of the emission coefficients are not necessarily unique. This was demonstrated for the x-ray images of tokamak sawtooth oscillations by Hanada et al. (1990), who found that they could discriminate between two possible model distributions only by using three detector arrays. Even more caution must be exercised in the interpretation of three-dimensional tomography in the absence of greatly overdetermined parameters of the possible source distributions. In medical tomography this over-determination is usually achieved by rotating the x-ray source, or in microtomography (Flannery et al. 1987) by rotating the object to be studied by x-ray backlighting. Note finally that optically thin emission or weak absorption are entirely equivalent in the present context. If finite optical depths are involved or if the emission is anisotropic as in the case of hard x-ray emission (von Goeler et al. 1995), the comparison with modeling results is generally the preferred, i.e., simpler and more experimentally oriented method. However, image reconstruction and radiative transfer equations, etc., may also be solved iteratively as in the analysis of the solar atmosphere.

# 9

# Radiation losses

In this and the following chapters, various applications of plasma spectroscopy will be discussed. Their selection is necessarily somewhat arbitrary, but they will hopefully serve as useful demonstrations of the general methods and principles described in the preceding chapters. A very broad class of applications is concerned with the energy loss or gain of plasmas because of emission or absorption of electromagnetic radiation. As usual, the need for comprehensive calculations of these processes is shared with astrophysics. Here the requirement of energy conservation within a stellar atmosphere not subject to any significant nonradiative energy transport must be imposed by having zero divergence of the spectrally integrated radiative flux which, in turn, is obtained from the radiative energy transfer equations of the preceding chapter. In many laboratory plasmas such a general approach is not necessary, because most of the emission normally comes from optically thin layers and because radiative heating, except for radio-frequency (Golant and Fedorov 1989) and microwave heating (Bekefi 1966), is not involved.

Very notable exceptions to the last point are laser-produced plasmas, in which the absorption of the, typically, visible laser light is indeed essential (Kruer 1988). Other exceptions are x-ray heated plasmas produced, e.g., for the measurement of absorption coefficients of hot and dense low, medium and high Z materials (Davidson et al. 1988, Foster et al. 1991, Perry et al. 1991, Springer et al. 1992, 1994, Schwanda and Eidmann 1992, DaSilva et al. 1992, Eidmann et al. 1994, Winhart et al. 1995). Experiments in the latter category are especially valuable because of the relatively uniform deposition of x-ray energy in the samples used. They thus come close to the ideal of spatially uniform plasmas, albeit at a great expense in heating efficiency. Because of their fairly high densities, they serve the important purpose of verifying LTE opacity calculations, as discussed in sections 8.1 and 8.3.

Before discussing in the following sections the contributions to radiation losses of bremsstrahlung associated with free-free transitions, of recombination radiation from free-bound transitions and of line radiation, an important contribution to radiation losses from strongly magnetized plasmas, namely, electron-cyclotron and cyclotron harmonic radiation, must be mentioned, or electron synchrotron radiation in the case of fully relativistic electrons. Although the classical theory of radiation as reviewed in sections 1.1 and 1.2 is nearly always sufficient in these cases, with the basic theoretical relations developed by Schott (1912), there are rather intricate problems in accounting for transport and dispersion effects (see, e.g., Bekefi 1966, Bornatici et al. 1983, Griem 1983, Hutchinson 1987) and for nonthermal electron velocity distribution functions. Naturally, the radiation associated with electrons spiraling in magnetic fields is also strongly polarized. This polarization is actually taken to be the signature of synchrotron radiation in astronomy, because it always occurs jointly with its anisotropy. Finally, a number of cyclotron harmonics from tokamaks tend to be optically thick. They therefore may approach blackbody radiation over frequency intervals usually determined by Doppler shifts, but also related to magnetic field gradients along the line of sight. These result in frequency shifts analogous to the Sobolev (1957) effect for expanding or contracting plasmas, which was mentioned in section 8.4. Nevertheless, the corresponding radiation loss tends to be small compared to that from high Z line radiation, basically because even for the highest optically thick electron cyclotron harmonic the ratio of photon and electron thermal energy is usually extremely small.

## 9.1 Bremsstrahlung losses

In fully ionized, high temperature plasmas, electron-ion collisions, and in relativistic plasmas also electron-electron collisions, give rise to electromagnetic radiation because of the deceleration during the collisions. This bremsstrahlung (German for "braking radiation") is therefore a universal and irreducible energy loss process, e.g., for magnetic confinement fusion plasmas which are optically thin for almost the entire spectrum. From the coefficient for inverse bremsstrahlung absorption, in plasma physics terminology, according to the last term in (5.12) and correcting for induced emission by multiplying with the factor $1 - \exp(-\hbar\omega/kT)$, this optical depth is for Maxwellian electron distributions

$$
\begin{aligned}
\tau_b \;\approx\; & 256\alpha\pi \left(\frac{\pi}{3}\right)^{3/2} g_z(\omega, T) \left[1 - \exp\left(-\frac{\hbar\omega}{kT}\right)\right] \\
& \times \left(\frac{E_H}{kT}\right)^{1/2} \left(\frac{E_H}{\hbar\omega}\right)^3 a_0^5 z'^2 N_i N_e d,
\end{aligned}
\tag{9.1}
$$

if we neglect the high density effects discussed in section 5.5. Here $z'$ is an effective charge of the ions which equals the nuclear charge $Z$ in case of completely stripped ions, $N_i$ is the ion density, $T$ the electron temperature, and $d$ the geometrical depth or characteristic size of the plasma. In case of ionic mixtures, $z'^2 N_i$ must be replaced by an appropriate average, and in case of incompletely stripped ions, the calculations mentioned in section 5.3 should be used to estimate $z'$, which is then generally smaller than the nuclear charge because of screening by the remaining bound electrons. However, these bound electrons may also add to the absorption by dynamical polarization effects (Tsytovich and Oiringel 1992).

For our present purpose, it is almost always safe to approximate the stimulated emission factor by $[\cdots] \approx \hbar\omega/kT$ and to use (5.27) for the Gaunt factor $g_z(\omega, T)$, leading to

$$\tau_b \approx 128\alpha \frac{\pi^{3/2}}{3}\{\cdots\} \left(\frac{E_H}{kT}\right)^{3/2} \left(\frac{E_H}{\hbar\omega}\right)^2 z'^2 N_i N_e a_o^5 d \qquad (9.2)$$

with the curly bracket as in (5.27). With $\tau_b \approx 1$, the frequency below which bremsstrahlung is significantly self-absorbed is thus approximately given by

$$\frac{\hbar\omega_1}{E_H} \approx 1.3\{\cdots\}^{1/2} \left(\frac{E_H}{kT}\right)^{3/4} \left(z'^2 N_i N_e\right)^{1/2} a_0^3 \left(\frac{d}{a_0}\right)^{1/2}. \qquad (9.3)$$

For high temperature laboratory hydrogen plasmas as encountered in fusion research, one would find, e.g., $\hbar\omega_1/E_H \lesssim 10^{-7}$ for large tokamaks and $\lesssim 10$ for high density laser-fusion plasmas. In the first case, these $\omega_1$ values are well below the plasma frequency, which would therefore provide an effective low frequency limit were it not for the fact that electron cyclotron and plasma frequencies are usually close to each other in such plasmas. If higher cyclotron harmonics are strong, they then tend to dominate the spectrum well beyond $\omega_1$, and inverse bremsstrahlung is not important. In the second case, say, at $N_e \approx 10^{25} \, \text{cm}^{-3}$, $\omega_1$ and the electron plasma frequency have similar magnitudes, and we should multiply (9.1) by $[1-(\omega_p/\omega)^2]^{-1/2}$, as discussed near the end of section 2.5, and make some further corrections (Dawson and Oberman 1962, Basov and Krokhin 1964, Johnston and Dawson 1973). However, the conclusion remains that bremsstrahlung somewhat above the plasma frequency would escape from $z = 1$ high temperature laboratory plasmas rather freely. Since these plasmas also have $\hbar\omega_p < kT$, their total bremsstrahlung power loss can usually be evaluated neglecting absorption.

On the other hand, for low temperature, high density plasmas, $\hbar\omega_1$ values may well be comparable to $kT$ or even larger. Then a significant portion of the bremsstrahlung spectrum could be greatly reduced by radiative transfer and refractive effects. An example for such situations are capillary-confined copper-wire plasmas (DeSilva and Kunze 1994) used for electrical conductivity measurements in strongly coupled plasmas ($kT/E_H \approx 0.1$, $N_e \approx 10^{22}$ cm$^{-3}$). For such conditions and assuming $z' \approx 2$, $d \approx 100\,\mu$m, (9.3) with $\{\cdots\} \approx 1$ suggests $\hbar\omega_1 \approx 2$ keV, at which photon energy one should include photoionization of bound electrons according to sections 5.1 and 5.2. In any event, radiation from such plasmas is dominated by transport effects and must be analyzed as discussed in the preceding chapter.

Assuming, nevertheless, radiation transport effects not to be important, we will hence take the frequency- and solid-angle integrated bremsstrahlung emission power density from the last term in (5.21), i.e.,

$$P_{ff} \approx \frac{64}{3} \left(\frac{\pi}{3}\right)^{1/2} (\alpha a_0)^3 \left(\frac{kT}{E_H}\right)^{1/2} \frac{E_H^2}{\hbar} z'^2 N_i N_e, \qquad (9.4)$$

as the reference for further discussions of radiation losses. We omit Gaunt or Biberman factors, which were discussed below (5.21), although their spectrally averaged values may be as large as $\sim 4$. Using for $z'$ the ionic charge of the target ions and omitting the $\xi_{ff}$ factor is appropriate only if contributing photon energies are smaller than ionization energies of remaining bound electrons. Since mostly photon energies $\hbar\omega \lesssim kT$ are important here, and ionization energies are larger than $kT$, these simplifications are justified for many plasmas near steady state ionization-recombination balance, in which ionization energies are larger than $kT$. High Z plasmas are an exception even if they are in steady state, not to mention rapidly ionizing plasmas. In extreme cases, i.e., at electron temperatures corresponding to $K$-shell ionization energies, $z'$ would come close to the nuclear charge, even if most of the atoms are still neutral or partially ionized.

In mixtures, the product $z'^2 N_i$ in (9.4) must be replaced by an appropriate average; and heavy, easily ionized impurity atoms are seen to be particularly effective in causing bremsstrahlung losses. Assuming such an admixture not to be important, we finally estimate a rate constant $R_{ff}$ for electron cooling due to this radiative process by taking the ratio of $P_{ff}$ and the energy density $(3/2)N_e kT$ of the electrons, namely

$$
\begin{aligned}
R_{ff} &\approx \frac{128}{9} \left(\frac{\pi}{3}\right)^{1/2} (\alpha a_0)^3 \left(\frac{E_H}{kT}\right)^{1/2} \frac{E_H}{\hbar} z'^2 N_i \\
&\approx 1.73 \times 10^{-14} \left(\frac{E_H}{kT}\right)^{1/2} z'^2 N_i \text{ sec}^{-1}.
\end{aligned} \qquad (9.5)
$$

Here $N_i$ is in cm$^{-3}$ units in the numerical expression, which must still be multiplied by Gaunt or Biberman factors for more accurate estimates.

As will become clear from the discussions in the following sections, cooling rates associated with recombination and line radiation are frequently much larger than the bremsstrahlung rate; but the requirement of balancing the unavoidable (in the laboratory) $z' = 1$ bremsstrahlung losses by the nuclear fusion power from D-T reactions gives the minimum temperature for a self-sustaining fusion reactor (Glasstone and Lovberg 1960). At high energies, electron-electron collisions result in additional bremsstrahlung (Joseph and Rohrlich 1958, Maxon and Corman 1967). In such cases, velocity distributions are generally quite different from Maxwellians, often following an inverse power law.

## 9.2    Recombination radiation losses

Both radiative and dielectronic recombinations are accompanied by the emission of electromagnetic radiation, with continuum or cascading line spectra, respectively. The former, free-bound continuum contribution to the radiative loss power density may be estimated from (5.16) and (5.19) by integrating over frequencies and solid angles,

$$P_{fb} \approx \frac{64}{3} \left(\frac{\pi}{3}\right)^{1/2} (\alpha a_0)^3 \left(\frac{E_H}{kT}\right)^{1/2} \frac{E_\infty E_H}{\hbar} z^2 N_i N_e. \qquad (9.6)$$

Here we neglected reductions of ionization energies and advances of series limits in dense plasmas, and we again omit Gaunt or Biberman factors. For recombination with bare ions, it is better not to use (5.19), but rather the first term in (5.17). This amounts to replacing the ionization energy $E_\infty$ of the recombined ion by $2z^2 E_H \sum g_n/n^3$, the sum being over principal quantum numbers of the final states and the $g_n$ being the Gaunt factors discussed in section 5.2.

From (9.4) and (9.6), the ratio of recombination continuum to bremsstrahlung emission is seen to be estimated by $E_\infty/kT$, which can be significantly larger than one. Only for $kT > E_\infty$ would bremsstrahlung become dominant, especially on incompletely stripped ions, because of the increased effective charges $z'$ in such situations, which were discussed in the preceding section. However, $E_\infty/kT$ is rarely very small for any important ionization stage, and line radiation would typically be more important in such situations anyway. The radiative cooling rate constant for electrons corresponding to the sum of (9.4) and (9.6) is finally analogous to (9.5), i.e., with $N_i$ again in cm$^{-3}$,

$$
\begin{aligned}
R_{\text{cont}} &\approx \frac{128}{9} \left(\frac{\pi}{3}\right)^{1/2} (\alpha a_0)^3 \left(\frac{E_H}{kT}\right)^{1/2} \frac{E_H}{\hbar} \left(1 + \frac{E_\infty}{kT}\right) z^2 N_i \\
&\approx 1.73 \times 10^{-14} \left(\frac{E_H}{kT}\right)^{1/2} \left(1 + \frac{E_\infty}{kT}\right) z^2 N_i \ \sec^{-1}, \qquad (9.7)
\end{aligned}
$$

if we drop the distinction between $z'$ and $z$, besides the Gaunt or Biberman factors. The reader should also remember the need for appropriate averages in multispecies plasmas, in which high $Z$ ions are particularly important.

These high $Z$ ions are even more important for radiation losses associated with dielectronic recombination. This contribution to the electron loss power density may be estimated by multiplying the individual contributions to the effective dielectronic rate coefficient (6.47) by the corresponding excitation energies $E_{m'm}$, and the resulting expression by the product of ion and electron densities. The radiated energy, including that associated with the deexcitation of the bound electron, is somewhat larger than $E_{m'm}$, say, by a factor of 1.5 on the average. In this manner and using (6.43), we estimate

$$P_{dr} \approx \frac{6\pi^{5/2}}{\alpha} \left(\frac{E_H}{kT}\right)^{3/2} \sum_{m'} \left[\left(\frac{3\bar{g}'}{z}\right)^{1/3} \ell_m'^{4/3} \frac{E_{m'm}^2}{\hbar} f_{m'n}\right.$$
$$\left. \times \exp\left(-\frac{\bar{E}_{m'm}}{kT}\right)\left(1 - 0.73 x_1' - \frac{0.37}{x_2'^2}\right)\right](\alpha a_0)^3 N_i N_e. \quad (9.8)$$

Remember also the choice of $\ell_m' \approx 3n_m$ or $\ell_m' \approx 1.5n_m$ for $\Delta n = 0$ or $\Delta n = 1$ excitations and $\bar{g} \approx 1$ or 0.2, respectively, for the effective Gaunt factors.

According to this very approximate expression and to (9.6), the ratio of dielectronic and radiative recombination energy loss rates may be written as

$$\frac{P_{dr}}{P_{fb}} \approx \frac{3^{5/2}\pi^2 E_H}{32\alpha z^2 kT} \sum_{m'} [\cdots], \quad (9.9)$$

with the square bracket as in (9.8) except for the replacement of $E_{m'm}^2/\hbar$ by $E_{m'm}^2/E_\infty E_H$, namely

$$[\cdots] = \left(\frac{3\bar{g}'}{z}\right)^{1/3} \ell_m'^{4/3} \frac{E_{m'm}^2}{E_\infty E_H} f_{m'm}$$
$$\times \exp\left(-\frac{\bar{E}_{m'm}}{kT}\right)\left(1 - 0.73 x_1' - \frac{0.37}{x_2'^2}\right). \quad (9.10)$$

Numerical values of $[\cdots]$ for temperatures approaching $E_{m'm}$ values for $\Delta n = 1$ transitions are typically of order $2z^2/9$, which results in

$$\frac{P_{dr}}{P_{fb}} \approx \frac{\sqrt{3}\pi^2 E_H}{16\alpha kT}$$
$$\approx 1.5 \times 10^2 \frac{E_H}{kT}. \quad (9.11)$$

This extremely crude estimate is in qualitative agreement with detailed calculations discussed in the final section of this chapter, e.g., with corona model calculations for iron (Davis et al. 1977, Jacobs et al. 1977a), but naturally cannot reflect the temperature modulation in this ratio associated with closed shells. Needless to say, (9.11) must not be used for completely stripped ions.

Another mechanism enhancing radiation loss, often called charge exchange recombination, involves charge transfer collisions as discussed in section 6.5, which preferentially populate excited states. Corresponding line radiation losses will be discussed further in section 9.4.

## 9.3    Line radiation losses

Near the end of the preceding section we have already begun to estimate some contributions of bound-bound transitions to the loss power density, namely those by transitions that stabilize the doubly-excited states reached by the radiationless capture reactions. This represents the first step in the complex dielectronic recombination process discussed in section 6.4. However, the corresponding satellites to normal lines of the ions that have not yet recombined tend to be weaker than these lines involving only singly-excited states, although they tend to be more numerous and are often unresolved from the parent line. The evaluation of the normal line contributions to the loss power density is therefore most critical, e.g., for the determination of permissible levels of heavy atom impurities in high temperature D-T plasmas for fusion reactor development.

Such line radiation loss power calculations, using detailed atomic physics as reviewed in chapter 6, will be discussed in the following section. Here we obtain only a first estimate for comparison with the other radiation loss processes by using the approximate excitation rate coefficient (6.27), multiplied by the excitation energies and the product of ion and electron density,

$$
P_{bb} \approx 32\pi \left(\frac{\pi}{3}\right)^{1/2} \left(\frac{E_H}{kT}\right)^{1/2} \frac{E_H^2}{\hbar}
$$

$$
\times \left[\sum_m \bar{g} f_{mn} \exp\left(-\frac{E_{mn}}{kT}\right)\right] a_0^3 N_i N_e. \tag{9.12}
$$

As in (9.8), the effective Gaunt factors are $\bar{g} \approx 1$ or 0.2 for $\Delta n = 0$ or $\Delta n \geq 1$ transitions. For temperatures approaching $E_{mn}/k$ for $\Delta n = 1$ transitions, the square bracketed factor is typically of order $n_e/3$, if $n_e$ is the number of equivalent electrons in the outer shell of the ion. Making this assumption and using (9.6) allows us to write the ratio of line and recombination radiation losses as

$$\frac{P_{bb}}{P_{fb}} \approx \frac{\pi n_e E_H}{2\alpha^3 z^2 E_\infty}, \tag{9.13}$$

$E_\infty$ being the ionization energy of the recombined ion and $n_e = 0$ for fully stripped ions. This ratio is very large for medium-heavy and incompletely stripped ions, namely $\sim 140$ for FeXXIII. As was our estimate (9.11) for dielectronic recombination radiation, this is consistent with the results of detailed calculations (Davis et al. 1977, Jacobs et al. 1977a). Taking ratios of (9.13) and (9.11) we can also estimate

$$\begin{aligned} \frac{P_{bb}}{P_{dr}} &\approx \frac{8 n_e}{\sqrt{3}\pi\alpha^2 z^2} \frac{kT}{E_\infty} \\ &= \left(\frac{166}{z}\right)^2 n_e \frac{kT}{E_\infty} \end{aligned} \tag{9.14}$$

which, together with $kT \lesssim E_\infty$ for any abundant high charge state, indicates that the dielectronic contribution is relatively small, except for very heavy and highly charged ions. For ions with many equivalent electrons, (9.11) may well underestimate $P_{dr}$, corresponding to some cancellation of the $n_e$ factor in (9.14).

Absolute values of the radiative loss power density for the various atoms and ions in a given plasma can be estimated from (9.4), (9.6), (9.8) and (9.12), provided the ionic abundances for a given chemical composition have been measured or calculated from one of the kinetic models discussed in section 6.1. One frequently finds that line losses dominate even in the case of relatively small admixtures of heavy impurities, where the cooling rate constant becomes instead of (9.7)

$$\begin{aligned} R_{\text{lines}} &\approx 6\pi \left(\frac{\pi}{3}\right)^{3/2} \left(\frac{E_H}{kT}\right)^{3/2} \frac{E_H}{\hbar} \left[\sum_m \bar{g} f_{mn} \exp\left(-\frac{E_{mn}}{kT}\right)\right] a_0^3 N_i \\ &\approx 6.2 \times 10^{-8} \left(\frac{E_H}{kT}\right)^{3/2} [\cdots] N_i \ \text{sec}^{-1}, \end{aligned} \tag{9.15}$$

obtained by division of (9.12) by $(3/2)N_e kT$, and with $N_i$ in $\text{cm}^{-3}$ units.

Before applying such estimates to any experimental situation, it is very important to verify the implicit assumption that the plasma is optically thin, even for the resonance lines. Using (8.4), (8.5) and (8.17), this optical depth is

$$\tau(\omega) \approx 2\pi^2 r_0 c f_{mn} L_{mn}(\omega) N_i \left(1 - \frac{g_n}{g_m} \frac{N_m}{N_n}\right) x, \tag{9.16}$$

if $x$ is a characteristic dimension of the plasma, and $N_i \approx N_n$ the ground-state density averaged along the line of sight. The induced emission correction $g_n N_m/g_m N_n$ is normally negligible except, perhaps, for $\Delta n = 0$

transitions. This leaves us with the line shape function $L(\omega)$, which for Doppler broadening in a Maxwellian plasma is given by (4.1) and (4.2), i.e., has a peak value of

$$
\begin{aligned}
L(\omega_{mn}) &= \frac{1}{\pi^{1/2}\omega_D} \\
&= \left(\frac{Mc^2}{2\pi k T_i}\right)^{1/2} \frac{1}{\omega_{mn}}
\end{aligned} \tag{9.17}
$$

in terms of the ion temperature $T_i$. Combining (9.16) and (9.17) we obtain

$$
\tau \lesssim 4\pi^{3/2} \left(\frac{E_H}{k T_i}\right)^{1/2} \frac{f_{mn} E_H}{\hbar \omega_{mn}} a_0^2 N_i x, \tag{9.18}
$$

omitting the correction for induced emission. Additional line broadening, e.g., from electron and ion collisions as discussed in chapter 4, or line splitting due to fine structure or even hyperfine structure not allowed for in the line loss calculations, would result in smaller optical depths.

Another general point to make concerns the somewhat indirect connection between electron energy loss rates and radiated power. This difficulty became obvious in our estimate of the loss related to dielectronic recombination near the end of the preceding section, but also occurs with normal line radiation in cases where the corona model is only marginally valid.

## 9.4 Numerical calculations of radiation losses

Because of their crucial role in magnetic fusion research (Isler 1984), and also in the physics of the solar corona, a number of much more detailed calculations of radiation losses from low density, high temperature plasmas have been undertaken based on atomic cross sections and kinetic models as introduced in chapter 6. A complete review of the results of such extensive calculations is not possible here, but enough examples can be given for comparison with the various estimates and some experiments in this category. For detailed references the reader should consult recent issues of the International Bulletin on Atomic and Molecular Data for Fusion of the International Atomic Energy Agency (Vienna).

Radiative cooling rates of steady state optically thin plasmas in corona equilibrium were calculated by Post et al. (1977) for half the elements with nuclear charges ranging from $Z = 2$ to 92, using an average-atom model for the ions. Such a model, first introduced by Strömgren (1932), replaces the various charge states at a given temperature by an average ion, which therefore has fractional populations in the outer shells. These fractional populations are obtained by balancing the rates at which electrons are added to a given principal quantum number shell, or removed from it, by radiative and collisional processes characterized by rate coefficients such as those introduced in chapters 2 and 6. The energy levels were taken to be hydrogenic, with relativistic corrections and using effective nuclear charges $Z_n$ for the various shells. These $Z_n$'s depend on all shell populations because of so-called inner and outer screening. They must therefore be determined self-consistently with the solution of the population balance equations.

The various contributions to the radiative loss rate could then be calculated similarly to the estimates in the preceding sections, but separately for the different $n$ to $n'$ transitions and using more involved expressions for the dipole excitation rate coefficients. Transitions with $\Delta n = 0$ were allowed for after augmenting the average ion model by corresponding level splittings, oscillator strengths, etc. These quantities were obtained from detailed atomic structure calculations by Post et al. (1977), assuming $LS$ coupling or $jj$ coupling for low and high $Z$, respectively. As to the radiation associated with dielectronic recombination, a factor $E_\infty + \langle E \rangle$ was applied to the dielectronic recombination coefficient, instead of the estimate 1.5 $E_{m'm}$ used in (9.8). In this case, $E_\infty$ is the ionization energy of the recombined ion, while $\langle E \rangle$ is an average of the excitation energies $E_{m'm}$.

Although Post et al. (1977) find for $Z \leq 26$ excellent agreement with other, less comprehensive but more detailed radiative loss power calculations (Breton, DeMichelis and Mattioli 1976, Merts, Cowan and Magee 1976, Tarter 1977), their estimated uncertainties correspond to a factor of 2 in the radiated power at given plasma conditions, and perhaps more than that for high-Z elements. Moreover, since only dipole ($\Lambda = 1$) excitations were included in the line loss calculation, there might have been a systematic underestimate, especially of the $\Delta n = 1$ contributions. As already discussed in section 6.3, for some of these transitions, monopole ($\Lambda = 0$) excitations could actually be very important.

The magnitude of this systematic error may be estimated by using a relation for monopole collision strengths $\Omega_0$ deduced by Hagelstein and Dalhed (1988), namely

$$\Omega_0 = \xi_{n\ell} \eta_{n\ell} n_e \frac{E_H}{E_{J'J}} \qquad (9.19)$$

for the near-threshold excitation of one of the $n_e$ electrons in the outer shell to $n + 1$, $\ell' = \ell$. The factors $\xi_{n\ell}$ are about 0.3, 0.5 and 0.6 for He-, Ne-, and Ni-like ions, while $\eta_{n\ell}$ ranges between 1 and 0.5 for excitation energies up to 2.5 keV for the dominant transition. Since the collision strength is related to the excitation cross section by (Hebb and Menzel

1940, Cowan 1981)

$$\Omega = \frac{E_{J'J}}{E_H}(2J + 1)\frac{\sigma}{\pi a_0^2}, \tag{9.20}$$

the effective Gaunt factor approximation (6.26) for dipole ($\Lambda = 1$) excitations corresponds to

$$\Omega_1 \approx \frac{8\pi}{\sqrt{3}}(2J + 1)\sum_{J'} f_{J'J}\ \overline{g}_{eff}\frac{E_H}{E}. \tag{9.21}$$

Here $f_{J'J}$ is the absorption oscillator strength as in (2.48), $\overline{g}_{eff}$ the effective Gaunt factor discussed in section 6.3, and $E \geq E_{J'J}$ is the energy of the incoming electron. From (9.19) and (9.21) we find for the ratio of monopole to dipole collision strengths (or cross sections)

$$\frac{\Omega_0}{\Omega_1} \approx \frac{\sqrt{3}}{8\pi}\frac{\xi_{n\ell}n_e}{(2J + 1)\sum_{J'} f_{J'J}\ g_{eff}}\frac{E}{E_{J'J}}. \tag{9.22}$$

Threshold effective Gaunt factors are roughly 0.2. The quantity $(2J + 1) \times \sum_{J'} f_{J'J}$, i.e., essentially the $\Delta n = 1$ oscillator strengths summed over the final angular momentum states, are about equal to $n_e$; so that we estimate $\Omega_0/\Omega_1 \approx \xi_{n\ell}/3$ at threshold, e.g., 0.10, 0.16 and 0.20 for He-, Ne-, and Ni-like ions, respectively. Radiation losses due to $\Lambda = 1$ excitations are therefore indeed dominant, as assumed by Post et al. (1977). This is, perhaps, different from what might have been expected from the significant role of $\Lambda = 0$ excitations in populating long-lived upper laser states, thus producing population inversions in collisionally pumped x-ray lasers (Elton 1990) as mentioned in section 6.3. In any event, the systematic errors due to omission of $\Lambda = 0$ excitations, as well as excitations via $\Lambda \geq 2$ interactions, are not significant relative to errors in rate coefficients and kinetic models used for radiation loss calculations. This is similar to much of the experience with electron collisional line broadening calculations as discussed in section 4.7.

The situation is not changed much by the use of distorted wave code results (Davis, Kepple and Blaha 1976, Davis et al. 1977) for the dipole excitation cross sections instead of the effective Gaunt factor approximation (6.26). This is especially true if $g_{eff}$ is chosen (Post et al. 1977) as a function of electron energy, etc., to reflect the results of ab initio cross section calculations. As shown in figure 9.1, these two detailed calculations for the radiation energy loss coefficient of iron plasmas in coronal equilibrium exhibit the same shell structure, although those of Post et al. (1977) have some of their maxima and minima at higher temperatures than those of Davis et al. (1977). This is a consequence of reduced dielectronic recombination coefficients in the latter calculation, in which the branching ratio, e.g., in (6.42) would contain an additional term in the denominator, which allows for $\Delta n = 0$ radiative decay of the doubly-excited states into autoionizing levels. The shift of ionic abundances versus temperature is probably also the main reason for the smaller Fe loss coefficients obtained by Davis et al. (1977) compared to those of Post et al. (1977), e.g., by a factor of 2 at $kT_e = 1\,\text{keV}$. The differences in the expressions used for the radiation loss associated with dielectronic recombination are less important. For comparison with some earlier calculations, see p. 416 of Post et al. (1977).

The total radiation loss coefficients, when multiplied with electron and total ion densities, yield the power densities due to the various processes estimated in the preceding sections of this chapter. These estimates according to (9.4), (9.6), (9.11) and (9.12), with $n_e [\cdots] = 0.6$, for $kT = 1\,\text{keV}$ and assuming Fe XXI to be the dominant species, are also indicated in figure 9.1 for comparison with the detailed calculations. They are found to be in reasonable agreement with the individual contributions given for only one of the detailed calculations (Davis et al. 1977). Of these contributions, bound-bound transitions comprise over 95% of the estimated total radiation loss at this temperature.

As already indicated at the end of section 9.2, charge transfer, e.g., from energetic hydrogen atoms into excited states of originally fully stripped impurity ions, can add to the line radiation losses discussed so far, although in this case certainly not at the expense of the energy of free electrons. Estimating the corresponding cross section to $\pi R_n^2/z$ with $1/z$ as probability for the transfer to occur (Janev, Presnyakov and Shevelko 1985) and $R_n \lesssim za_0$ according to the discussion following (6.50), and the energy radiated by the captured electron to $z_B^2 E_H/n_g^2$, $n_g$ being the principal quantum number of the ground state of the recombined ion, the resulting loss power density is

$$P_{cx} \lesssim \pi a_0^2 z v_H (z^2 E_H/n_g^2) N_i N_H. \tag{9.23}$$

For monoenergetic neutral beams, i.e., $v_H = (2E/M_H)^{1/2}$, this contribution to line radiation is best compared with normal line radiation from electron impact excitation. Using (9.12) and the same estimate for the factor in the square bracket as in (9.13), the corresponding ratio becomes

$$\frac{P_{cx}}{P_{bb}} \lesssim \frac{\pi}{16} \left(\frac{3}{\pi}\right)^{3/2} \left(\frac{mkTE}{M_H E_H^2}\right)^{1/2} \frac{z^3}{n_e n_g^2} \frac{N_H}{N_e}. \tag{9.24}$$

If the recombining ions are hydrogen- or helium-like, i.e., for $n_e n_g^2 = 1$ or 2, respectively, and for, e.g., $z \approx 10$, $kT$ and $E \approx 10\,\text{keV}$, the factor of $N_H/N_e$ turns out to be $\sim 3 \times 10^4$ and $1.5 \times 10^4$, respectively. Relatively small neutral atom concentrations thus can cause a significant increase in

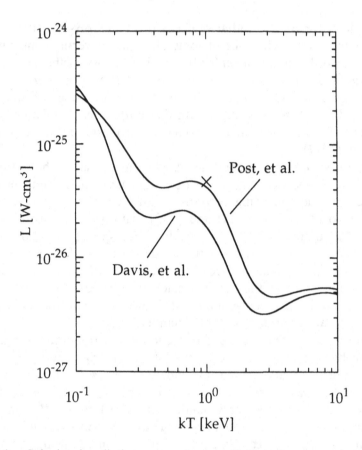

Fig. 9.1.  Calculated radiation energy loss coefficients for corona equilibrium plasmas containing iron ions according to Post et al. (1977) and to Davis et al. (1977). Also indicated, by the cross at $kT = 1\,\text{keV}$, is an estimate obtained as discussed in the text.

line radiation, although better cross sections (Janev et al. 1985), and an average over impurity species and charge states would be required for a more realistic assessment of the total charge exchange radiation.

In applications, some of which will be discussed further in the final chapter, additional uncertainties may arise. For example, the coronal model may become inappropriate because of rapid ion transport in spatially inhomogeneous plasmas, or because of rapid variations in plasma conditions. At increasing electron densities, one may, especially for low $Z$ ions, also have to allow for two-step collisional excitation and ionization (Sasaki et al. 1994), and for reductions in dielectronic recombination rates as discussed in section 6.4.

There are very few measurements of radiation losses from high temperature plasmas with known electron and ion densities. In one of the experiments (Zhu et al. 1983), the absolute density of Fe X ions was determined (Wang and Griem 1983) from the absolute intensity of the coronal Fe X line due to a forbidden transition between fine-structure levels of the groundstate configuration, whose relative populations could be assumed to be as in LTE. The relative densities of Fe VII, VIII, IX and XI ions were inferred from solutions of the time-dependent rate equations using measured electron temperatures and densities from Thomson scattering as input. Together with the Fe X density, this gave a total iron ion-concentration amounting to 2% of the electron density. At a measured temperature of 45 eV, the actual charge state distribution was thus found to be equivalent to coronal equilibrium at 65 eV, using the relative abundances of Post et al. (1977). Scaling their loss coefficient at 65 eV to 45 eV, according to (6.27), their prediction becomes $2.5 \times 10^{-25} \mathrm{W} \, \mathrm{cm}^3$. An analogous scaling of the results of Davis *et al.* (1977) gives a loss power coefficient of $3.6 \times 10^{-25} \mathrm{W} \, \mathrm{cm}^3$, while the measured value of Zhu et al. (1983) is $1.6 \times 10^{-25} \mathrm{W} \, \mathrm{cm}^3$, with estimated errors of $+70\%$, $-40\%$. The larger error in the positive direction was assigned to account for lines that may have escaped detection; but it still does not accommodate the larger of the two theoretical values, which is about a factor of 3 below a maximum value reported for M-shell emission (Davis and Jacobs 1980). This apparent discrepancy is probably not significant because of the theoretical and experimental uncertainties. This conclusion is supported by comparison with a third calculation (Uchikawa, Griem and Düchs 1980) listing specific contributions to the line loss power coefficient. It yielded a loss coefficient of $2.3 \times 10^{-25} \mathrm{W} \, \mathrm{cm}^3$ at the conditions of the experiment.

Another experiment (Isler, Neidigh and Cowan 1977) on tungsten radiation from a tokamak plasma near 1 keV also gave loss powers consistent with calculations as discussed here at tungsten concentrations of 0.17% of the electron density. This concentration was inferred from absolute line intensities using approximate excitation rate coefficients. It would probably be fruitful to resume such measurements and compare them with calculations based on atomic data as reviewed in a recent reference book (Drake and Hedgecock 1996) or as those employed for astrophysical opacity calculations discussed in the preceding chapter. For high Z impurities, a non-LTE version of the supertransition array method (Bar'-Shalom et al. 1995) should be particularly useful. This statistical method for the evaluation of complex spectra had been mentioned already in section 8.3 because of its use in opacity calculations.

# 10

# Spectroscopic density measurements

Depending on various plasma conditions, such as size, composition, densities and temperatures, the electron density and related quantities can be measured using a number of techniques. Of these, Langmuir probes (Tonks and Langmuir 1929, Hutchinson 1987) provided the first means to infer local values of the electron density, mostly at relatively low densities. Much more recently, Thomson scattering of laser light has become a method of choice for localized electron density measurements (Kunze 1968, Evans and Katzenstein 1969, DeSilva and Goldenbaum 1970, Sheffield 1975) over a range of about $10^{11}$ to $10^{21}$ cm$^{-3}$. Then there are the usually inherently most accurate interferometric techniques, mostly laser-based as well (Jahoda and Sawyer 1971, Hauer and Baldis 1988), which span a similar range, recently extended to $N_e \approx 3 \times 10^{21}$ cm$^{-3}$ (DaSilva et al. 1995), but provide localized values of the electron density only indirectly. Using two or more wavelengths, one can, however, separate the free-electron contribution (1.37) to the refractive index from any boundstate contribution (2.103). In single-species, partially ionized plasmas, this boundstate contribution is a direct measure of the neutral atom density. Interferometric local values of the density can, in principle, be determined by methods analogous to those discussed for emission or absorption measurements in section 8.5. These methods must also be used to infer local density values from spectroscopic density measurements, which will be the only ones discussed in any detail.

The purely spectroscopic density determinations are based on the interpretation of measurements of at least one of the following quantities: spectral line widths or profiles, absolute continuum intensities, absolute line intensities, or relative line intensities. In all cases, this interpretation depends on some knowledge of the temperature. In general, an iterative procedure using also the methods of temperature determination as discussed in the following chapter is thus called for. In practice, this iterative procedure is frequently replaced by comparison between measured and synthetic spectra, the latter being calculated for sets of assumed plasma conditions until a satisfactory fit is obtained. Such synthetic spectra are calculated using the procedures described in chapters 6, 7 and 8 and normally require a large set of atomic data. Plasma conditions are either taken from a preliminary analysis of a subset of the spectroscopic measurements, or from theoretical predictions, e.g., of conditions in laser-produced plasmas obtained with the help of a hydrodynamic code. In either case, redundancy is desirable to avoid the inherent ambiguity of such inversion problems.

In the following sections, we will discuss the four above-mentioned spectroscopic density determinations independently.

## 10.1  Densities from spectral line widths and profiles

Under favorable circumstances the widths, shifts and profiles of suitable spectral lines are very insensitive to both electron and ion temperatures, at least according to theoretical considerations discussed in chapter 4. The proviso here is that there are relatively few lines for which this temperature dependence, or rather the lack of it, has been checked experimentally. In any case, it is perhaps worth noting that by using suitable lines electron densities have been inferred ranging from about $N_e = 0.1 \, \text{cm}^{-3}$ (Payne, Anantharamaiah and Erickson 1994) in interstellar molecular clouds to over $N_e = 1 \times 10^{24} \, \text{cm}^{-3}$ (Keane et al. 1993, 1994, Hammel et al. 1994, Hooper et al. 1995, Nishimura et al. 1995) in inertially-confined laser-produced plasmas. In the first case, radio-frequency $\Delta n = 1$ transitions, i.e., $n - \alpha$ lines of C I were used, with principal quantum numbers $n \lesssim 740$. The high density measurement involved x-ray (K-shell) lines of helium- and hydrogen-like argon and (L-shell) lines of, e.g., neon-like krypton or xenon. Not quite as impressive as the factor of over $10^{25}$ between the electron densities in these experiments is the factor of about $10^{11}$ in the photon energies of the transitions used. Also, the temperature differences are fairly extreme, namely near 50 K for the molecular clouds and over $1 \times 10^7$ K for the laser-compressed plasmas.

As with some of the other spectroscopic methods for the determination of electron densities, line broadening does not require an accurate knowledge of the plasma composition. This fortunate situation is in the present case due to the insensitivity of Stark broadening calculations, as discussed in section 4.7, to variations in the ionic composition at fixed electron density. Although multiply-charged ions are contributing more to the broadening at low densities but constant electron density than what would correspond to a linear scaling with ionic charge, this effect is often compensated more or less by the decrease of average ion-microfields caused by ion-ion correlations (see section 4.3).

The most important experimental considerations for choosing spectral lines for plasma density measurements are their strengths and separations from neighboring lines; and also the wavelength range, i.e., the relative ease of obtaining sufficient resolution and signal. On the theoretical side, it is important that broadening mechanisms other than Stark effects can be ruled out, or at least accounted for. In regard to broadening caused by interactions with neutral atoms, one normally has to proceed by estimates according to section 4.8; and Doppler broadening can be calculated from (4.1) and (4.2), if the atom or ion kinetic temperatures are known. In practice, they are more often than not inferred from line widths and profiles of lines which are relatively insensitive to Stark broadening, not to mention interactions with neutrals (see section 11.8). Since these conditions are more likely to be fulfilled for lines with relatively low-lying excited states than for those with highly excited states, great care has to be exercised in regard to self-absorption and to instrumental resolution. The former effect must normally be estimated according to chapter 8, while instrumental broadening should be measured separately. Both effects are best accounted for by including them in any simulated spectra. If this is not possible, they should be allowed for in error estimates.

The principal advantages in using synthetic spectra for comparison with measured data before inferring electron densities are the possibility of fitting entire profiles, rather than only comparing measured and calculated line widths; and the ease of accounting for variations of plasma conditions along the line of sight. These variations are particularly important in the present context, because for purely Stark-broadened lines low density layers tend to contribute quite substantially to the central regions of observed profiles. This effect may well cause the observed line to be narrower by factors 2 to 3 than a line that would be emitted from a homogeneous plasma at the peak density along the line of sight.

If no information concerning electron density and excited-state densities is available, and if high accuracy is required, inversion or reconstruction methods as discussed in section 8.5 must be applied to spectra measured along a suitable array of lines of sight to extract values of local emission coefficients and line profiles. If some absorption and re-emission occurs, this also should be allowed for. However, it is preferable to use optically thin lines having high intensity and good sensitivity to Stark effects. A good example in this category is the $H_\beta$ line of hydrogen. For it the basic theory has been well verified, most recently by comparison with Abel-inverted data (Helbig 1991, Thomson and Helbig 1991, Kelleher et al. 1993) in experiments discussed in section 4.9 for which other density diagnostics were available. For the present purposes, the line profile measurements would be done similarly, but the density would be determined by comparison with a family of calculated profiles for a range of assumed electron densities. This procedure is, at temperatures near 1 eV, applicable for electron densities from about $10^{14}$ to $10^{17} \, \text{cm}^{-3}$. At, say, 10 eV, only densities above $10^{15} \, \text{cm}^{-3}$ can be determined reliably because of greater Doppler broadening.

Above $N_e = 10^{17} \, \text{cm}^{-3}$, it is preferable to use the less sensitive $H_\alpha$ line, because the $H_\beta$ line becomes too broad for a clean separation from the underlying continuum and from the $H_\gamma$ line. However, $H_\alpha$ will often be optically thick, and remaining uncertainties in calculated profiles seem considerable, as already mentioned in section 4.9. Nevertheless, as shown by the measurements of Böddeker et al. (1993), again with independent density diagnostics, $H_\alpha$ profiles can be used up to $N_e \approx 10^{19} \, \text{cm}^{-3}$. Higher Balmer-series lines can be used to extend this density diagnostics to lower electron densities even at relatively high electron temperatures. The characteristic $\alpha$- and $\beta$-type, i.e., single-peaked vs. double-peaked, structures prevail in calculated profiles (Bengtson, Tannich and Kepple 1970), which were based on quasistatic broadening by ions and electron-impact broadening. However, measured profiles in that work were found to be all nearly Lorentzian at $N_e \approx 2 \times 10^{13} \, \text{cm}^{-3}$, $kT \approx 0.2$ eV. This is also true for the profiles observed (Welch et al. 1995) in tokamak divertor and plasma edge regions shown in figure 10.1. Fits of calculated Stark profiles, etc., to observed spectra, an example of which is shown in figure 10.2, suggest electron densities $N_e \lesssim 5 \times 10^{14} \, \text{cm}^{-3}$ at $kT \lesssim 4$ eV. This is consistent with Langmuir probe data, if one allows for some smoothing of the central profile structures by Doppler, Zeeman and ion-dynamical effects and, more importantly, for spatial variations.

In all of these examples, the dependence on electron temperature is rather weak; and this would also be the case for the members of other spectral series of hydrogen, which are rarely used for laboratory density measurements. This is different for ionized helium, for which the $n = 4 \rightarrow 3$ "Paschen-$\alpha$" line is the most studied example, as discussed in section 4.9. Similar to the situation for the hydrogen $H_\alpha$ line, there is as yet no completely satisfactory calculation for this line. This can be seen from figure 4.10, in which measurements (Büscher et al. 1996) at independently determined electron densities are compared with various calculations and numerous measurements.

For other He II lines the situation may be similar, and the relatively few measurements on overlapping lines from more highly charged ions could not confirm the various calculations to much better than a factor of 2, as already discussed in section 4.9. Moreover, two additional

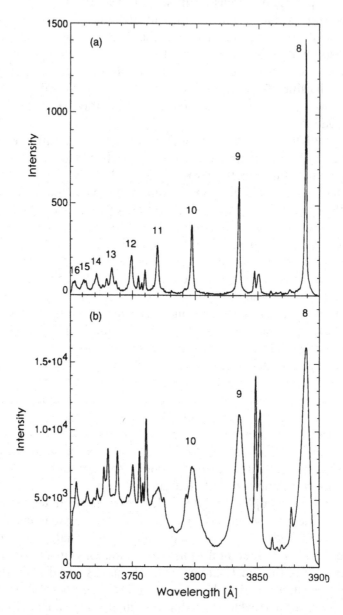

Fig. 10.1. High-*n* Balmer lines representative of tokamak divertor and plasma edge regions (Welch et al. 1995). (a) Low density spectrum showing distinct lines including $n = 16$ as upper level. (b) High density spectrum showing lines with increasing widths up to $n = 10$. The higher lines have merged into the continuum, leading to an advance of the Balmer edge from 3648 Å to about 3775 Å. The unmarked lines are F II and, mainly, O III impurity lines.

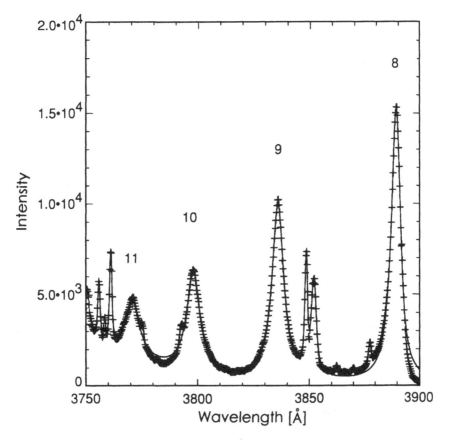

Fig. 10.2. Fit of a calculated spectrum, assuming $N_e = 5 \times 10^{14}\,\mathrm{cm}^{-3}$, $kT_e = 4$ eV, to a high density tokamak spectrum similar to that shown on figure 10.1(b). The fit includes the major impurity lines.

complications may arise. Depending on plasma conditions and their time-evolution, dielectronic satellite lines, partially blended with the low-photon-energy wings of the main lines, may be important. They should therefore be included in the calculation of synthetic spectra (Woltz et al. 1991, Mancini et al. 1992, Keane et al. 1993) to avoid an overestimate of the electron density from overall widths. Another complication arises from the electron-produced line shifts discussed in sections 4.7 and 4.10, which so far have been neglected in almost all overlapping line profile calculations. On the other hand, a recent experiment (Leng et al. 1995) on C VI $L_\alpha$, $L_\beta$, and $L_\gamma$ suggests that consistency between all three measured profiles and synthetic spectra at $N_e \le 10^{22}\,\mathrm{cm}^{-3}$ can be obtained only if the calculated shifts (Nguyen et al. 1986) are included, especially for $L_\gamma$. This is demonstrated in figure 10.3, where the calcula-

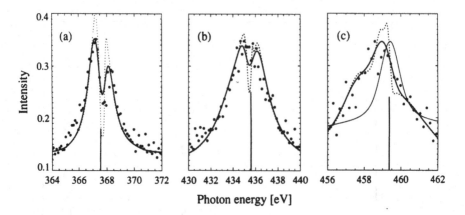

Fig. 10.3. Comparison of measured (solid dots) profiles of C VI Lyman-$\alpha$, $\beta$ and $\gamma$ lines (Leng et al. 1995) from a 10 psec KrF laser-produced plasma with calculated profiles before (dotted lines) and after (solid lines) allowing also for instrumental broadening. The vertical markers indicate the unperturbed line positions and the thin solid curve for Lyman-$\gamma$ represents the calculated profile without inclusion of electron-collisional shifts. The Lyman-$\alpha$ self-absorption dip is slightly shifted in the other direction due to Doppler shifts associated with the radial expansion.

tions without shifts have $L_\gamma$ profiles that are much too narrow and too symmetrical. The apparent broadening stems from the superposition of lines emitted from layers having different densities and, therefore, also different line positions. In any event, had the shifts been ignored and only the $L_\gamma$ width been used, one would have overestimated the density by a factor $\sim 3$.

Neither of these complications is allowed for in the interpretation of the measurements of profiles of the $n = 3 \rightarrow 1$ lines of helium-like argon (Hammel et al. 1993, 1994, Keane et al. 1993, 1995, Hooper et al. 1995, Haynes et al. 1996), which are very similar to $L_\beta$ of hydrogen-like chlorine. From the various fits shown in figure 10.4, electron densities up to $N_e = 2 \times 10^{24}\,\mathrm{cm}^{-3}$ at $kT \approx 1$ keV were inferred, although the observed profile is distinctly asymmetric and shifted (see figure 6 of Hammel et al. 1994) at the highest density. Calculated electron-produced line shifts (Griem, Blaha and Kepple 1990) amount to $\sim 10$ eV at the highest density, i.e., may well have contributed some to the observed FWHM width of $\sim 40$ eV because of density gradients and some time-averaging. However, significant contributions from unresolved Li-like, etc., dielectronic satellites cannot be ruled out either. Note also that the line profiles in figures 10.4 and 10.5 below are volume averaged, weighted according to upper level densities. Since the comparisons are with calculated profiles for homogeneous plasmas, maximum electron densities may therefore be higher than $N_e = 2 \times 10^{24}\,cm^{-3}$.

In experiments (Hooper et al. 1989) with argon-filled microballoons, time-resolved profiles of the hydrogen- and helium-like resonance lines, which are both Stark- and opacity-broadened, suggest electron densities up to $8 \times 10^{24}\,cm^{-3}$. The helium-like line and its satellites from inner-shell transitions in lower charge-states could also be observed in absorption from outer, i.e., cooler layers, with the central core providing the back-lighting continuum. These absorption spectra can be interpreted (Mancini, Hooper and Coldwell 1994) in terms of a three-layer model, with most of the absorption at $N_e \approx 1 \times 10^{24}\,cm^{-3}$, $kT \approx 250$ eV. So far, isolated lines from multiply-ionized ions have scarcely been used for density diagnostics. One reason is their relatively low sensitivity to Stark and related effects; and then there are the very considerable uncertainties in experiment-theory comparisons discussed in section 4.9. However, measurements of neon-like, etc., L-shell lines of krypton have yielded densities in compressed microballoon inertial fusion experiments which seem consistent with the results obtained using overlapping lines (Hooper et al. 1990). This is demonstrated by the argon and krypton spectra on figures 10.5(a) and 10.5(b), which were obtained on the same laser shot. Improved fits to the central profile region of the helium-like argon $1s^2$-$1s3p$ line are obtained by allowing for both ion-dynamics and opacity (Hooper et al. 1995, Haynes et al. 1996). Ion-dynamical effects in the 10% range had been predicted for the hydrogen-like $\beta$ and $\gamma$ lines of argon at $N_e \lesssim 10^{24}$ $cm^{-3}$ by Cauble and Griem (1983). For lines with forbidden components, e.g., lithium-like lines of multiply-charged ions, the experiment-theory comparisons have also improved (Godbert et al. 1994b, Glenzer et al. 1994c), making them better candidates for density measurements.

Returning to lower density and temperature plasmas, neutral helium lines with or without forbidden components can be used very nicely for density determinations, depending on the line, over about the same density range as hydrogen lines, but for higher temperatures. Recommended He I lines are the lines at 4471Å (Hey and Griem 1975, Uzelac, Stefanović and Konjević 1991) and at 7281Å (Pérez et al. 1995). For further suitable He I lines, see Mijatović et al. (1995). This use of He I lines again comes down to an application of the results discussed in sections 4.7 and 4.9, as is also the case for the very large set of isolated lines from neutral atoms and, mostly singly-charged, ions. Other things being equal, their sensitivity to electron density is relatively small. They are therefore useful for electron densities above or near $10^{16}\,cm^{-3}$ in case of atom lines, $10^{17}\,cm^{-3}$ for first ions, etc. This was the practical reason for tabulating (Griem 1974) calculated electron impact HWHM

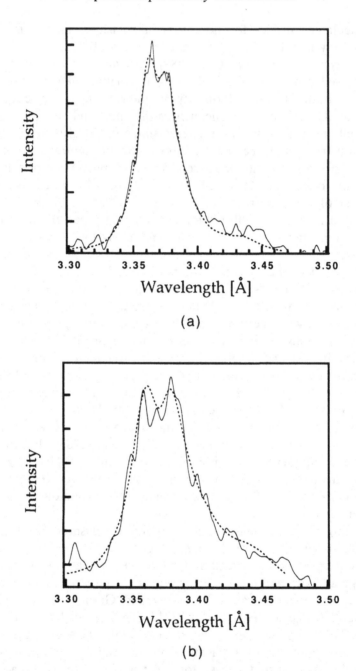

Fig. 10.4.   Profiles (Hammel et al. 1994) of helium-like argon $(1s^2)^1$S-$(1s3\ell)^1$S,P,D lines obtained from x-ray-compressed deuterium plasmas (solid lines) containing about 0.1% of argon for two laser-drive conditions; (a) 1 n-sec pulse, (b) 3:1 contrast, pulse-shaped drive. The calculated profiles (dashed lines) are for $N_e = 1.2 \times 10^{24}\,\mathrm{cm}^{-3}$, $kT = 1.2$ keV and $2.0 \times 10^{24}\,\mathrm{cm}^{-3}$, 1.0 keV, respectively.

(a)

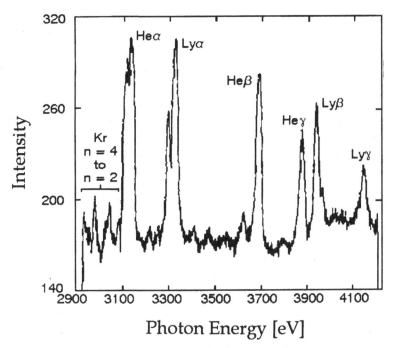

(b)

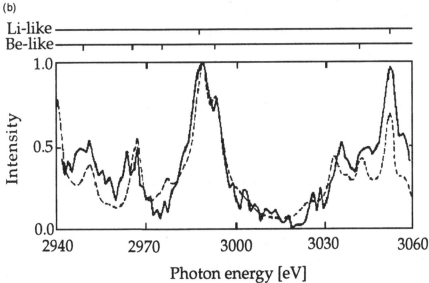

Fig. 10.5. Comparison of He- and H-like argon line profiles with profiles of Li- and Be-like krypton lines due to $n = 4$ to 2 transitions (Hooper et al. 1990). Using $N_e = 1.5 \times 10^{23}$ cm$^{-3}$, $kT_e = 1.1$ keV inferred from the argon lines shown in (a), the krypton spectrum shown on an expanded energy scale in (b) was calculated (dashed line) and found to be in satisfactory agreement with the observed spectrum (solid line).

(half widths at half maximum) widths at these densities, with the ion broadening and Debye shielding corrections already mentioned in section 4.7. Most of the more recent calculations discussed in that section contain FWHM widths for a set of densities to allow for shielding effects. They treat ion-atom or ion-ion collisions in the impact approximation, which is appropriate only at relatively low electron densities. However, allowing for the almost linear electron density dependence and the factor of 2 between HWHM and FWHM widths, there is normally no significant difference between the calculated widths, exceeding, say, 20%. The comparisons for the usually relatively small shifts are less favorable. This is discussed in section 4.10, together with the profile asymmetries especially of neutral atom lines due to quasistatic ion-produced fields. It is therefore recommended not to use shifts and asymmetries of isolated lines for density diagnostics, except for preferably small corrections of the widths. The latter, after extracting local and instantaneous profiles that are nearly Lorentzian and after correcting for Doppler and instrumental broadening, if necessary, can be used to determine electron densities with errors corresponding to a theoretical uncertainty of about $\pm\ 20\%$.

This theoretical uncertainty is not as small as that attained for hydrogen lines, especially for $H_\beta$ at intermediate electron densities, say $10^{15}$–$10^{17}\,cm^{-3}$, but clearly substantially lower than that which can safely be assumed for lines from highly-charged ions, both overlapping lines or, especially, isolated lines. In these cases more experiments with independently measured electron densities are required to reduce the approximate factor-of-2 theoretical uncertainty. Similarly, electron densities inferred from dip structures caused by wave-produced fields as discussed near the end of section 4.11 cannot be expected to be any more accurate than this, until the outstanding experimental and theoretical problems are better understood.

## 10.2   Densities from absolute continuum intensities

The reader will have gleaned that more often than not the accuracy of electron density measurements using the Stark broadening method may still be limited by theoretical uncertainties in calculated line profiles. In spite of its very large range of applicability, the Stark-broadening method must therefore be supplemented by methods requiring only some more basic theoretical underpinnings. At least for plasmas containing only atoms and ions with a small number, or even no, bound electrons, a prime example of such a spectroscopic method is one based on absolute continuum intensity measurements.

The underlying theoretical formula for plasmas containing only fully stripped ions is

$$
\int_\lambda^{\lambda+\Delta\lambda} I_c(\lambda)d\lambda \approx \frac{2^6\sqrt{\pi}E_H c}{3^{3/2}\lambda^2}\left(\frac{E_H}{kT}\right)^{3/2}\Delta\lambda
$$

$$
\times \int dx \sum_z \left[ z^4 \sum_{n_{\min}}^{n_{\max}} \frac{g_n^z(\lambda)}{n^3}\exp\left(\frac{z^2 E_H}{n^2 kT}\right) \right.
$$

$$
\left. +z^2\frac{kT}{2E_H}g_f^z(\lambda, T)\exp\left(\frac{\Delta E_L}{kT}\right)\right](\alpha a_0)^3 N_z N_e \exp\left(-\frac{\hbar\omega+\Delta E_z}{kT}\right),
$$

$$(10.1)$$

as obtained from (5.16) and (5.17) by multiplication with $|\Delta\omega| \approx (2\pi c/\lambda^2)\Delta\lambda$ and by integrating over $x$ along the line of sight. The $x$-dependence of temperatures and densities is implied, and emission is assumed to be optically thin. The Gaunt factors $g_n^z$ and $g_f^z$ are as shown in figure 5.3, and the temperature dependence of $g_f$ is seen to be rather weak. From the theoretical point of view, and bearing in mind (5.13) or (5.24a or 5.24b) for $n_{\max}$, and (7.30) and (5.18) for $\Delta E_z$ and $\Delta E_L$, respectively, one can see that measurements at wavelengths corresponding to photon energies well below $kT$ are preferable for accuracy. Moreover, wavelengths should be long enough such that radiative recombination is possible only into states such that $z^2 E_H/n^2$ is also less than $kT$. Using the first version of (5.20), we also find

$$
\hbar\omega+\Delta E_L \gtrsim \frac{z^2 E_H}{n_{\min}^2}
$$

$$(10.2)$$

for the lowest principal quantum number for the inner summation in (10.1) and notice that there is some cancellation of the seemingly exponential temperature dependence. This leaves us, more or less, with the $T^{-3/2}$ and $T^{-1/2}$ explicit temperature dependencies of recombination and bremsstrahlung contributions, respectively. Long wavelengths measurements are therefore indeed preferable, except that one would want to avoid wavelengths where the refractive index factor $[1-(\omega_{pe}/\omega)^2]^{1/2}$ should have been included in (10.1), as discussed in section 5.5.

Although we have assumed complete stripping of all bound electrons, it is still important to choose a wavelength interval $\Delta\lambda$ that is free from line and molecular or even $H^-$ emission, and of impurity radiation, say from edge regions of a hot plasma. This may require preliminary high resolution measurements, before undertaking the absolute intensity measurements in intervals $\Delta\lambda$ that had been found free of lines or other radiation not due to radiative recombination and bremsstrahlung with or on the completely-stripped ions considered so far. This is not to say that (10.1) cannot be generalized to ions already possessing some bound electrons. For each species of such ions, (10.1) would then according to (5.21) have to be replaced by

$$
\int_{\lambda}^{\lambda+\Delta\lambda} I_z(\lambda)d\lambda = \frac{2^5\sqrt{\pi}E_H cz^2}{3^{3/2}\lambda^2} \left(\frac{E_H}{kT}\right)^{1/2} \Delta\lambda \int dx
$$
$$
\times \left\{ \xi_{fb}(\lambda, T) \left[1 - \exp\left(-\frac{\hbar\omega}{kT}\right)\right] + \xi_{ff}(\lambda, T)\exp\left(-\frac{\hbar\omega}{kT}\right) \right\}
$$
$$
\times \exp\left(\frac{\Delta E_L - \Delta E_z}{kT}\right)(\alpha a_0)^3 N_z N_e, \tag{10.3}
$$

where $\xi_{fb}$ and $\xi_{ff}$ are the Biberman factors introduced in section 5.2 and discussed further in section 5.4. Since they are either derived from numerical calculations of photoionization and bremsstrahlung cross sections, or from measurements as discussed in section 5.6, they are generally known only approximately.

For both (10.1) and (10.3), a Maxwell distribution of the electrons was assumed, whereas only in case of (10.3) there is also an implied assumption of LTE populations for the various bound states of the target ion. Their populations are important because especially recombination cross sections are state-sensitive. This LTE assumption for target ion states is again less critical for long-wavelength radiation, because the recombination is then essentially into Rydberg levels, which are not affected much by the state of the target ion. Nevertheless, both cross section uncertainties and any remaining dependence on initial level populations should be allowed for in estimating errors of electron and ion densities from absolute continuum intensities.

The ideal case for an application of the present method are highly ionized and pure hydrogen plasmas. Quasi-neutrality then gives $N_z = N_e$, and what is actually measured is the line integral of $N_e^2$, or its local values if inversion methods as described in section 8.5 are used. Cross section errors can be excluded, and relativistic electron-electron bremsstrahlung, already mentioned in section 9.1, and deviations from a Maxwell distribution and molecular radiation tend to be negligible in steady-state high-temperature hydrogen plasmas. The relative error in $N_e$ is therefore about half the experimental error in the intensity measurement, plus a usually small error associated with the rather weak temperature dependence.

As a matter of fact, the inherent accuracy of the method is so good that any difference between the $N_e N_z$ product thus obtained, or from, e.g., Thomson scattering or interferometric determinations of $N_e$ and from quasi-neutrality, is often interpreted as a measure of the impurity ion concentration (von Goeler et al. 1975, Röhr and Steuer 1988). Since at high temperatures and long wavelengths both (10.1) and (10.3) suggest an intensity scaling proportional to $z^2 N_z = z N_e$, the ratio of measured and predicted intensity, for $z = 1$, is then interpreted as a measure of the effective charge $z_{\text{eff}}$ (von Goeler et al. 1975, Kadota, Otsuka and Fujita 1980, Röhr and Steuer 1988). Because of changes in Gaunt factors and deviations of the Bibermann factors $\zeta_{\text{ff}}$ from the free-free Gaunt factors, this $z_{\text{eff}}$ does not precisely correspond to the factor appropriate, i.e., for electrical conductivities (Spitzer 1962). This and other methods for $z_{\text{eff}}$ determinations are discussed further in section 12.6.

Returning briefly to electron density determinations in plasmas containing significant fractions of different ions, these fractions must be determined separately, e.g., from relative line intensity and temperature measurements. One can then write the intensities as a weighted sum of the appropriate versions of (10.1) and (10.3) and, again using quasi-neutrality, express the result in terms of $N_e^2$. In such situations, additional errors in the $N_e$-determination would arise from any errors in the ion fractions.

## 10.3 Densities from absolute line intensities

If continuum measurements are performed at shorter wavelengths and, in particular, on both sides of an absorption edge, the edge position may define a particular ion or atom species, if electron densities are low enough for the edge shifts discussed in section 5.5 not to be large. The intensity ratio across an edge also can help in estimating any underlying stray light. From the absolute value of the additional intensity on the short-wavelength side of the edge, which corresponds to one additional term in (10.1) or to a corresponding jump of the free-bound Biberman factor $\zeta_{fb}(\omega)$ in (10.3), one can then infer a value for the line integral of $N_z N_e$ for this particular ion. However, mostly well-defined edges are observed only at relatively short wavelengths, for which the temperature dependence is quite strong (see section 11.4).

More fruitful is usually the absolute measurement of spectrally integrated line intensities, in particular of lines from sufficiently highly-excited levels such that PLTE relations like (6.10) and (7.24) are valid, i.e., if the principal quantum number of the upper level of the line is above the critical quantum number according to (7.77) or to more detailed estimates as discussed in section 7.6. Primarily, the spectrally integrated intensity of an optically thin line yields according to (3.1) and (8.10) the line-of-sight integrated density $N_m$ of emitting atoms or ions in the upper level, i.e.,

$$\int I_{nm}(\lambda) d\lambda = \frac{(2\pi c)^2 \hbar r_0}{\lambda_{mn}^3} f_{mn} \frac{g_n}{g_m} \int N_m dx, \qquad (10.4)$$

if we again use wavelengths, absorption oscillator strengths and the ratio of lower to upper state statistical weight factors. For the oscillator strengths, the reader is reminded of chapter 3, especially also of the references in the introduction to that chapter. Moreover, the present application is basically the inverse of one of the methods for the measurement of transition probabilities, etc., briefly discussed in section 3.6.

An important aspect of both types of experiments is the need to correct for any underlying continua and, possibly, for blends with other lines. Both needs are served by having moderate resolution and the ability to monitor the background continuum in a suitable wavelength interval which is essentially free of line radiation. In many cases it is not practical to extend the intensity measurements sufficiently far from the line core; or if a monochromator is employed, to use a sufficiently broad exit slit so as to include almost all of the integrated line intensity. For a dispersion profile as in (4.3), but in wavelength-units, the corresponding fractional correction is in terms of the HWHM width $\Delta\lambda_h$ and the largest separation $\Delta\lambda_m$ from the peak position

$$2 \int_{\Delta\lambda_m}^{\infty} \frac{(\Delta\lambda_h/\pi)d\Delta\lambda}{(\Delta\lambda)^2 + (\Delta\lambda_h)^2} \approx \frac{2\Delta\lambda_h}{\pi\Delta\lambda_m}. \tag{10.5}$$

This assumes a reasonably symmetrical interval and a relatively small correction, e.g., $\sim 6\%$ for $\Delta\lambda_m = 10\Delta\lambda_h$. Such corrections are therefore surprisingly important for Lorentz and similar profiles, but are much less of a problem for Doppler profiles. This can be seen, e.g., from figure 4.1.

With these wing and background corrections the accuracy of total line intensity measurements tends to be limited by the absolute intensity calibration. A more severe limit on the accuracy of upper state densities is often imposed by uncertainties in oscillator strengths. However, in the case of hydrogen plasmas the line integral of $N_m$ can be determined to the accuracy of the calibration, and according to (7.24) therefore also $\int N_e N_i dx = \int N_e^2 dx$, if we ignore temperature gradients. In any event, we find that absolute line and continuum intensities can be used very similarly. Even the temperature dependence is not too different, if one chooses lines from high-lying levels.

A more distinct application is to use the upper level density via (7.24) and a separately determined electron density and temperature to obtain groundstate densities of the next ion of the emitting species, provided the PLTE criterion corresponding to (7.77) is fulfilled and that the data can be inverted to obtain local values. Ionic abundances inferred in this manner can be relatively accurate, with major errors usually from oscillator strengths and from assuming PLTE.

A unique method based on the observation of $\Delta n = 1$ transitions from high-lying levels of multiply-ionized ions, corresponding to dielectronic satellite lines between doubly-excited states of the preceding ion, has been proposed by König et al. (1989). If the principal quantum numbers are above the critical quantum number for PLTE, Saha equations can be used to relate the upper level densities of the satellites and the parent transitions to the populations of the corresponding states of the next ion, namely of its lowest excited state and its ground state.

An alternative method for the determination of atomic and ionic abundances is to use absolute intensities of lines from low-lying levels. If the corresponding level populations are in LTE with respect to their ground state populations, i.e., fulfill the conditions discussed in section 7.6, and if these usually strong lines are optically thin, then one can infer the groundstate densities from the upper level densities via Boltzmann factors, i.e., (7.21). It is relatively rare that all of these conditions are fulfilled, and the determination of atomic ground state densities from emission measurements remains a difficult task. If deviations from LTE are the major problem, one may employ one of the kinetic models in section 6.1 or those discussed in the following section. Otherwise, it may be possible to measure the line absorption coefficient and then to use (8.4) to infer the groundstate density. More powerful is generally a tunable-laser-based method (Mertens and Bogen 1987, Bogen et al. 1992, Demtröder 1996) utilizing resonance fluorescence as discussed theoretically in section 2.9 or two-photon polarization spectroscopy (Dux et al. 1995). These methods also provide spatial resolution, as do some other methods discussed in the final chapter of this book.

## 10.4 Electron densities from relative line intensities

Returning from the spectroscopic determination of atom and ion densities once more to that of electron densities, but allowing now explicitly for significant deviations from LTE and even from PLTE, a number of special diagnostic methods have been proposed on the basis of detailed rate-equation modeling. Naturally, one would normally expect to infer temperatures from relative line intensities, as will be discussed in the next chapter. However, for highly transient plasmas or in the presence of strong particle flows, but also in steady-state non-LTE plasmas, it may turn out that some relative intensities depend more strongly on the electron density. Temperatures must then be assumed to be known from other measurements, spectroscopic or otherwise, or be obtained from plasma modeling calculations.

An early example of a non-LTE, but possibly steady-state plasma spectrum suitable for electron density determinations is that of helium-like ions, e.g., of O VII (Elton and Köppendörfer 1967) or of C V (Kunze,

Gabriel and Griem 1968b). A schematic energy level diagram containing the ground state and the singly excited $n = 2$ levels is shown in figure 10.6. There are three line transitions from these excited levels to the ground state, the resonance line from $(1s\,2p)^1P$, the intercombination line from $(1s2p)^3P$, and the highly-forbidden singlet-triplet line from $(1s\,2s)^3S$. The $(1s2s)^1S$ state decays radiatively only by two-photon transitions, which give a weak contribution to the continuum.

Because the radiative lifetimes of these four excited states are so different from each other, at least for low and intermediate (nuclear charge) $Z$-values along the isoelectronic sequence, their relative populations behave quite differently as a function of electron density. At very low densities, the metastable levels are vastly over-populated compared with the $(1s2p)^1P$ population, because the electron-collisional near-threshold excitation cross sections are of the same order of magnitude, as already mentioned in section 6.3. Perhaps surprisingly, the three total line intensities and also the spectrally integrated two-photon continuum all have comparable intensities in the validity region of a simple corona model corresponding to (6.6).

As the electron density increases, collisional transfer from the excited states begins to compete first with the weaker radiative decays, thus reducing the relative intensities of the transitions to the ground state according to a suitable branching ratio of radiative decay to the sum of radiative and collisional decays, including ionization. If the plasma is not rapidly ionizing, but is in steady state or even decaying, recombination into the excited levels must be included as well. These refinements of the modified corona model have been pursued by a number of authors (Gabriel and Jordan 1972, Vinogradov et al. 1975, Skobelev et al. 1978, Fujimoto and Kato 1990, Masai 1994), the latter emphasizing the importance of excitations from Li-like ions in rapidly ionizing plasmas. Calculated steady-state line intensity ratios for He-like oxygen and iron are shown in figure 10.7. They give an indication of the ranges of density sensitivity, but differences between various calculations are quite large. Besides this difficulty, there is also the possible experimental problem, especially in laser-produced plasmas, of self-absorption of the resonance line. The increases of the intercombination line intensities above $10^{10}$ and $10^{16}\,cm^{-3}$ are due to collisional excitation energy transfer from the $(1s2s)^3S$ level, at the expense of the forbidden line.

A similar problem occurs in argon micro-pinches of about 40 $\mu m$ diameter produced in plasma focus discharges (Koshelev et al. 1994). The resonance line is optically thick also in this case, but the optical depth during the implosion or expansion phases is smaller than one might think because of Doppler shifts associated with radial velocity gradients. If both Li-like satellites and these Doppler shifts are included in the

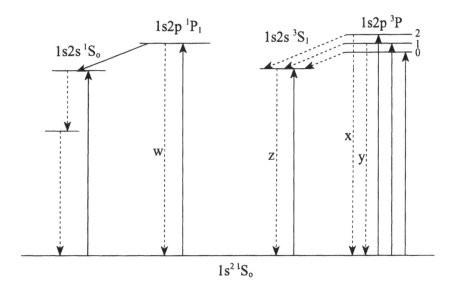

Fig. 10.6. Schematic energy level diagram for helium-like ions showing the ground state and all $n = 2$ singly excited states. (See figure 6.1 for doubly excited states of lithium-like ions.)

modeling (Schulz and Koshelev 1995), reasonable fits to measured spectra are obtained up to $N_e \approx 10^{23}$ cm$^{-3}$, including asymmetries caused by one line of the group pumping the other.

Other than He-like ionic species have been considered for very similar electron density determinations, e.g., Be-like ions (Kato, Masai and Mizuno 1983, Kato, Lang and Berrington 1990) with $(2s2p)^3$P as the metastable level. Such ions and also B-like ions were discussed in the present context by Masai (1994) and, for the case of $\Delta n = 1$ transitions, by Feldman, Seely and Bhatia (1985), who considered Ne-like ions as well. The latter authors used the fact that the $2s^n2p^m$ - $2s^n$ $2p^{m-1}3p$ monopole excitation rates are comparable to $2s^n2p^m$ - $2s^n2p^{m-1}3s$ or 3d dipole rates, while the radiative decay of 3p levels to 3s is relatively slow. They pointed out that therefore at higher densities than appropriate for the corona model, lines from 3p would increase with electron density less than those from 3s or 3d, until the usual saturation of line intensity ratios sets in near the LTE limit.

Feldman et al. (1985) also solved, at a temperature of half the ionization energies, steady-state rate equations for 20 levels of B-like ions and Be-like ions ranging from argon to krypton, and for 27 levels of Ne-like ions from silicon to krypton. Their figures 3, 6 and 7 show corresponding density-dependent line ratios, which should be useful from $N_e \approx 10^{16}$ cm$^{-3}$ to $N_e \approx 10^{22}$ cm$^{-3}$, depending on the ions and possible effects of self-absorp-

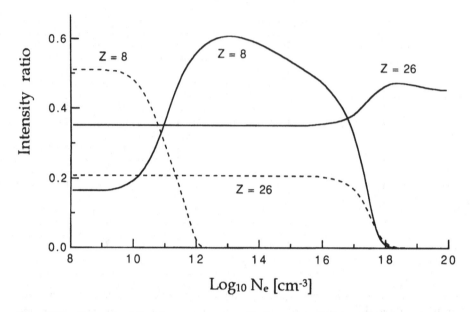

Fig. 10.7.   Calculated line intensity ratios (Fujimoto and Kato 1990, Masai 1994) of the intercombination line ($1s^2 \, {}^1S$-$1s2p\,{}^3P$, solid lines) and forbidden line ($1s^2 \, {}^1S$-$1s2s\,{}^3S$, dashed line) to the resonance line ($1s^2 \, {}^1S$-$1s2p\,{}^1P$) of helium-like oxygen ($Z = 8, kT = 300$ eV) and iron ($Z = 26, kT = 3$ keV), respectively.

tion. Masai (1994) presents results for Be-like oxygen that can be used to infer electron densities near $N_e = 10^{13}$ cm$^{-3}$ and for B-like iron, which should be useful near $N_e = 10^{14}$ cm$^{-3}$. In both papers, there is also a discussion of the temperature dependence, which is weak but not negligible. However, Masai (1994) considers only $\Delta n = 0$ transitions, so that no detailed comparison with the earlier calculation is possible, and the appropriate densities are very different. Model calculations for F- and Ne-like lines of xenon (Keane et al. 1994) have been used to infer electron densities and temperatures in the fuel-regions of indirectly driven laser fusion targets at densities near $10^{24}$ cm$^{-3}$.

Intensity ratios of dielectronic satellite lines, e.g., lines from doubly-excited Li-like ions near the resonance line of the helium-like ion, can also be used for electron density determinations. This is possible because for some of these lines angular-momentum changing electron-ion collisions are important (Jacobs and Blaha 1980, Kononov et al. 1980) through their effects on the population of levels that are not directly populated by dielectronic recombination. This method was used for $N_e \gtrsim 10^{22}$ cm$^{-3}$ by Lunney (1983) for gas-filled microballoon targets and near $N_e = 10^{23}$ cm$^{-3}$ by Zigler et al. 1992 in short-pulse, laser-produced silicon and aluminum plasmas, with calculated satellite line ratios presented in their figures 2 and 3. Although these ratios are insensitive to temperature changes, they may be affected by rapid changes in the charge state (Peyrusse et al. 1993, Mancini et al. 1994a, 1995) and by Stark broadening (Woltz et al. 1991). One can also use satellites involving, e.g., $n = 3$ electrons (Audebert et al. 1984), but then the LTE limit is reached already near $N_e = 10^{22}\,cm^{-3}$ for $A\ell$ and Si.

Returning to much lower density plasmas, e.g., the $N_e \approx 10^{12}$ to $10^{15}\,cm^{-3}$ edge region plasmas of tokamaks or stellarators, an interesting possibility for electron density determinations from relative intensities of hydrogen lines has been pointed out by Fujimoto, Miyachi and Sawada (1988). Atomic hydrogen in these regions of relatively high electron temperature may mostly be considered as in a rapidly ionizing phase. As already discussed in section 7.6, populations of states with principal quantum number $n$ above the critical quantum number $n_{cr}$ according to (7.77) are then proportional to $n^{-4}$. On the other hand, for lower states, the usual corona model suggests populations proportional to $n^{3/2}$. For each pair of lines, there is therefore a transition region between low- and high-electron density plateaus for the intensity ratio. This transition region is located such that $n_{cr}$ lies between the two upper level principal quantum numbers. It has been evaluated by Fujimoto et al. (1988) using a collisional-radiative model and two sets of cross sections. As can be seen from their figures 5a and 5b, electron densities ranging from $N_e \approx 10^{11}$ to $10^{14}\,cm^{-3}$ can be inferred from various Lyman or Balmer line ratios.

For $N_e \lesssim 10^{13}\,cm^{-3}$ the method, modified for contributions due to excitation by $H_2$ molecular dissociation (Fujimoto, Sawada and Takahata 1989a, see also Sawada, Eriguchi and Fujimoto 1993) has been compared with laser-interferometry results (Fujimoto et al. 1989b). The molecular dissociative-excitation rate coefficients are approximately proportional to $n^{-6}$ rather than to $n^{-3}$ as in case of the excitation of hydrogen atoms. Inclusion of the molecular process into the collisional-radiative model for a rapidly ionizing plasma, with the molecular to atomic hydrogen ratio as parameter, then gives line intensity ratios as function of electron density, e.g., $H_\beta/\,H_\alpha$ and $H_\gamma/H_\alpha$. These two line ratios were used by Fujimoto et al. (1989b) to measure radial distributions of both the electron density and the molecule-to-atom ratio.

Measurements of the intensity ratios of resolved fine-structure components may also be used for electron density measurements, e.g., the intensity ratio of optically thin high $Z$ Lyman-$\alpha$ fine-structure components ($j = 1/2$ and $3/2$) is not necessarily equal to $1/2$ in high temperature tokamak plasmas, but can be as large as 0.8. This is due to a delicate balance between collisional and radiative processes (Ashbourn and Ljepojević 1995), suggesting large errors in electron densities inferred from measured density ratios.

Relative intensities of neutral helium lines have recently been used to infer electron densities in edge region and similar low density plasmas (Schweer et al. 1992, Sasaki et al. 1995). Their interpretation requires extensions of Fujimoto's (1979a) collisional-radiative model and self- consistent determinations of electron density and temperature (Sasaki et al. 1995).

# 11

# Spectroscopic temperature measurements

Next to the qualitative determination of the chemical composition of a plasma from emission and absorption line identifications, the measurement of electron and ion or atom temperatures is the oldest application of spectroscopic methods to plasma and gaseous electrical discharge physics, not to mention astronomy. It continues to play an important role, e.g., in fusion research (DeMichelis and Mattioli 1981, 1984 and Kauffman 1991). In the laboratory, independent methods based on laser light scattering and Langmuir probes are available, as already mentioned in the introduction to the preceding chapter. However, in astronomy spectroscopic methods normally must stand alone. Another important distinction is the usually dominant role of radiative transfer (see chapter 8) in astronomical applications, compared with the relatively small optical depth in some useful portion of the spectrum of most laboratory plasmas.

In many cases, it is necessary to distinguish between kinetic temperatures of electrons, ions and atoms, say, $T_e$, $T_z$, and $T_a$. These temperatures may differ from each other even if the individual velocity distributions are close to Maxwellians, because, e.g., electron-electron energy transfer rates are much larger than electron-ion collision rates, as are ion-ion rates (Spitzer 1962). In most applications, at least the electrons do have a Maxwellian distribution, and we will assume this here. If a more general distribution must be considered, one may determine the corresponding parameters by measuring and interpreting more features than would be required normally (Behringer and Fantz 1994), or measure the continuum emission over a wide photon energy range (section 11.4). Depending on which of the collisional interactions discussed in section 6.5 dominate in a particular experiment, the relationship between ion-atom and atom-atom rates, not to mention rates of molecular processes, could either favor or counteract the establishment of a Maxwell distribution also for the neutral atoms. It is therefore especially important not only to measure $T_a$, often called the gas temperature, but also to measure the velocity distribution. A spectroscopic method for this purpose, as for the corresponding measurement of ion distribution functions, will be discussed in the last section of this chapter.

Many laboratory plasmas emitting enough light for spectroscopic measurements are sufficiently dense and long-lived, and also have only relatively smooth spatial variations, such that the electron distributions are indeed close to Maxwellian both locally and instantaneously. Important exceptions are some laser-produced plasmas, for which the energy deposition is so localized and in an ultra-short pulse that a nonlocal description as a function of time is called for. Assuming this not to be the case, the concept of a local electron temperature, with the time as parameter, is normally valid and useful. The same cannot be said of related effective temperatures, e.g., excitation and ionization temperatures. Instead of introducing such effective temperatures, the underlying, often non-LTE, physics will be emphasized here. Perhaps surprisingly, it will turn out that spectroscopic $T_e$ measurements remain possible even in extreme non-LTE situations.

As a corollary to the corresponding situation mentioned in the introduction to chapter 10, most of the spectroscopic methods for the determination of $T_e$ also imply some knowledge of the electron density $N_e$. Again, in principle we must either use the observed quantities to obtain self-consistent solutions for $T_e$ and $N_e$ or, often more practical, fit synthetic spectra to the data. This approach is also suitable for the inclusion of any radiative transfer effects, spatial and time variations, etc. In the following, we will, however, assume that such effects are either not important or already allowed for in the data reduction, e.g., according to section 8.5.

Most of the spectroscopic temperature measurements primarily yield $T_e$. They are based on relative line intensities either of the same atom or ion, of neighboring ionization stages, or of successive isoelectronic ions; on relative continuum intensities, on ratios of line and continuum intensities, on relative and absolute intensities from optically thick plasmas, or on time histories of emission from transient plasmas. The following sections address these methods. Finally, there is one spectroscopic method for the determination of ion or neutral atom temperatures, $T_i$ or $T_a$. This is the time-honored interpretation of measured spectral line profiles in terms of one-dimensional distribution functions for the velocity component along the line of sight. The basic assumption here is that the nonrelativistic, free-streaming Doppler effects dominate over the various other line broadening and, even, narrowing mechanisms discussed in chapter 4.

## 11.1   Relative intensities of lines of the same atom or ion

In optically thin plasmas of length $\ell$ along the line of sight, spectrally integrated emission line intensities are given by

$$
\begin{aligned}
i_{nm} &= \int I_{nm}(\omega)d\omega \\
&= \frac{\hbar\omega_{mn}}{4\pi} A_{nm} \int_0^{\ell} N_m dx \\
&\approx \frac{\hbar\omega_{mn}}{4\pi} A_{nm} N_m \ell,
\end{aligned}
\tag{11.1}
$$

assuming a homogeneous plasma in the final step. As already discussed in section 3.6 in connection with the measurement of oscillator strengths and summarized by (10.5), it is very important to allow for the contribution of extended line wings to the frequency integral and to consider continuum corrections and any blends with neighboring lines. All this and errors in relative intensity calibrations rarely permit measurements of the ratio of two integrated line intensities to be better than 10%.

For PLTE as discussed in section 7.6 down to the upper levels $m_1$ and $m_2$ of two lines of the same ion or atom, one can use (7.21) for the ratio of the two populations. One thus obtains

$$
\begin{aligned}
R &= \frac{i_{n_1 m_1}}{i_{n_2 m_2}} \\
&\approx \frac{\omega_{m_1 n_1} A_{n_1 m_1} g_{m_1}}{\omega_{m_2 n_2} A_{n_2 m_2} g_{m_2}} \exp\left(-\frac{E_{m_1} - E_{m_2}}{kT}\right)
\end{aligned}
\tag{11.2}
$$

or, using (3.1), an equivalent expression in terms of the two oscillator strengths. The temperature corresponding to a given line ratio,

$$
kT = \frac{E_{m_2} - E_{m_1}}{\ell n(\omega_{m_2 n_2} g_{m_2} A_{n_2 m_2} R / \omega_{m_1 n_1} g_{m_1} A_{n_1 m_1})},
\tag{11.3}
$$

depends only logarithmically on $R$ and the other quantities in the argument $X$ of the logarithm. The relative error in the temperature is related to any error $\Delta X$ of this argument through

$$
\left|\frac{\Delta T}{T}\right| = \frac{kT}{E_{m_2} - E_{m_1}} \left|\frac{\Delta X}{X}\right|.
\tag{11.4}
$$

As already mentioned, experimental errors in $\Delta X / X$ are rarely below 0.1, and the prefactor of this quantity is typically of order 1. This suggests $\sim \pm 10\%$ error in electron temperatures from such line ratios, even under favorable conditions concerning the validity of PLTE conditions and the availability of accurate $A$-values. Any errors from these two sources must be added, often resulting in relatively poor accuracies of inferred electron temperatures.

To some extent, this accuracy can be improved by measuring line ratios along some spectral series and then plotting the logarithms of the relative intensities, each multiplied with the appropriate factors occurring in (11.3), as function of the excitation energies of the upper levels. If the corresponding data on such a Boltzmann plot fall close to a straight line, one may obtain a more accurate value for $kT$ from the reciprocal of the slope of this line. Any significant deviations from such linear fit could signal deviations from PLTE or errors in the $A$-values, not to mention in the various corrections required to extract total line intensities from observed spectra. Such corrections are especially uncertain for lines from highly excited levels that begin to merge, as discussed in section 5.5. Using reduced probabilities for the occupation of these levels (Gündel 1970 and 1971, Däppen et al. 1987) may improve the consistency of a Boltzmann plot, but is not likely to improve the accuracy of the temperature measurement.

For plasmas for which PLTE cannot be assumed for the upper levels of otherwise suitable lines, the above relations should not be used. However, depending on the availability of collisional rate coefficients and of sufficiently realistic models as discussed in chapter 6, one may still determine temperatures from measured line ratios, either by fitting synthetic spectra or by considering limiting, i.e., simple cases. An example for such simple situations are resonance lines and higher series members of lithium-like ions (Heroux 1963) for plasmas in the corona model regime as discussed in section 6.1. In such situations, (6.6) leads to line intensity ratios

$$
\begin{aligned}
R &= \frac{i_{n_g m_1}}{i_{n_g m_2}} \\
&\approx \frac{\omega_{m_1 n_g} X_{m_1 g}}{\omega_{m_2 n_g} X_{m_2 g}} \\
&\approx \frac{(\bar{g}f)_{m_1 g}}{(\bar{g}f)_{m_2 g}} \exp\left(-\frac{E_{m_1} - E_{m_2}}{kT}\right).
\end{aligned}
\tag{11.5}
$$

where the effective Gaunt factor approximation (6.27) for the excitation rate coefficients was used to obtain the final version and radiative transitions to the ground state were assumed to dominate. To the extent that actual excitation cross sections can be represented by constant Gaunt factors and absorption oscillator strengths, the temperature dependence of line ratios is therefore the same as for PLTE line ratios, although one has the additional uncertainty from the collisional cross sections. Moreover, the model tends to be inappropriate for lines from highly excited levels, because for them population through recombination and cascading may be important. Similarly, it would be best to avoid lines from long-lived levels, because their populations are quite susceptible to further collisional effects as already discussed in section 10.4.

. A possible exception to this are low density plasmas, say, with $N_e \lesssim 10^{13} \, \text{cm}^{-3}$, containing neutral helium. In that case the collisional excitation from the ground state is dominant for triplet levels, and triplet to singlet line ratios are functions mainly of the electron temperature (Fujimoto 1979a, Behringer 1992, Sasaki et al. 1995). This is due to the different threshold behaviors of the cross sections, the ratios descreasing by a factor of approximately 5 from $kT_e = 5$ eV to 50 eV. Note, however, that extensions of this method to helium-like ions, e.g., of titanium (Bitter et al. 1985, Kato et al. 1987) and of nickel (Zastrow, Källne and Summers 1990) did not yield very satisfactory results. This may in part be due to suprathermal electrons, but also to charge exchange recombination (Kato et al. 1987, Kato and Masai 1988) as discussed in section 6.5.

## 11.2 Relative line intensities of subsequent ionization stages of the same element

Because of the relatively small differences between excitation energies, experimental errors, and theoretical uncertainties, line ratios within a given atom or ion normally do not lend themselves to very precise temperature measurements. In dense LTE plasmas of known electron density, this situation can be much improved by measuring the relative integrated intensities $i$ and $i'$ of, say, a line from a neutral atom and the following ion. From (7.21) and (7.24) their ratio is

$$
R = \frac{i'}{i} = \frac{\omega' A' g'}{\sqrt{\pi} \omega A g} (4\pi a_0^3 N_e)^{-1} \left( \frac{kT}{E_H} \right)^{3/2}
$$
$$
\times \exp \left( -\frac{E' + E_\infty - E - \Delta E_\infty}{kT} \right) \tag{11.6}
$$

in terms of the two sets of $\omega, A, g$ (upper level statistical weight) and $E$ (excitation energy) values, the ionization energy $E_\infty$ of the "atom", and its reduction $\Delta E_\infty$ according, e.g., to (7.30). The point here is that the combination of the various energies appearing in the exponent is often substantially larger than $kT$. In an error estimate analogous to (11.4), the prefactor on the right-hand side is therefore typically well below 1, say $\sim 0.2$. This reduction may well compensate for any additional errors associated with the ionization equilibrium under LTE or near-LTE conditions and with the measurement of the electron density.

If electron densities are not sufficient to ensure LTE for ionization and excitation up to level $E'$, a combination of PLTE for the two upper level populations relative to the groundstate populations of the following charge states according to (6.10) or (7.24) and of corona ionization equilibrium according to (6.2) may be appropriate. In that case (11.6) should be replaced by

$$R = \frac{i'}{i} = \frac{\omega' A' g'}{\omega A g} \exp\left(\frac{E'_\infty - E' - E_\infty + E}{kT}\right) \frac{g_i S}{g'_i \alpha}, \tag{11.7}$$

the ionization and recombination coefficients being those for the "ion", e.g., for the ionization of $He^+$ or leading into $He^+$ from $He^{++}$ for the recombination. The $g_i$, $g'_i$ are groundstate statistical weights of the "next" ions, i.e., $He^+$ and $He^{++}$, respectively. If the $S$ and $\alpha$ are taken from a suitable collisional-radiative (C-R) model as discussed in section 6.1, then this relation can also be used at intermediate electron densities between the corona and LTE regimes. Examples of such calculations for the ratio of the He II 4686 Å and He I 5876 Å lines based on the C-R model of Burgess and Summers (1976) are shown in figure 11.1, together with LTE results from (11.6) for $N_e = 10^{18}\,cm^{-3}$. At this density, LTE relations may still be accurate to $\sim 10\%$ accuracy, provided the He II resonance line is optically thick. (See section 7.6 for a discussion of this point.)

From the results presented on figure 11.1, we may infer that the He II/ He I ratio can be used to infer fairly accurate temperatures between $kT \approx 3$ to 7 eV at $N_e \lesssim 10^{18}\,cm^{-3}$, and at somewhat higher temperatures if electron densities are larger. Using highly charged pairs of ions, this method is readily extended toward higher temperatures. However, for the upper levels of strong lines, PLTE will soon begin to fail except at very high densities, and there may be relaxation effects in these usually rather transient plasmas. All this requires more complete modeling of synthetic spectra, as was done by Keane et al. (1993, 1994) in order to measure temperatures from intensities of H- and He-like lines of argon and of F- and Ne-like lines of xenon up to $\sim 2$ keV in inertial fusion targets. Analogous measurements using argon lines were made by Nishimura et al. (1995) and, with spatial resolution, by Uschmann et al. (1995).

A closely related method (Gabriel 1972) for electron temperature measurements in high temperature plasmas is based on observing relative intensities of dielectronic satellite lines, with respect to the resonance line of the following ion in the ionization sequence. Assuming, e.g., a helium-like ion, this reference line would be due to the $(1s^2)^1S$-$(1s2p)^1P$ transition, with dielectronic satellites of the lithium-like ion, e.g., due to $(1s^2 2\ell)\,^2\ell$-$(1s2p2\ell)\,^2L$ transitions. Both of the upper levels can be excited by electron collisions with $(1s^2)^1S$ groundstate ions, $(1s2p)^1P$ directly by

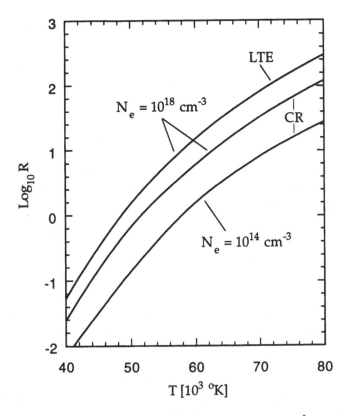

Fig. 11.1. Logarithm of the intensity ratio of the He II 4686 Å and He I 5876 Å lines as a function of temperature (see text). If the He II resonance line is optically thick, the CR (collisional-radiative model) curve for $N_e = 10^{18}$ cm$^{-3}$ would be closer to the corresponding LTE result.

electrons with above-threshold energies, $(1s2p2\ell)^2L$ by below-threshold excitation involving capture of the colliding electron into $2\ell$. Inner-shell excitation of the lithium-like ions must also be considered for accurate calculations.

Assuming corona model conditions, the resonance line intensity $i_r$ is proportional to its excitation coefficient $X$, estimated by (6.27), while the sum of $n = 2$ dielectronic satellite intensities, $\sum i_d$, is proportional to the dielectronic capture rate coefficient estimated by (6.37), but summed over $\ell$ and multiplied by the branching ratio $[1 + (n_s/n)^3]^{-1}$ for radiative decay used in (6.42). In this ratio $n_s$ corresponds to that principal quantum number $n$ of the captured electron for which radiative decay and autoionization are equally likely.

The transition probabilities, oscillator strengths, effective Gaunt factors and frequencies are essentially the same for both transitions. The resonance-to-satellite line ratio is therefore simply estimated by

$$R = \frac{i_r}{\sum i_d} \approx \frac{X}{\sum d_{n\ell}}[1 + (n_s/n)^3]$$

$$\approx \frac{1}{2}[1 + (n_s/n)^3]\frac{n^2 kT}{z^2 E_H}\exp\left(-\frac{z^2 E_H}{n^2 kT}\right). \tag{11.8}$$

This estimate, using (6.41) with $\bar{g} = 0.2$ and $n_1 = \ell_m = 2$, is compared in figure 11.2 with more detailed calculations of the quantities $q$ and $C$ in their equation (12) by Vainshtein and Safronova (1978), who have also calculated corresponding data for $n = 3$ (1980). These detailed calculations (see also Itikawa, Kato and Sakimoto 1995, Kato et al. 1995) yield relative intensities of individual satellites, the full use of which is in avoiding those satellites whose relative intensities are dependent also on electron density, as discussed in section 10.4. Moreover, the calculated wavelength separations from the parent line are required for line identification and to assess blends.

The temperature dependence of these ratios is much weaker than that of the line ratios in figure 11.1, at least in the temperature range where the various ions are likely to be abundant and radiating. However, this is offset by having to measure only relative intensities in the immediate vicinity of the resonance line. Agreement with experiments, e.g., for helium-like nickel between measured and calculated satellite-to-resonance line ratios is good (Zastrow, Källne and Summers 1990), both for $n = 2$ and $n = 3$ satellites.

The reader should not be disturbed by the vertical displacements between the argon and iron curves on figure 11.2 corresponding to (11.8) and those according to Vainshtein and Safronova (1978). These displacements are due to the simplifications made in section 6.4. However, the main point here is to elucidate the temperature dependence, which is practically the same if $z$ in (11.8) is taken as the ionic charge of the excited and re-combining ion, i.e., $z = Z - 1$ or $Z - 2$ in case of hydrogen- or helium-like ions of nuclear charge $Z$. The basic virtue of the method is that it does not even require a relative intensity calibration, but only a linear response of the detector and sufficient range to measure both the usually rather weak satellites and the resonance line. Perhaps the greatest experimental problem is to ensure that the resonance lines are indeed optically thin, or else to correct for radiative transfer and photo-excitation (Duston and Davis 1980). Some of these difficulties are alleviated by using satellites to the $(1s^2)^1S$ - $(1s3p)^1P$ transition, say, of Ar XVII. In that case the tempera-ture sensitivity of satellite line ratios is reduced and Stark broadening and non-LTE level kinetics must be considered in a realistic model (Mancini et al. 1992). A thorough experimental and theoretical study (Beiersdorfer

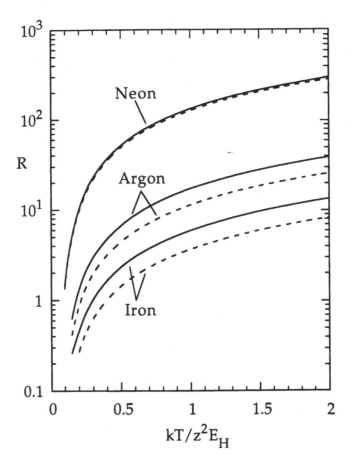

Fig. 11.2.  Intensity ratios of resonance lines and dielectronic satellites of helium-like ions according to Vainshtein and Safranova (1978), solid curves, and according to (11.8), dashed curves.

et al. 1995) at low densities ($N_e \lesssim 2 \times 10^{13}$ cm$^{-3}$) indicates good agreement for the stronger dielectronic satellites but deviations by factors of $\sim 2$ for weaker satellites.

## 11.3  Lines from isoelectronic transitions of different elements

The methods for spectroscopic temperature measurements discussed so far depend more or less critically on atomic parameters and the validity of LTE, PLTE or of the particular kinetic model used. As pointed out by Marjoribanks et al. (1992) and further discussed by Marjoribanks et al. (1995) and by Shepard et al. (1996), much of this dependence can be removed by measuring and interpreting relative intensities of corresponding lines from isoelectronic ions of different elements. One first observes that corona model ionization and excitation calculations as discussed in section 6.1 yield steady-state populations, e.g., for hydrogen- and helium-like ions of the elements K and Cl, which are very nearly identical functions of a scaled electron temperature, $kT/Z^2E_H$. The rise and fall of the excited level populations as function of actual temperature is thus shifted in temperature by an amount of the order

$$\Delta(kT) \approx 0.2\overline{Z}\Delta Z E_H, \qquad (11.9)$$

if one takes $kT \approx 0.1Z^2E_H$ as a typical value and introduces a mean nuclear charge $\overline{Z}$.

By measuring isoelectronic line intensity ratios, one basically determines this shift to within some fraction $f$, say one half, of its value. One may therefore expect a relative error in the inferred temperature of

$$\begin{aligned} \frac{\Delta T}{T} &\approx 2f\frac{\Delta Z}{Z} \\ &\approx \frac{\Delta Z}{\overline{Z}}, \end{aligned} \qquad (11.10)$$

e.g., $\Delta T/T \approx 0.05$ for $\Delta Z = 1$, $\overline{Z} \approx 20$. As can be seen from figure 11.3, this is reasonably consistent with detailed calculations, using a refined version of the RATION code (Lee, Whitten and Strout 1984), of the intensity ratio of optically thin $1s^2\,{}^1S$ - $1s3p^1P$ helium-like ion lines of vanadium and titanium for $kT \approx 700$ eV. This assumes $\sim 5\%$ accuracy in the measurement and neglects any other errors. At temperatures below $\sim 1$ keV these calculated ratios (Marjoribanks et al. 1995) are, moreover, only very weakly dependent on electron density, as expected in the validity regime of the basic corona model. Near $kT = 2$ keV the density dependence is more pronounced, presumably because of the increasing importance of hydrogen-like ions and excited state populations. Still, as long as the electron density is known within a factor of 2, there is no significant deterioration in the expected accuracy.

So far, this discussion implied steady-state ionization and excitation. For highly transient laser-produced plasmas, this assumption should be checked by estimating especially the ionization relaxation times using, e.g., the approximate ionization rate coefficients discussed in section 6.2. At high densities, these times would be upper limits, since then two-step ionization and also photo-excitation, etc., would tend to diminish relaxation effects in ionizing plasmas. In recombining plasmas, a similar role might be played by three-body recombination and collisional deexcitation. In both cases, the situation is rather analogous to the re-

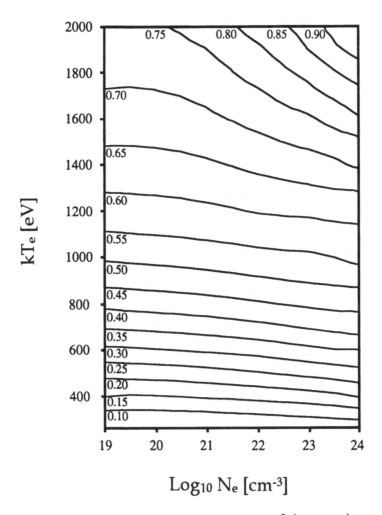

Fig. 11.3.   Contour plot of the intensity ratio of $1s^2$ $^1S$ - $1s3p^1P$ vanadium and titanium lines as a function of electron temperature and electron density for optically thin plasmas in corona equilibrium (Marjoribanks et al. 1995).

laxation to LTE conditions discussed in section 7.6. Nevertheless, as pointed out by Marjoribanks et al. (1995), a fully time-dependent model is required for quantitative temperature measurements if relaxation effects are important. Such a model was used in the interpretation of isoelectronic spectra from plasmas in laser-heated hohlraum and related laser targets, for which the line ratio technique not only remained useful for highly transient plasmas, but also could be shown to be rather insensitive to the effect of the hohlraum background radiation (Shepard et al. 1995,1996).

## 11.4    Relative continuum intensities

A general point should be made in regard to all spectroscopic temperature measurements, namely, that in case of deviations from Maxwellian electron velocity distributions one essentially obtains an effective temperature characterizing that part of the distribution which is mostly responsible for the observed quantity. By measuring quantities depending on different portions of the distribution, one can thus infer the actual shape also of a non-Maxwellian velocity distribution (Behringer and Fantz 1994). An especially transparent case is continuum emission due to free-free or free-bound transitions, provided one uses accurate cross sections (Jung 1994). The simplest situation occurs in fully ionized, high-temperature plasmas. In that case only electrons with energies exceeding $\hbar\omega$ contribute to the bremsstrahlung intensity $I_{ff}(\omega)$ according to the last term in (5.17), which dominates any free-bound emission for $kT \gg z^2 E_H$. (Remember that the ionic charge $z$ of the scattering or recombining ion equals the nuclear charge $Z$ in this case.) Except for the rather weak variation of the Gaunt factor with frequency, one therefore has from (5.16)

$$I_{ff}(\omega) \sim \exp(-\hbar\omega/kT). \tag{11.11}$$

Measuring relative intensities at photon energies exceeding $kT$ by some factor $F$ will then yield temperatures with relative errors corresponding to $\Delta I/I \cdot F$, if $\Delta I/I$ is the error in the relative intensity measurement. In practice, $\Delta I/I \cdot F$ has a minimum at some frequency beyond which the intensity becomes too small, or the electron-ion (dipole) bremsstrahlung is no longer much more intense than electron-electron (quadrupole) bremsstrahlung (Joseph and Rohrlich 1958, Maxon and Corman 1967). In any case, $F \lesssim 3$ may be considered typical, and errors in corresponding intensity ratios will rarely be below, say, 15%, resulting in $\sim 5\%$ accuracy in $T_e$ if proper Gaunt factors are used. In case of incompletely stripped target ions, the Gaunt factor implied in (11.11) should be replaced by the appropriate Biberman factor or the corresponding quantity obtained from more accurate bremsstrahlung cross sections as discussed in sections 5.3 and 5.4.

While the relative continuum intensity measurements just discussed depend on calibration of the relative instrumental response over a broad spectral range and on avoiding any line contributions and absorption edges, continuum intensity measurements on both sides of an edge basically only depend on the linearity of the detector response and on extrapolating the continuum levels between the various lines toward the ideal limit of the spectral series. Such extrapolation is normally required because of the various high density effects discussed in section 5.5. Except for any accidental blends, the continuum intensity on the high photon energy side of an edge can be measured more directly. In any case, it is desirable to obtain well-resolved spectra to locate lines before using, e.g., filter-detector combinations for broad-band measurements.

Calculated ratios of the intensities $I_+$ and $I_-$ on the high and low frequency sides of the Lyman ($n = 1$) and Balmer ($n = 2$) limits of hydrogen and one-electron ions obtained from (5.16) and (5.17) are shown on figure 11.4. The sensitivity of the ratios extends up to $kT \approx \frac{1}{2}Z^2 E_H$, but the temperature accuracy is typically inferior to that achievable by bremsstrahlung measurements. Moreover, it is more difficult to generalize it to plasmas containing various ions.

## 11.5  Ratios of line and continuum intensities

The method just discussed does not involve the assumption of LTE boundstate populations. In contrast, we now turn to a method which is also mostly useful for pure hydrogen, helium, etc. plasmas, but rests on assuming PLTE for the upper level of the line, whose spectrally integrated intensity is measured relative to that in an adjacent or underlying continuum band.

Using (7.24) and the frequency integral of (8.10) for the line emission and (5.16) multiplied with the frequency interval $\Delta\omega$ for the continuum emission, the line to continuum intensity ratio can be written as

$$\frac{i_\ell}{i_c} = \frac{\pi(3)^{3/2} g_m A_{nm}}{32\alpha^3 Z^4 \Delta\omega} \frac{\hbar\omega}{E_H} \frac{\exp(E_{ni}/kT)}{[\cdots]} \tag{11.12}$$

with $[\cdots]$ as in (5.17) and $g_i = 1$ for the completely stripped ions assumed here. The reduction of the ionization energy appearing in (5.16) is cancelled by the corresponding correction to be applied to (7.24), and we used the fact that $E_{mi}$, the ionization energy of the upper level, added to the photon energy $\hbar\omega$, results in $E_{ni} = Z^2 E_H/n^2$ for the one-electron atoms and ions being considered.

After expressing $g_m A_{nm}$ in terms of $g_n f_{mn}$ as in (3.1) and substituting $\Delta\omega = 2\pi c\Delta\lambda/\lambda^2$, (11.12) is seen to agree with (13.8) of Griem (1964), namely,

$$\frac{i_\ell}{i_c} = \frac{\pi^3 3^{3/2}(a_0/\alpha)^2 g_n f_{mn} \exp(E_{ni}/kT)}{2\lambda\Delta\lambda g_i[\cdots]}, \tag{11.13}$$

which may be more convenient for applications to other than hydrogen or hydrogenic ion plasmas. The important point is that the sum in equation (5.17) for $[\cdots]$ over final states in the recombination continuum begins with $n_{min} = n + 1$ so that the exponential in the numerator dominates the temperature dependence of the intensity ratios, e.g., for pure hydrogen

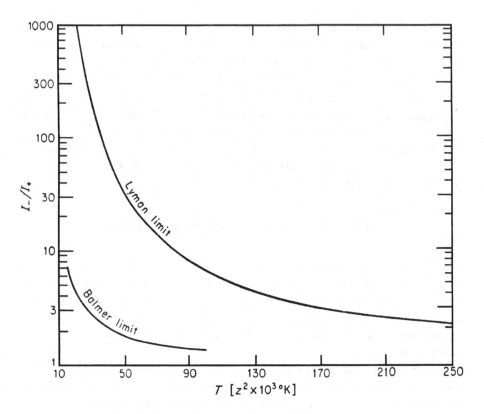

Fig. 11.4.  Ratio of continuum intensities on the high and low photon energy sides of Lyman and Balmer limits (see text) as a function of temperature in plasmas containing essentially only bare ions.

plasmas as shown in figure 11.5. Suitable neutral helium line-to-continuum ratios, calculated neglecting doubly ionized helium, are shown in figure 11.6. Ratios for isoelectronic ions could be estimated by scaling $A_{nm}$ and $\omega$, $\Delta\omega$ and $E_{ni}$, etc., in (11.12) with $Z^4$ or $Z^2$, respectively; but one would usually find that the PLTE assumption for the upper levels of strong lines is questionable according to the criteria in section 7.6. At relatively low temperatures, one should also consider, e.g., $H^-$ emission, bremsstrahlung on helium atoms, etc. (see chapter 5).

## 11.6  Intensities from optically thick layers

So far we assumed that lines, not to mention continua, used for temperature measurements were not subject to reabsorption or induced emission. On the other hand, in the rather academic case of emission from an optically thick and homogeneous, but sharply bounded, plasma, the emitted

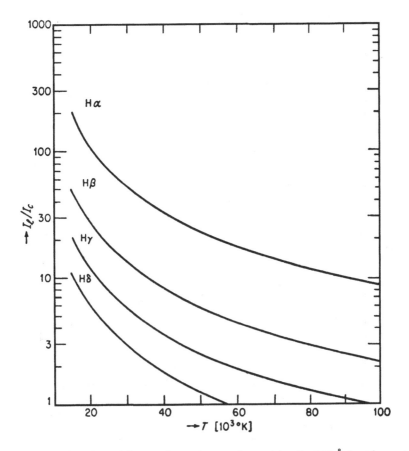

Fig. 11.5. Ratio of total line and continuum intensities (in 100 Å bands centered at the lines) as a function of temperature in hydrogen plasmas.

intensity is given by Planck's law, (2.54), for blackbody radiation. Such an ideal situation would be indicated by a continuous spectrum of the same spectral shape as (2.54); and this shape could be used to infer the electron temperature if one assumes that electrons do indeed control level populations and the state of ionization. Other methods for the temperature determination using Planck's law would involve the absolute intensity at a certain frequency, or the spectrally integrated intensity.

Having a continuous spectrum corresponding to the Planck function over a significant frequency range, rather than the usual and more characteristic recombination and bremsstrahlung continua with super-imposed emission or absorption lines, is a rare exception. In other words, if at all, laboratory plasmas are normally optically thick only in the vicinity of strong lines or in the continuum at very low frequencies, i.e., long wavelengths. In the latter case, which was already mentioned in the intro-

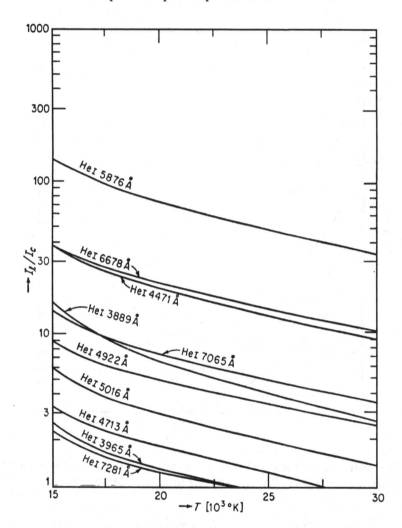

Fig. 11.6.  Ratio of total line and continuum intensities (in 100 Å bands centered at the lines) as a function of temperature in helium plasmas.

duction to chapter 5, Planck's law can almost always be approximated by the Rayleigh-Jeans law, i.e., the emitted intensity is only a linear function of the temperature. It increases then with frequency as $\omega^2$, or transformed to wavelengths, decreases with $\lambda^{-4}$. The absolute intensities in frequency or wavelength ranges corresponding to these power laws can be used with some confidence to deduce temperatures via (2.54), although it is generally a good idea to first consider the absorption mechanism and to either estimate or ascertain that the optical depth of the plasma is large at the frequencies or wavelengths used.

Optically thick lines, whose upper to lower level population ratio is near the LTE value at the electron temperature, can be used similarly. In our idealized, sharply bounded but otherwise homogeneous plasma, their emission profiles would have a flat-top at a level corresponding to (2.54) evaluated at the transition frequency. The temperature could then be deduced from the measured absolute intensity in this saturation region. As is well known, observed and sufficiently well-resolved line profiles show typically much more structure than this idealized model would imply. Quite common for near LTE conditions is the appearance of a self-absorption dip near the line center caused by a decrease of the upper-to-lower level population ratio towards the plasma boundary. As discussed in chapter 8, the corresponding radiative transfer problem is much more involved in non-LTE situations, which may well occur in the plasma edge region even if in the interior LTE population ratios can be expected according to the various criteria in section 7.6.

Assuming LTE and cylindrical symmetry, a method due to Bartels (1950 a,b), which has been reviewed by various authors (Zwicker 1968, Preobrazhenskii 1971, Karabourniotis 1986) can be used. It is based on the measurement of self-reversed, pressure-broadened line profiles as a function of chord height. Basically it relies on the experience from modeling of optically thick emission for the most likely case of radially decreasing temperature, and therefore also of the LTE source function $S$ defined by (8.18), that the two maxima of the self-reversed profiles correspond to $\tau \approx 2$ and that their intensities reach typically 80% of the blackbody intensity at the temperature in the $\tau \approx 2$ layer. A mapping of $\tau$ vs. radius is to some extent possible by an analysis of the chordal measurements.

Closely related to Bartels' method is the generalized line reversal method (Kohn 1932, Faizullov, Sobolev and Kudryavtsev 1960, Lochte-Holtgreven 1968) which is, however, directly applicable only for homogeneous plasmas and requires a background source of comparable brightness. One measures the emerging intensities both with and without the additional source. Its contribution is represented by the last term in (8.19), which for the present case becomes

$$I(\omega) = I_T(\omega)(1 - e^{-\tau_\omega}) + I_S(\omega)e^{-\tau_\omega}, \qquad (11.14)$$

if we replace the source function by its LTE value (2.54). By measuring the emerging intensity $I(\omega)$ both with and without the source, and having it and therefore $I_S(\omega)$ calibrated absolutely, one obtains, at a given frequency, both $I_T$, and therefore the temperature, and also the optical depth $\tau_\omega$. For reasonable sensitivity $\hbar\omega > kT$ and $\tau_\omega \approx 1$ are preferable, as are measurements at various frequencies in order to validate the rather restrictive assumptions. The reader will notice that this line-reversal

method is closely related to measurements of LTE absorption coefficients
or opacities mentioned in the introduction to chapter 9.

## 11.7    Line intensities in rapidly ionizing plasmas

Inverting the interpretation (Kunze 1972) of measured time histories of
lines from successive ionization stages in rapidly heated, transient plasmas
in terms of effective ionization coefficients mentioned briefly in section 6.2,
electron temperatures can be inferred from such measurements, provided
the electron density has been determined independently. The charge
state distribution in plasmas of interest here normally can be described
by a time-dependent corona model generalizing (6.1) by including time-
derivatives and transport terms. A typical time-lag between a step-function
temperature pulse and the appearance of some ion is then of the order of
$(SN_e)^{-1}$, $S$ here being the ionization rate coefficient of the preceding ion
and $N_e$ the electron density.

   If the observed line is primarily excited from the ground state of the
resulting ion, no significant further delay is to be expected, because the
corresponding relaxation time is either determined by the inverse of the
spontaneous transition probability or even shorter. If a line from some
highly excited level of principal quantum number $n$ is used, e.g., a line
involving a $\Delta n = 1$ transition, its time history could be closer to that
of the ground state of the following ion, provided the electron density is
within some factor $\lesssim 10$ of the critical density corresponding to (7.77)
or larger. In such a case, one would have to use the ionization rate
coefficient of the ground state of the observed ion in estimating the
relaxation time. In intermediate situations, it would be preferable to use
a time-dependent collisional-radiative model to obtain synthetic spectra
for a variety of assumed temperature histories and measured electron
densities, varying the assumed temperatures versus time until satisfactory
agreement is reached.

   This method is applicable as such only if the relevant $(SN_e)^{-1}$ values
are longer than the characteristic times for significant temperature variations. Since the ionization rate coefficients vary according to section 6.2
approximately as $\exp(-E_\infty/kT)$ for ionization energies $E_\infty$ substantially
larger than $kT$, and since $E_\infty \gtrsim 5\ kT$ is quite common, temperatures can
be estimated to within a factor, say, 1.5, even if one assumes a combined
uncertainty of a factor of 10 in rate coefficients, etc.

   For impurity ions in the edge region of a magnetically-confined plasma,
another application of time-dependent rate equations is to relate the
observed maximum charge state to the impurity ion confinement time $\tau$,
again equating it with $(SN_e)^{-1}$. An additional assumption in this case

is that radially outward transport of such ions from higher temperature regions is not important or can be accounted for.

Before leaving the subject of spectroscopic measurements of the electron temperature, it is good to remember that most of the methods used are somewhat indirect. They should therefore be supplemented, e.g., by Thomson scattering measurements (Kunze 1968, Evans and Katzenstein 1969, DeSilva and Goldenbaum 1970, Sheffield 1975) if the results are crucial for the interpretation of the experiment. A recent example for such cross-checks for laser-produced plasmas is the work of La Fontaine et al. (1994).

## 11.8  Doppler profiles

Although the gas temperature, i.e., the kinetic temperature of neutral atoms in very partially ionized gases of known total pressure, is often inferred from spectroscopic or other density measurements, the only direct emission or absorption spectroscopic method for the determination of ion and atom temperatures remains the profile measurement of primarily Doppler-broadened spectral lines. For thermal distributions of nonrelativistic emitting or absorbing atoms or ions, these profiles and their widths are given by (4.1) and (4.2), respectively. The very first experimental test to make in selecting lines for the present purpose is therefore to verify that their profiles are reasonably consistent with (4.1), i.e., that they are indeed Gaussian over a significant intensity range, say, down to at least 20% of peak intensity.

However, a positive result of such test may well be misleading. One reason could be that other physical processes together with instrumental broadening effects might mimic a Gaussian profile. They and also non-Gaussian line broadening effects, including any self-absorption, etc., should therefore be considered, e.g., according to chapters 4 and 8, before accepting Doppler temperatures from a small number of measured line profiles. A more experimental approach is to use lines whose sensitivity to competing, e.g., Stark, effects can safely be assumed to vary substantially. Only if all of these lines yielded the same temperature, say, within $\pm 20\%$, would it be safe to accept the corresponding mean value.

Another source of possible misinterpretation of Gaussian line shapes is the rather common occurrence of hydrodynamic turbulence in high-temperature plasmas. Since velocity components of the corresponding eddies may also have Gaussian distributions for fully developed turbulence, and since emission or absorption measurements are averaged over these eddies, they basically measure a convolution of thermal and fluid velocity distributions, which is also Gaussian or at least very nearly so.

An important example for such complications may be encountered in the measurement of impurity ion temperatures in hydrogen plasmas. Because of their larger masses, impurity ion thermal velocities tend to be much smaller than those of hydrogen atoms or ions. For equal temperatures, thermal Doppler widths of impurity lines thus tend to be small and are easily masked by turbulent motion with velocities approaching thermal velocities of hydrogen, i.e., by subsonic turbulence. To guard against systematic temperature errors due to such effects, it is a good procedure to first deduce rms velocities from measured line widths for several or all impurity ions in their various charge states. Should these rms velocities approach some common value for heavier impurity ions, or effective ion temperatures begin to scale with the square root of the ion mass, turbulence is likely to be important. Depending on its cause, it may not be equally developed in all directions of velocity space, a situation that could be gleaned from profile measurements from various directions.

In magnetized plasmas, Zeeman effects (Condon and Shortley 1951, Cowan 1981, Sobel'man 1992) may have to be accounted for — over and above any other broadening not due to Doppler effects, including natural broadening of high $Z$ lines, before deducing temperatures from measured line widths. Simply checking the appearance of near-Gaussian line shapes does not suffice; merged Zeeman and fine-structure components of lines with relatively complex spectra may well combine to form a composite Gaussian profile. Some indication of such a situation would be a dependence of profiles and widths on polarization, but in any event modeling (Hey 1994) of combined Zeeman and Doppler effects should be used to estimate corrections (Hey et al. 1993) of preliminary ion temperatures inferred from total line widths. For lines from high $Z$, two- and more-electron ions, one usually deals with the anomalous Zeeman effect; while for one-electron spectra and high fields the magnetic interactions may be comparable to or larger than the fine-structure splitting. The corresponding Paschen-Back effect must then be accounted for exactly (Hey et al. 1993, 1994).

Zeeman shifts or splittings are always of the order of the electron cyclotron frequency $\omega_{ce} = eB/m_e$, so that Doppler widths according to (4.2) should be, say, five times larger than this characteristic frequency. For Zeeman effects not to compromise the temperature measurement, the ion or atom temperature should therefore fulfill

$$kT \geq (\ell n2)^{-1} \left( \frac{5a_0^2}{c\mu_0 e} \right)^2 \left( \frac{B\lambda}{a_0/\alpha} \right)^2 \frac{M}{m_e} E_H, \qquad (11.15)$$

if $\lambda$ is the wavelength of the line used and $M/m$ the radiating-ion to electron mass ratio. Inserting numerical values for $a_0/\alpha = 7.25$ nm $=$

72.5 Å and $(5a_0^2/c\mu_0 e)^2 = 5.38 \times 10^{-8}$ SI units, this condition is seen to give similar values as Hey's (1994) equation (5) for $n \approx 10$ as the principal quantum number of the upper level of the line. For example, for the Si XII 4792 Å line and $B = 2T$, a temperature above 1.5 keV is required according to Hey (1995), rather than the 1.0 keV suggested by (11.15). However, (11.15) does not account for any incomplete Paschen-Back transition, i.e., for residual fine-structure effects; while the factor $n^2$ in Hey's (1994) equation (5) is probably not representative of the general case. To summarize, short wavelength lines are relatively insensitive to Zeeman effects; but their Doppler widths are often not much larger than instrumental widths. Another possible complication can arise from unresolved dielectronic satellite lines (Bitter et al. 1979).

As with all spectroscopic measurements, the observed profiles should be reduced to local profiles of emission coefficients as discussed in section 8.5 in order to obtain temperatures, e.g., as a function of minor radius in a tokamak (Miyachi et al. 1989). For more direct measurements of local quantities, see the first two sections of the following, final chapter.

# 12

# Other diagnostic applications
# of plasma spectroscopy

In addition to the most general applications of spectroscopic methods to density and temperature measurements, which were discussed in the two preceding chapters, there are numerous special applications. Some of these will be discussed in this concluding chapter, without any prejudice against any well-established or recently developed methods which are omitted. If there is any common thread, it is in the strong role played by atomic collision theory and by the physics of atoms and ions in electric and magnetic fields. As in the other chapters, no attempt will be made to describe the often very sophisticated instrumentation or other experimental details, which the interested reader should be able to find with the help of the references in the original papers.

The first two special applications to be discussed, namely charge exchange recombination and beam emission spectroscopy, obviate the essential difficulty in emission or absorption spectroscopy in obtaining spatial resolution along the line of sight. This is accomplished by injecting heating or diagnostic neutral beams into magnetically confined plasmas either to preferentially populate some excited states of plasma ions by charge exchange recombination, as discussed in section 6.5, or by having the atoms in the beam ionized and excited by the plasma electrons, as discussed in sections 6.2 and 6.3, or even by protons and other plasma ions (Mandl et al. 1993), see section 6.5. Either way, by observing the various emissions obliquely to the beam, spatial resolution is provided as, e.g., in case of 90° Thomson scattering or laser-induced fluorescence (LIF).

The subject of polarization spectroscopy is introduced in the third section of this chapter, with one application being in the measurement of electric fields due to plasma waves, i.e., of fields that cannot be attributed to charged particles within a Debye radius from the radiating atom or ion. The reader will remember that these more-or-less stochastic particle-produced electric fields were the subject of section 4.3, while some wave (oscillating) field effects on spectral line shapes were discussed in section

4.11. Another application of polarization spectroscopy is concerned with any deviation from isotropy of the electrons responsible for excitation, which would then be reflected in the polarization of the line radiation.

The fourth section deals with the measurements of magnetic fields, which may or may not involve diagnostic beams. In any case, it is advantageous to measure Zeeman patterns as a function of polarization and possibly also of the direction of emission.

The challenging subject of the measurement of macroscopic electric fields, in contrast to microfields and plasma wave fields, is taken up in the fifth section. No well-established methods seem to exist, but there could well be promising developments through experimental research and computer modeling.

A final application to be discussed in the last section are spectroscopic methods for the determination of the effective charge of plasma ions in magnetic fusion experiments.

## 12.1 Charge exchange recombination spectroscopy

Because of the state-selectivity of charge exchange, say, from groundstate hydrogen atoms into specific excited states of the recombined ion, line spectroscopy (CHERS) may provide diagnostic signatures exceeding the continuous background and any impurity line radiation even in high temperature plasmas consisting mostly of hydrogen and its isotopes. This advantage and the spatial resolution already mentioned, which is possible if the charge exchange is dominated by beam atoms, facilitate diagnostic measurements, improving the usual emission spectroscopy considerably.

For quantitative spectroscopy, much more accurate cross sections than the estimates in section 6.5 and even than those obtained from earlier detailed calculations as reviewed by Janev, Presnyakov and Shevelko (1985) are required. Examples of such atomic data for magnetic fusion applications are effective line excitation cross sections of von Hellermann et al. (1991), which account for mixing of the angular momentum sublevels of the recombined, originally fully-stripped, ion. A very useful feature of these cross sections for fusion $\alpha$-particle diagnostics (Post et al. 1981, von Hellermann et al. 1990) is their strong peak at small relative velocities between $\alpha$-particles and beam atoms. For a tokamak diagnostic arrangement (Stratton et al. 1994) as shown in figure 12.1 and for a 100 keV deuterium neutral beam, $\alpha$-particles moving at about 200 keV more-or-less parallel to the beam thus become preferred targets for recombination into $n = 4$ levels of He II. The radiative branching ratio for decay into $n = 3$ is 30%, also $\ell$-averaged, and the corresponding He II 4686 Å ($4 \rightarrow 3$) radiation is Doppler-shifted by about $-48$ Å $\cos\phi$, $\phi$ being the angle between a sight line and the neutral beam.

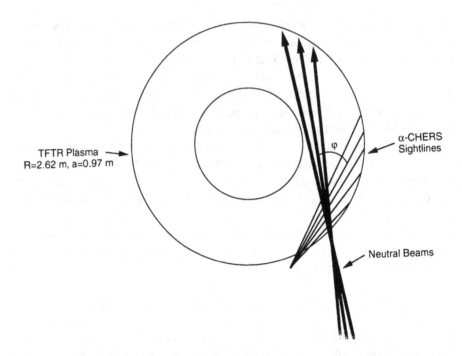

Fig. 12.1.   Schematic of experimental setup for charge exchange recombination spectroscopy (Stratton et al. 1994) on a tokamak.

Other $\alpha$-particles than those favored by the peak cross section contribute also to the He II line emission, but with considerably reduced efficiency. For example, 200 keV $\alpha$-particles moving anti-parallel to the 100 keV neutral deuterium beam are detected only with negligible efficiency compared with the parallel case (von Hellermann et al. 1991). Line profiles calculated by averaging over the $\alpha$-particle velocity distribution are thus asymmetric, except for $\phi = 90\%$. This asymmetry permits the separation of contributions from collisions with thermalized atoms, which primarily provide $n = 4$ electrons to thermalized $\alpha$-particles ($\alpha$-$s$). The remaining density and energy distribution of the fast fusion $\alpha$-$s$, in turn, can be inferred from the absolute intensity of the asymmetric red wing (McKee et al. 1995 a,b). This data analysis involves considerable modeling of neutral beams [including excitation before ionization (Boley et al. 1984)] and of the charge exchange collisions of the remaining beam atoms with the fast $\alpha$-$s$ (Post et al. 1981, von Hellermann et al. 1991, Stratton et al. 1994). Besides the desired line radiation from prompt charge exchange, recombined $\alpha$-s may also drift out of the beam volume and be excited by further collisions (Fonck et al. 1984).

In addition to this relatively recent application to fusion-$\alpha$ diagnostics, the new method of space-resolved and state-selective quantitative spectroscopy has found much use for other ions and in other magnetic fusion experiments since the early measurement (Isler et al. 1981) of the density of completely stripped oxygen ions in a tokamak by neutral hydrogen beam, charge-exchange recombination spectroscopy. The general method has been reviewed by Fonck in 1985, and the various associated line shape considerations were presented by the same author in 1990. Contributions of charge exchange recombination, e.g., to C VI Lyman-series lines were discussed by Mattioli et al. (1989). This mechanism is also of interest to x-ray laser pumping (Elton 1990).

A further application of charge exchange recombination is in the measurement of space-resolved ion temperatures and plasma rotation (Fonck et al. 1984, Fonck 1985, Danielsson et al. 1992, Schorn et al. 1992, Hey 1994) in high temperature plasmas, especially also in the edge plasma region or even near a limiter (Bogen et al. 1995) or a divertor plate. To facilitate such measurements on low to medium-$Z$ impurities, which are mostly fully stripped, recombination into, say, $n = 8$ to 12 bound states followed by a $\Delta n = 1$ ($n$-$\alpha$) radiative transition is generally preferable. (The $\alpha$ here refers to an $\alpha$-type hydrogen or hydrogenic ion line, not to an $\alpha$-particle.) The spectral resolution at these relatively long wavelengths is usually quite sufficient for resolving the Doppler width, and other line broadening mechanisms are normally not too important. However, fine-structure and other inherent level splittings and Zeeman effects must be considered very carefully, as already discussed in section 11.8. Moreover, some of the spatial resolution may be compromised due to plasma edge emission (Ida and Hidekuma 1989, Synakowski et al. 1995).

The number of Zeeman levels for these hydrogenic $n$-$\alpha$ lines is large (Hey et al. 1993), and fine-structure splittings (Bethe and Salpeter 1957, Klarsfeld and Maquet 1973) are typically small but not negligible. The Zeeman effects must therefore often be calculated for intermediate field-strengths (Kiess and Shortley 1949, Condon and Shortley 1951) in order to account for the Paschen-Back effect. For not fully stripped recombining ions, it is also necessary to include $\ell$-dependent core-polarization corrections (Edlén 1964) to account, e.g., for the distortion of the shielded Coulomb potential due to the presence of $1s^2$ electrons. In spite of these complications, at least the long-wavelength-side of mostly Doppler-broadened lines of this kind remains close to Gaussian (Hey et al. 1993). These authors also emphasize that their calculated correction factors for ion temperatures inferred from measured line widths are not valid if the density is too low for collisions to ensure statistical populations for different $\ell$-levels (Fonck, Darrow and Jaehnig 1984, Schorn et al. 1992). Since charge exchange favors intermediate $\ell$-levels in these cases, this is not a moot point. Another important consideration is a possible overlap of, e.g., C VI and O VI $n$-$\alpha$ lines. Corresponding calculated profiles (Bogen et al. 1995) for $n = 8$ are shown in figure 12.2. The O VI line has a hump on the blue wing, which is associated with low-$\ell$ levels, and its peak is blue-shifted with respect to the C VI line by about 80 mÅ, mostly because of the larger Rydberg constant associated with the larger reduced mass of the active electron in case of O VI.

The major effect of some unknown admixture of oxygen to the dominant carbon impurity is simply an underestimate of the temperature due to the use of too small an ion mass in (4.2) for the Doppler width. While having good wavelength resolution and calibration may help to decide between extreme cases, it will not be of much use in determining carbon and oxygen line ratios. However, high resolution turned out to be extremely important for the detection and subtraction of $D_2$ molecular lines (Bogen et al. 1995). Had these lines not been recognized, the C VI line would have appeared to be significantly broader and have indicated almost twice the actual temperature. In the experiment just discussed, recycling hydrogen atoms provided most of the electrons. Similar measurements were made using neutral H, D or Li beams (Fonck et al. 1984, Fonck 1985, Schorn et al. 1992, Hey et al. 1993). If absolute cross sections, etc., are available, impurity (e.g., carbon) densities can be inferred from absolute intensities of C VI lines emitted mostly due to Li-beam charge exchange recombination (Schorn et al. 1991), analogously to the work of Isler et al. (1981).

For further discussions of charge exchange recombination spectroscopy involving Li-beams, the reader is referred to an early review by Winter (1982), and for recent measurements of corresponding emission cross sections to Wolfrum et al. (1992) and to Hoekstra (1995), who also discusses the important role of metastables in He beams. Some of these cross sections are about two orders of magnitudes larger than corresponding cross sections for neutral hydrogen beams (Janev, Bransden and Gallagher 1983), provided these hydrogen beams have only a very small fraction of 2s metastable or other excited atoms (Isler and Olson 1988). As might be expected, Li(2s) and H(2s) cross sections for charge transfer are about the same (Winter 1982) at given relative velocities, and estimates of the metastable fraction in hydrogen beams or recycled hydrogen atoms are therefore very important.

In most applications of CHERS, the beam atom velocities are much higher than those of the target ions. The population rate of the upper level of the diagnostic line is then nearly independent of the ion velocity. This allows one to infer ion velocity distributions from the measured emission line profiles, i.e., not only temperatures can be obtained but also any flow velocity of the tokamak plasma (Fonck et al. 1984, Fonck 1985). By

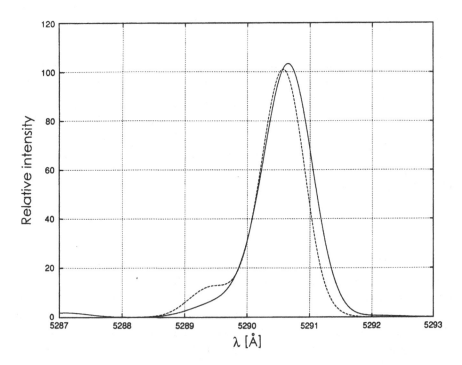

Fig. 12.2. Comparison of calculated line profiles, including Gaussian instrumental broadening (FWHM 0.11 Å), of the $n = 8 \rightarrow n' = 7$ transitions in C VI (solid line) and O VI (dashed line), for an ion temperature of 50 eV and a magnetic field of 2.25 T. The observation direction is perpendicular to the magnetic field, and $\pi$ components only have been selected (340 $\pi$ components, 268 allowed, 72 $\Delta J$-forbidden). The differences in profile shape arise from incomplete screening of the nucleus by the $1s^2$ core electrons in the case of O VI, as a result of which s, p and d components do not contribute to the main line. Practically the full strength of the C VI line occurs within the wavelength range indicated, but about 10% of the strength of the O VI line occurs elsewhere. Statistical populations among the upper fine-structure levels were assumed. Appreciably narrower and differently shaped lines could be produced, were this requirement not fulfilled. [Calculations by J.D. Hey and J. Wienbeck for the experiment of Bogen et al. (1995).]

employing high-throughput filter spectrometers, time-and-space resolution can be improved to such an extent that both ion temperature and parallel velocity fluctuations and their spatial correlations can be measured and analyzed (Evensen et al. 1995). This is a significant complement to measurements, e.g., of ion density fluctuations to be discussed in the following section.

## 12.2    Beam emission spectroscopy (BES)

Diagnostic neutral beams are not only providing line emission through charge exchange recombination as discussed in the preceding section, but also through emission by beam atoms (see, e.g., Mandl et al. 1993) which are excited and ionized, mostly but not only by plasma electrons, as the beam penetrates into regions of higher density. Since the ionized beam atoms are deflected by the Lorentz force and slowed down very effectively by ion-ion collisions, the ionization determines the attenuation of the beam particle current $J$ according to

$$v_B \frac{dJ}{dx} = -S_{av} N_e J, \qquad (12.1)$$

$v_B$ being the beam velocity and $S_{av}$ the average ionization rate coefficient of the beam atoms. The beam often may be assumed to be in a quasi-stationary state, so that there is no $\partial J/\partial t$ term in (12.1), and that populations of the atomic states are in steady-state. Since understanding the beam attenuation is not only important for the following diagnostic applications, but also for those discussed in the preceding section, and because the ensuing line emissions give much of the diagnostic information, we now consider the beam propagation in some detail.

For simplicity, we will use a two-state model (Hey 1994), i.e., write the population rate equation for the excited state as

$$\frac{\partial N_2}{\partial t} \approx 0 = X_{21} N_e N_1 - [A_{12} + (X_{12} + S_2)N_e]N_2, \qquad (12.2)$$

since relaxation times for excited state populations are relatively short. This is not only because the deexcitation coefficient $X_{12}$, which is related to the excitation coefficient $X_{21}$ by the principle of detailed balancing expressed by (6.11), can be sufficiently large for the deexcitation rate to be significantly enhanced above the spontaneous radiative decay rate $A_{12}$, but also because the ionization coefficient $S_2$ for excited atoms tends to be comparable to $X_{12}$. Moreover, the excited fraction, namely

$$\frac{N_2}{N_1 + N_2} = \frac{X_{21} N_e}{A_{12} + (X_{12} + S_2)N_e} \left[1 + \frac{X_{21} N_e}{A_{12} + (X_{12} + S_2)N_e}\right]^{-1} \qquad (12.3)$$

according to the quasi-stationary solution of (12.2), usually remains small. Neglecting $\partial N_2/\partial t$ therefore should cause smaller errors than neglecting $\partial N_1/\partial t$ would in an analogous equation for this quantity. This equation contains the groundstate ionization coefficient $S_1$, which is normally much smaller than $X_{21}$.

From (12.3) and the corresponding relation for the groundstate fraction,

$$\frac{N_1}{N_1 + N_2} = \left[\ \cdots\ \right]^{-1}, \tag{12.4}$$

with the square bracket as in (12.3), we have for the state-averaged ionization coefficient

$$S_{av} = S_1 \left[1 + \frac{X_{21}N_e S_2/S_1}{A_{12} + (X_{12} + S_2)N_e}\right] \left[\ \cdots\ \right]^{-1}, \tag{12.5}$$

called the total ionization rate coefficient by Hey (1994). [See his equation (16) and the following numerical examples.] The spatial decay constant of the beam current is according to (12.1)

$$k_s = \frac{S_{av}N_e}{v_B}. \tag{12.6}$$

Its measurement therefore yields a local value of the electron density, provided the electron temperature and therefore $S_{av}$ can be determined. As should be clear from (12.5), this electron density measurement actually requires some iteration, which is not an essential difficulty as long as ionization out of the excited state is not dominant. Otherwise, one would have to use a more complete atomic model.

Instead of determining the ionization rate $S_{av}N_e$ from the attenuation of the beam, one actually measures the accompanying absolute line emission. To extract from this quantity the ionization rate, one first considers the ratio of ionization and photon emission rates,

$$\begin{aligned} R &= \frac{(N_1 + N_2)S_{av}N_e V}{N_2 A_{12} V} \equiv \frac{dN_i/dt}{dN_{ph}/dt} \\ &= \frac{S_2 N_e}{A_{12}} + \frac{S_1[A_{12} + (X_{12} + S_2)N_e]}{A_{12}X_{21}}, \end{aligned} \tag{12.7}$$

with $V$ being the volume defined by the intersection of the beam with the observation optics. The photon rate $dN_{ph}/dt$ is measured absolutely with allowance for the solid angle ratio, while $(N_1 + N_2)$, the effective volume and also $R$ and $S_{av}$ must be calculated. In case of $N_1 + N_2$, this calculation would begin with the measured beam current entering the plasma, followed by consideration of the losses according to (12.1), etc. (Winter 1982) and also allowing for any spread in beam velocities and changes in this velocity distribution along the beam (Guenther et al. 1989).

The inferred electron density from such measurements can finally be written from (12.5) and (12.7) as

$$
\begin{aligned}
N_e &= \frac{R}{(N_1 + N_2)S_{av}V} \frac{dN_{ph}}{dt} \\
&= \frac{1 + (X_{12} + S_2)N_e/A_{12}}{N_1 V X_{21}} \frac{dN_{ph}}{dt},
\end{aligned}
\tag{12.8}
$$

to be solved by iteration as discussed below. Again, the electron density dependence of the right-hand side should be small, or the two-level model would not be sufficient. For his Li-beam example, and for $T_e = 20\,\text{eV}$ and $N_e = 5 \times 10^{12}\,\text{cm}^{-3}$, Hey (1994) estimates $(X_{12} + S_2)N_e/A_{12} \approx 0.07$ and an excited state fraction as in (12.3) of 0.09, both well within the validity limit of the model. Electron densities significantly over $10^{13}\,\text{cm}^{-3}$, on the other hand, would be subject to additional theoretical errors unless one uses a more complete numerical model (Mandl et al. 1993), which also includes beam-atom, plasma-ion collisions resulting in excitation and ionization.

To implement (12.8), the local value of the groundstate density $N_1$ of the beam atoms is required. In the case of a mono-energetic and highly collimated beam implied by our simplified discussion, this quantity is given by

$$
N_1 = \frac{J}{v_B[\cdots]A_B}
\tag{12.9}
$$

in terms of the local value $J$ of the total beam current $J$, the groundstate fraction (12.4) and the cross-sectional area $A_B$ of the beam. While $A_B$ can be determined by imaging of the beam emission, $J$ must be calculated from the attenuation of the entering beam according to (12.1), etc., i.e., basically also from measurements of the beam emission as a function of the position along the beam. Substitution of corresponding expressions into (12.8) or its generalizations leads to rather involved relations (see, e.g., Guenther et al. 1989) for the electron density in terms of emitted intensities and beam parameters. Returning to (12.8), note that it can also be written, with (12.3) and (12.9), as

$$
N_e = [1 + (X_{12} + S_2)N_e/A_{12}][\cdots]\frac{v_B}{J(z)}\frac{dN_{ph}/dt}{X_{21}\Delta z},
\tag{12.10}
$$

if $\Delta z$ is the element of beam length viewed by the optical system.

Other than Li-beams, e.g., H- or He-beams may also be used for space-resolved $N_e$ determinations, especially to probe higher density plasmas. However, while in case of Li one directly measures the 2s-2p, 6708 Å resonance line, i.e., a line whose upper level is indeed the excited level "2" in the two-state model, one usually prefers not to measure the absolute H and He resonance line intensities because of the experimental difficulties in the vacuum-ultraviolet. Rather, one typically uses the $H_\alpha$ line or analogous He I lines, i.e., lines whose upper levels are more highly excited than level "2", which then corresponds to the final state of the observed transitions. To interpret such measurements, the two-level model remains appropriate for the inclusion of ionization from $n = 2$ excited levels, but depending on conditions, metastable level populations may require some consideration of nonlocal effects because of their long relaxation times. This must be kept in mind when using He I triplet-to-singlet line ratios (Fujimoto 1979a) for electron temperature determinations, say, in the 5-50 eV range, as discussed at the end of section 11.1.

Besides for electron density and temperature measurements, beam emission spectroscopy can also be used for detailed measurements of density fluctuations, including their frequency and wavenumber spectra or time- and two-dimensional space correlations (Fonck, Duperrex and Paul 1990). The basic experimental arrangement is similar to that used for charge exchange recombination spectroscopy (see figure 12.1), except that one now uses a two-dimensional high-quantum efficiency detector array and interference filters tuned, e.g., to the Doppler-shifted $D_\alpha$ line emitted by neutral-beam atoms.

For a steady-state beam, the line intensity is modulated by any density fluctuation of the species most responsible for the excitation of the line, i.e., by the majority ions in case of high beam energies and possibly also by impurity ions. The impurity-ion density fluctuations should therefore be measured separately, as discussed in the preceding section, and their contribution to the observed intensity fluctuations be estimated theoretically. As already pointed out by Stern and Cheo (1969) in connection with their demonstration of modulations of line emission by high-frequency plasma oscillations using heterodyne down-conversion, one must also keep in mind that the frequency response is limited by the radiative transition probability in case of low density plasmas. At higher densities, collisions leading to depopulation of the upper level of the diagnostic line increase the time response (Fonck et al. 1990), but naturally degrade the amplitude of intensity fluctuations corresponding to a given density fluctuation (Gianakon et al. 1992, Fonck et al. 1993), typically by a factor of about 3.

A further complication in the determination of density fluctuations (Durst et al. 1992) arises from fluctuations in the neutral beam intensity caused by plasma fluctuations along the beam path, especially by the usually very large fluctuations in the plasma edge region. Both computer simulations of time- and space-dependent beam propagation effects (Gianakon et al. 1992) and multipair correlation analysis (Fonck et al. 1993) can be used to obtain corresponding corrections, e.g., to $\sim 0.5\%$ density fluctuations in the core region of the TFTR tokamak in the presence of $\sim 20\%$ edge fluctuations.

Another application of beam emission spectroscopy is in the measurement of local magnetic fieldstrengths via the motional Stark effect, i.e., the Stark effect caused by the $\mathbf{v}_B \times \mathbf{B}$ electric field in the beam-frame, $\mathbf{v}_B$ being the vector corresponding to the beam velocity in (12.1), etc. For hydrogen and its isotopes this effect is more important than the Zeeman effect already at moderate beam energies. The linear Stark effect of these lines results in frequency shifts of the order

$$\Delta\omega_{MS} \approx ea_0 n^2 |\mathbf{v}_B \times \mathbf{B}|/\hbar, \tag{12.11}$$

the first factor being a typical electric dipole matrix element of the upper levels with principal quantum number $n$. (Stark shifts of the lower levels of $H_\alpha$, etc., are not very important in the present context.) The Zeeman splitting $\Delta\omega_Z$ may again as in section 11.8 be estimated by $\omega_{ce} = eB/m_e$, the electron cyclotron frequency. The ratio of motional Stark effects and Zeeman effects is therefore of the order

$$\frac{\Delta\omega_{MS}}{\Delta\omega_Z} \approx \frac{n^2 v_\perp}{\alpha c}, \tag{12.12}$$

$v_\perp$ being the perpendicular component of the beam velocity and $\alpha^{-1} = 137$. Writing the beam velocity for nonrelativistic atoms in terms of their kinetic energy $E_k$ and rest energy $Mc^2$, the ratio of the two effects can be expressed as

$$\frac{\Delta\omega_{MS}}{\Delta\omega_Z} \approx \frac{n^2}{\alpha} \left(\frac{2E_k}{Mc^2}\right)^{1/2} \sin\delta, \tag{12.13}$$

where $\delta$ is the angle between beam and magnetic field directions.

For nearly perpendicular directions at $E_k = 100$ keV, the motional Stark effect of, e.g., the $H_\alpha$ line, is thus seen to be an order of magnitude larger than the Zeeman effect, allowing a fairly simple interpretation of the spectral line structure and its polarization. It is also important to consider the Doppler shift associated with the beam, i.e., $\Delta\omega_{DB} \approx (2\pi c/\lambda_0)(v_B/c)\cos\phi$, if $\lambda_0$ is the center wavelength and $\phi$ the angle between beam direction and the optical line of sight. The ratio of the motional Stark effect and beam-Doppler shift is

$$\frac{\Delta\omega_{MS}}{\Delta\omega_{DB}} \approx \frac{n^2}{\alpha} \frac{\lambda_0}{2\pi c} \Delta\omega_Z \frac{\sin\delta}{\cos\phi}, \tag{12.14}$$

again using (12.11) and assuming $\Delta\omega_Z \approx \omega_{ce}$. For $H_\alpha$ at $B = 5$ T, this gives $\Delta\omega_{MS}/\Delta\omega_{DB} \approx 0.42\sin\delta/\cos\phi$. Motional Stark effect and beam Doppler shift can accordingly be of the same order, depending on the angles $\delta$ and $\phi$ and on $B$. Thermal Doppler widths, on the other hand, are relatively small, as required for discrimination against $H_\alpha$ or $D_\alpha$ emission from the cooler edge regions in a tokamak plasma.

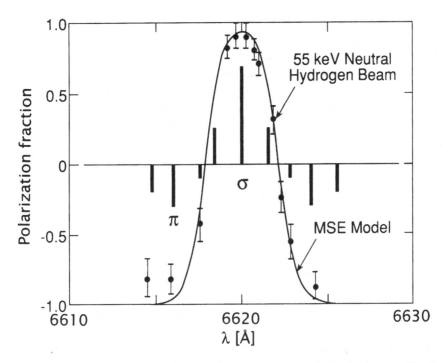

Fig. 12.3. Measured and calculated Stark pattern of the hydrogen $H_\alpha$ line in terms of the polarization fraction $(I_\parallel - I_\perp)/(I_\parallel + I_\perp)$ in a beam emission spectroscopy experiment on a tokamak (Levinton et al. 1989). The relative intensities of the components are indicated by bar lines, in the positive direction for $\sigma$-components, negative for $\pi$ components. The smooth curve is from a motional Stark effect model calculation.

If a neutral hydrogen beam is injected in the midplane of a tokamak as shown in figure 12.1, but now observed (Levinton et al. 1989, Wróblewski et al. 1990) in this plane along a major radius, the expected Stark effect pattern of the beam-shifted line is as shown in figure 12.3. The essential point is that the unshifted component and the two nearest shifted components are $\sigma$-components, i.e., linearly polarized perpendicular to the (vertical) $\mathbf{v}_B \times \mathbf{B}$ electric field or parallel to a purely toroidal magnetic field (Condon and Shortley 1951). The more widely displaced $\pi$ components are linearly polarized parallel to the electric field. Viewed parallel to the magnetic field direction, the emission is unpolarized. A measurement of the fractional polarization in the core of the beam-shifted line provides then a measure of the pitch angle of the actual magnetic field (Levinton et al. 1989, Wróblewski et al. 1990, Levinton 1992, Levinton et al. 1993, Mandl et al. 1993). As pointed out by Xu, Ida and Fujita (1995), there is actually some spectral asymmetry between blue- and red-shifted

$\pi$-components due to the different solid angles of the observation optics. These authors propose a pitch angle measurement of about 2% accuracy from $\sigma$ component measurements using two polarizers.

Neutral helium diagnostic beams may also be suitable for measurements of the motional Stark effect (Pohlmeyer, Dinklage and Kunze 1996), both because of their relatively modest attenuation by collisional ionization and because the Stark effect of the $nD$ levels is almost as large as the linear Stark effect for corresponding hydrogen levels. Additional attractive features are forbidden components, whose relative intensities provide a direct measure of the $\mathbf{v}_B \times \mathbf{B}$ electric field according, e.g., to (4.131), as long as this relative intensity is small.

### 12.3    Polarization spectroscopy

The hydrogen motional Stark effect experiment discussed in the preceding section was already an example for which deviations from the often implicitly assumed unpolarized nature of the radiation field are not only important but also useful for diagnostics. There are two general classes of polarization phenomena (Fujimoto and Kawachi 1995). The motional Stark effect associated with beam atoms belongs to the first class of polarization caused by external and macroscopic electric or magnetic fields. Such fields split spectral lines into various components, which are primarily characterized by the various magnetic quantum numbers, or parabolic quantum numbers in case of linear Stark effect (Bethe and Salpeter 1957), and these components are polarized (Condon and Shortley 1951, Cowan 1981, Sobel'man 1992) depending on the relative orientations of fields and viewing optics. Continuing first with beam emission spectroscopy on magnetically confined plasmas, but now considering lithium or neutral helium beams, the magnitudes of Zeeman and motional Stark effects tend to be inverted compared to the situation for hydrogen beams. This occurs because corresponding quadratic Stark effects are usually negligible, $nD$ levels being notable exceptions, not to mention $nF$, etc.

Neutral lithium beams were successfully employed to probe the magnetic field structure in moderate density tokamak plasmas by measurements of the polarization of the Zeeman components of the $2^2$S-$2^2$P resonance line at 6708 Å. For magnetic fields $\gtrsim 1\,\mathrm{T}$, the Zeeman splitting between the outer components is $\gtrsim 2 \times 10^{11}\,\mathrm{sec}^{-1}$, i.e., larger than the fine-structure splitting, $0.34\,\mathrm{cm}^{-1}$ or $6.4 \times 10^{10}\,\mathrm{sec}^{-1}$. One therefore deals with the normal Zeeman triplet, in case of observation perpendicular to the magnetic field, or the doublet for observation parallel to the field. Experiments beginning with McCormick et al.'s (1987) work are closest to the first case, and the unshifted $\pi$-component of the Zeeman triplet would be linearly polarized were the $B$-field entirely toroidal. However, a small poloidal field gives rise to some depolarization, which is measured to obtain the pitch angle of the actual field lines.

To increase the signal-to-noise ratio of such difficult measurements, West et al. (1987) directed a dye-laser tuned to the $\pi$-component of the Zeeman-triplet collinearly with the neutral lithium beam. The polarization vector of the laser beam was rotated at some angular frequency $\omega_m$, resulting in a corresponding modulation of the laser-induced fluorescence. Actually measured were the phase shifts of the optical signals of the laser polarization modulation, which can then be related to the desired pitch angle (West et al. 1986).

Turning briefly to passive, in contrast to beam-aided active, emission spectroscopy, passive emission spectroscopy may also be used to infer in this case the line-of-sight-averaged poloidal field from the relative difference of left- and right-handed circularly polarized, mostly Doppler-broadened, impurity line profiles. Using the Ti XVII 3834 Å magnetic dipole line, this was demonstrated by Wróblewski et al. (1988a). Details of the interpretation of such measurements and of the instrumentation can be found in Wróblewski, Huang and Moos (1988b).

The second class of polarization processes (Fujimoto and Kawachi 1995) is associated with possible alignment (Omont 1977) of atomic states, i.e., anisotropic distributions in space and therefore nonstatistical distributions over magnetic quantum numbers. Isotropic distributions had been assumed, e.g., in chapters 2, 3 and 4 in the definitions and calculations of radiative transition probabilities, oscillator strengths and spectral line profiles, but also implicitly in chapter 5 for the calculation of continuous spectra. For recombination radiation and for free-free emission or absorption, this assumption is equivalent to assuming isotropic velocity distributions for the free electrons. Returning to level populations, there are many possible mechanisms for forcing the distributions away from isotropy, beginning with photo-excitation by an ordinary light beam which cannot excite $\Delta m = 0$ transitions, if the beam axis is used as a quantization axis. Polarized light is even more selective; but polarization of the second kind can result also from collisional excitation (Oppenheimer 1927, Percival and Seaton 1958) if the corresponding velocity distributions are not isotropic, i.e., are beam-like or at least contain a corresponding admixture to the usual isotropic distribution. A comprehensive account of the various mechanisms has been given by Fujimoto et al. (1992). Not surprisingly, disalignment by electron collisions occurs at about the same rate as electron impact broadening (Hirabayashi et al. 1988b).

To describe the resulting polarization of atoms and ions quantitatively, a generalized collisional-radiative (CR) model based on the density-matrix formalism can be developed (Fujimoto and Kawachi 1995), in which the polarization creation rates are balanced with destruction rates estimated by the electron collisional line width, in angular frequency units. The latter estimate may have to be improved somewhat, because elastic $m \to m$ collisions do not contribute to disalignment. In any event, care should be exercised in interpreting polarization measurements in terms of field, i.e., class I effects, or one or the other kinds of class II phenomena.

Examples of experiments that may be interpreted in terms of class II phenomena are a number of x-ray line polarization measurements (Kieffer et al. 1992) in terms of a beam-like electron velocity distribution in a laser-produced plasma, and a proposed experiment (Lyaptsev and Kazantsev 1993) for the measurement of electric fields near tokamak divertor plates. Such fields would give rise to anisotropic electron velocity distributions, which in turn would result in alignment of excited atoms or ions and polarization of their line emissions. Polarization spectroscopy of low pressure discharges and its applications to remote sensing and studies of the solar atmosphere are subjects of a recent monograph (Kazantsev and Henoux 1995).

### 12.4  Magnetic field measurements

The only spectroscopic method for magnetic field measurements before the advent of the motional Stark effect method discussed near the end of section 12.2 was the more-or-less direct measurement of Zeeman splitting (Condon and Shortley 1951, Cowan 1981, Sobel'man 1992) of suitable spectral lines. As already discussed in section 11.8 and estimated by (11.15), more often than not even the thermal Doppler broadening turns out to be more important than Zeeman splitting, not to mention Stark broadening in dense plasmas. To avoid the possible complications arising from Stark broadening, it is important to select lines that are insensitive to this effect, e.g., lines from $n = 2$, $\Delta n = 0$ transitions in helium-like ions (Peacock and Norton 1975, McLean et al. 1984). If then essentially only Doppler broadening and Zeeman splitting remain, one can separate their contributions by measuring the resulting line profiles as a function of polarization as in Babcock's method (1953) for the measurement of solar magnetic fields. Only if magnetic fields are very large and ion temperatures relatively low can this procedure be side-stepped, as may be the case in some ultrashort, high intensity laser-produced plasmas or in edge regions of high-field magnetic fusion experiments. Early applications of suitable modifications of Babcock's method were to magnetically compressed plasmas (Jahoda, Ribe and Sawyer 1963, Hübner 1964, Eberhagen et al. 1965), followed by plasma focus (Peacock and Norton 1975) and z-pinch (Finken 1979) experiments and then a moderate intensity, nsec laser, blow-off plasma experiment (McLean et al. 1984).

This laser-plasma application will serve as a case study here. It involved the C V lines of the $2S^3S_1$-$2^3P_{0,1,2}$ multiplet rather than the much weaker, but single, $2^1S_0$-$2^1P_1$ line. The Zeeman pattern for the singlet line has simply the classical Lorentz pattern, if one correctly assumes the quadratic Zeeman effect (Garstang 1977) to be negligible. For hydrogen and hydrogenic ions, this is the case for fields (in teslas) obeying (Hey 1994)

$$B \ll 5 \times 10^4 \frac{z^2}{n^4},$$
(12.15)

if $n$ is the principal quantum number and $z$ the effective nuclear charge acting on the active electron. The quadratic Zeeman effect is associated with the square of the vector potential $(\mathbf{A})^2 \approx (Br)^2$ in the atom-electromagnetic-field Hamiltonian (2.1); whereas the usual linear Zeeman effect arises from the products of momentum and vector potential operators. The right-hand side of (12.15) is therefore basically the inverse of the square of the excited-state Bohr radius. For our C V lines, (12.15) requires $B \ll 10^5 T = 10^3$ MG, a relatively minor restriction in laboratory experiments.

The lower level of the C V triplet line, $2^3S_1$, splits into a normal $M' = \pm 1$ Zeeman pattern according to

$$\Delta E_{M'} = 2M'\mu_0 B = \pm 0.93372B \text{ cm}^{-1},$$
(12.16)

if the Bohr magneton, etc., as used, e.g., by Cowan (1981) is written so that magnetic fieldstrengths are in teslas. For the upper $2^3P_J$ levels, with $J = 0$, 1 and 2, no simple expression can be obtained. One or the other of the fine-structure intervals $E_2(0) - E_1(0) = 136$ cm$^{-1}$ or $E_1(0) - E_0(0) = -13$ cm$^{-1}$ is comparable to magnetic interaction matrix elements, which are of the order of (12.16), for fields ranging from about 1 T to $10^3$ T. Only below this range would one have a clear case of the anomalous Zeeman effect associated with the three individual lines of the multiplet, but the splittings would be very difficult to observe in the presence of Doppler broadening, etc. For fields above $10^3$ T, i.e., 10 MG, on the other hand, the Paschen-Back (1912, 1913) transition is practically complete, resulting in the classical Lorentz triplet centered at the center of gravity of the multiplet with a splitting of $\pm 0.46686B$ cm$^{-1}$ or $\pm 46.686B$ m$^{-1}$, which is equal to one half of the electron cyclotron frequency if multiplied by $2\pi c$ for conversion into angular frequency units.

This leaves most of the fields of practical interest in the intermediate region (McLean et al. 1984), and the field-mixing of different $J$, but the same $M$, must be taken into account. Within the $LS$ coupling approximation, this is accomplished by writing the field-dependent upper level wave functions as

$$|LSM\rangle = \sum_J C_{JM}|LSJM\rangle. \qquad (12.17)$$

The summation is over $J$ values $\geq |M|$, i.e., over three terms for $M = 0$ and two terms for $|M| = 1$, while the $|M| = 2$ functions remain pure $|LS, J = 2, \ M = \pm 2\rangle$ functions. Their Zeeman splitting is as in (12.16), except for a different Landé (1921) g-factor (Cowan 1981), namely $g = 3/2$ instead of 2, so that these $M = \pm 2$ level splittings are $\pm 1.40058B$ cm$^{-1}$. For the $M = 0$ and $M = \pm 1$ levels, one obtains diagonal and nonzero off-diagonal interaction energy matrix elements for $2^3P_J$ from

$$\langle JM|H_{\mathrm{mag}}|JM\rangle = \frac{3}{2}M\mu_0 B \qquad (12.18)$$

and

$$\langle JM|H_{\mathrm{mag}}|(J-1)M\rangle = -\frac{1}{2}\mu_0 B[(J^2 - M^2)(9 - J^2)/(4J^2 - 1)]^{1/2}, \quad (12.19)$$

respectively. The corresponding energy matrices must then be diagonalized and utilized to determine the expansion coefficients in (12.17). Normalizing the coefficients and using the usual rules for relative intensities of Zeeman components (Condon and Shortley 1951, Cowan 1981), one can finally calculate the relative intensities of the components as a function of fieldstrength and polarization. The minus sign in (12.19) did not occur in the early work but is consistent with Cowan's phase conventions. It has no effect on the results discussed here.

The assumption of pure *LS* coupling may be questioned, because the $2^3P_J$ fine-structure splittings do not agree with the Landé (1923 a,b) interval rule. However, judging by the small ratio of intercombination-to-resonance line transition probabilities, namely $\sim 3 \times 10^{-5}$ for C V (Elton 1967), the singlet admixture is too small to significantly affect the calculations of the Zeeman effect of the triplet levels.

For observation perpendicular to the magnetic field, such intensities can be found from equations (A.7) and (A.8) of McLean et al. (1984) for $\pi$ and $\sigma$ polarizations, respectively, or from figure 11 of this reference. Significant deviations from intensities for the weak field (anomalous) Zeeman effect case begin to occur above $B = 1$ T and affect most components strongly beyond 10 T.

To model experimental spectra, these relative intensities and Zeeman shifts must be combined with, e.g., thermal Doppler profiles and estimated Stark broadening. The result of such a calculation is shown in figure 12.4 for fieldstrengths expected at moderate laser intensities. The C V $J = 1$ and 0 components near 2278 Å are not resolved, their wavelength spacing

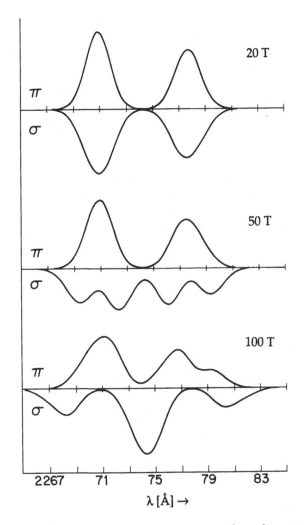

Fig. 12.4. Calculated Zeeman patterns of the C V 2s³S-2p³P multiplet for π- and σ-polarizations, for perpendicular observation and an assumed 2.1 Å line width of the individual components, at various magnetic fieldstrengths (McLean et al. 1984).

being only 0.6 Å compared to an assumed line width for the components of 2.1 Å. Although actual splitting becomes noticeable only at $B \geq 50\,T$ for σ-polarization, there is already about 15% difference between the peak intensities of the two polarizations near 2271 Å at $B = 20\,T$. The corresponding change in the line absorption coefficient was invoked by McLean et al. (1984) in their interpretation of measured σ- and π-profiles showing the σ-intensities to be almost 20% higher than π-intensities, i.e., deviations of intensity ratios in the opposite direction of expectations for emission lines.

This apparent anomaly was explained assuming a cooler outer absorbing layer for the radially expanding plasma, with the emission coming mostly from an intermediate layer moving radially at $\sim 1.5 \times 10^7$ cm/sec, judging by the overall $\sim 1.4$ Å blue shift of the entire multiplet. Line emission from the far side of the emitting layer, which would be red-shifted, was evidently absorbed by a dense but again relatively cool absorbing core region, corresponding to a minimum in the radial laser deposition profile. Comparisons of measured and modeled profiles are shown on figure 12.5. An azimuthal 20 T magnetic field was assumed, and calculations with various fieldstrengths suggested an error of about $\pm 7.5$ T. As already recognized by Finken (1979), since absorption can affect observed profiles in an exponential fashion, it actually enhances the sensitivity to magnetic fields although complicating the data analysis.

### 12.5    Electric field measurements

Laboratory plasma spectroscopy is used mostly in electrically neutral and not too tenuous plasmas. Spectroscopic measurements of any macroscopic electric fields, in contrast to the microfields discussed in section 4.3, are then difficult but not impossible, because these fields are usually weak and also localized in thin boundary layers. Even if use is made of the polarization dependence of the Stark effect, e.g., of $H_\alpha$ as shown on figure 12.3, in order to discriminate against Doppler and other line broadening effects, it would be difficult to see, say, an approximately $\pm 0.1$ Å Stark splitting of the $\pi$-components of this line, which is less than the 0.15 Å fine-structure effect associated with the lower level and would correspond to a field of about 1 kV/cm in a perpendicular direction to the line of sight. For still smaller fields, e.g., in tokamak edge regions (Groebner et al. 1990, Ida et al. 1990), also the fine-structure splitting of the $n = 3$ upper levels would have to be considered (Bethe and Salpeter 1957), while for fieldstrengths approaching 10 MV/cm higher order Stark effects are important in case of the $H_\alpha$ line. For higher members of the Balmer series, Stark splittings are larger, e.g., about $\pm 0.2$ Å for the outer components of $H_\gamma$ at fields of 1 kV/cm, but corrections are very important already near 1 MV/cm.

These examples suggest that measurements of electric fields via linear Stark effects in hydrogen and, perhaps, also highly excited atoms in general should be possible over a large range of fieldstrengths, especially if use is made of polarization to discriminate against Doppler and normal Stark broadening. In the case of magnetized plasmas, discrimination against

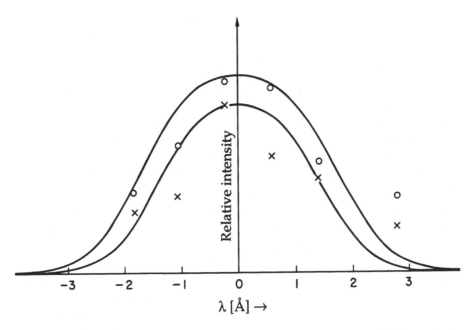

Fig. 12.5. Comparison of measured and modeled (at $B = 20\,\mathrm{T}$) profiles of the C V $2s^3S_1$-$2p^3P_2$ line (McLean et al. 1984). Circles are for $\sigma$-polarization, crosses for $\pi$-polarization.

Zeeman effects could be obtained by measuring several lines, the Zeeman splitting being about the same for all lines and the linear Stark effect splitting increasing as $n^2$, using wave number or frequency units. In high temperature plasmas, one would also have to contend with the motional Stark effects discussed in section 12.2. According to (12.13) and replacing the kinetic energy of the hydrogen atoms by kT, these additional Stark effects may be comparable to Zeeman effects for $n = 3$, increasing in importance along the Balmer series.

As in the case of motional-Stark-effect magnetic-field diagnostics, and given the difficulties in discriminating against other line broadening and splitting mechanisms, one may also use relative intensities of forbidden components, e.g., of neutral helium lines (Pohlmeyer et al. 1996). One would then use (4.131) to relate this relative intensity to the square of the electric fieldstrength, and ascertain that other mechanisms for breaking electric dipole selection rules are not likely under the experimental conditions. As already mentioned in section 4.11, an analogous method involving polar molecules (Moore, Davis and Gottscho 1984, Derouard and Sadeghi 1986, Maurmann and Kunze 1993, Maurmann et al. 1996) is more sensitive, down to fields $\lesssim 0.1\,\mathrm{kV/cm}$. Combined with laser excitation, it even provides good spatial resolution.

The reader may also be reminded of the discussions in section 4.11 of the effects of fluctuating electric fields on the structure of spectral line profiles (see also Oks 1995). Measurement of these satellite and dip structures typically allows the determination of plasma wavefields with rms fieldstrengths of the order of 10 kV/cm or larger. Increased sensitivity would be provided by laser-fluorescence or by using hydrogen lines, instead of the usual neutral helium lines. One could use polarization spectroscopy (Gavrilenko 1993, see also section 12.3 above) to determine any polarization of the fields. Again, great care must be exercised in the interpretation in order to properly account for competing line broadening mechanisms and for blends with other lines.

Returning to the measurement of more-or-less macroscopic electric fields, say, in the boundary or divertor regions of a tokamak, a method proposed by Lyaptsev and Kazantsev (1993), which was already briefly discussed in section 12.3, should be mentioned one more time. It would involve polarization spectroscopy, e.g., of $H_\alpha$, with the polarization caused by the anisotropy of the distribution of the exciting electrons near a plasma boundary. In the interpretation of such measurements in terms of electric fields or potentials, one would again have to consider Zeeman effects, Doppler broadening, etc., but could in this case probably neglect motional Stark effects.

To measure electric fields between high voltage electrodes Knyazev et al. (1991 a,b and 1992) have developed an active spectroscopy method involving the injection of neutral lithium into these vacuum regions. The lithium is excited by two lasers to 2p and then to 4d, 4f levels, the Stark pattern of the fluorescence yielding the fieldstrength.

## 12.6  Effective charge measurements

To characterize the high temperature plasmas produced and investigated in magnetic fusion experiments, it is convenient to account for the effects of multiply-charged impurity ions, e.g., on the electrical conductivity, by defining an effective ionic charge,

$$z_{\text{eff}} \equiv \sum_i z_i^2 N_i / \sum_i z_i N_i = \sum_i z_i^2 N_i / N_e, \qquad (12.20)$$

the sum being over charge states and chemical species. As already indicated near the end of section 10.2, the absolute continuum intensity of a fully ionized plasma containing a mixture of various ions is enhanced over that of a fully ionized hydrogen plasma according to (10.1) essentially by the factor $z_{\text{eff}}$, except for usually relatively small changes in the Gaunt or Biberman factors.

The established method for $z_{eff}$ determination is therefore to measure absolute continuum intensities in carefully selected regions free of line, negative ion, and molecular emissions, and to compare the corresponding volume emission coefficients with those of a pure hydrogen plasma at independently measured electron densities and temperatures (von Goeler et al. 1975, Kadota et al. 1980, Röhr et al. 1988). As should be clear from the discussion of corresponding cross sections in chapter 5, this may introduce systematic errors if some of the impurity ions are not completely ionized, especially if temperatures are low enough and wavelengths sufficiently short for recombination radiation to be important. As can be seen, e.g., from figures 5.1 and 5.2, at least some of the contributions would then only be very poorly represented by Gaunt factors for one-electron systems. Moreover, the bremsstrahlung on impurity ions may differ substantially from predictions based on simple shielding approximations (Tsytovich and Oiringel 1992).

Corresponding errors are particularly severe in boundary or edge regions of high temperature plasmas. Short of specific measurements of the various impurity ion densities, e.g., by laser-induced fluorescence as already indicated at the end of section 10.3, there is therefore a great need for some integral measurement of $z_{eff}$, which would yield at the same time sufficient spatial resolution to study, say, any variations along magnetic field lines in the divertor of a tokamak.

A method meeting these requirements had been proposed by Abramov and Lisitsa (1977) and applied by Bychkov, Ivanov and Stotskii (1987) using the $H_\alpha$ line. It requires either saturation spectroscopy (Szöke and Javan 1963, Bloembergen 1965, Letokov and Chebotayev 1977) of an optical transition by a narrowband laser and observation of the ensuing fluorescence, or, e.g., Doppler-free two-photon excited fluorescence. In the first case, the Lorentzian width of the Voigt profile of the fluorescence can be much enhanced by power broadening (Karplus and Schwinger 1948, Townes and Schawlow 1955) over that of the ion impact width of the emission line expected, e.g., at $N_e \approx 10^{13} \, cm^{-3}$, $kT \gtrsim 100 \, eV$, so that it might be measurable in the presence of Doppler broadening of the spontaneous emission background. In the second case, Doppler broadening is avoided by selecting only atoms in a very small velocity range (Demtröder 1996), at the expense of photon signal, and one requires very high spectral resolution.

Abramov and Lisitsa (1977) pointed out that most of the Lorentzian broadening for low-$n$ hydrogen lines would be due to ion-atom collisions for tokamak edge plasma conditions, and that one could use the impact approximation to describe their effects at these low densities. Using Lisitsa's (1977) adiabatic approximation, they estimated a FWHM collisional width of about 5 mÅ for the Balmer-$\alpha$ and 0.03 mÅ for the

Lyman-$\alpha$ lines, respectively, at the above plasma edge conditions, to be multiplied by $z_{eff}$ in the case of ionic mixtures. This is comparable to the natural, or radiative, widths of these lines of about 1.2 mÅ and 0.04 mÅ, respectively; and a large enhancement of the Lorentzian width through saturation pumping is essential, considering the Doppler widths near 4 Å or 0.8 Å, respectively, at $kT_i = 100$ eV. (Remember that kinetic temperatures of ions and atoms can be much higher than the electron temperatures in tokamak edge plasmas.)

Another complication was pointed out recently by Derevianko and Oks (1994), who reconsidered the validity of Lisitsa's (1977) adiabatic (phase-shift) approximation for the ion impact width, e.g., of Lyman-$\alpha$. They accounted for the Zeeman effect on the $n = 2$ levels. It causes line splittings of $\pm 35$ mÅ at $B = 5$ T and is therefore also important, being much larger than the 4 mÅ fine-structure splitting. Using polarization spectroscopy, one could therefore selectively measure the unshifted $\pi$-component of the Lorentz triplet. Its ion impact width is, according to the convergent semiclassical theory of Derevianko and Oks (1994, 1995) discussed in section 4.7, equal to the FWHM width estimated according to Abramov and Lisitsa (1977) and Lisitsa (1977), except that the $z_{eff}$ factor should for this component be replaced by $z_{eff} - 1 + 2^{-1/2} \approx z_{eff} - 0.29$. Derevianko and Oks emphasize that this near-agreement holds only for this particular component and does not carry over to other lines. Also the temperature dependence is simple in this case, namely $T_i^{-1/2}$, an important point in view of the fact that the temperatures do vary with charge state and chemical species in the edge region of tokamaks (Hey 1994). Any measured additional and power-broadening-enhanced Lorentzian width of the $\pi$-component of the Lyman-$\alpha$ line, observed perpendicularly to the magnetic field, therefore is really only a measure of $\sum z_i^2 N_i / T_i^{1/2}$; and space-resolved impurity and hydrogen ion temperatures must be obtained as discussed in section 11.8 and in section 1 of the present chapter, before $z_{eff}$ can be inferred from the Lyman-$\alpha$ ion impact width.

In addition to the great difficulties in measuring the small changes in the Lorentzian widths (Bychkov et al. 1987) and the problems caused by the requirement for reasonably accurate ion temperatures, and in the presence of a mostly Doppler-broadened background from spontaneous emission, motional Stark effects might be suspected to cause further complications. From estimates analogous to (12.13) and (12.14), respectively, one expects corresponding level shifts of only about 10% of the Zeeman splitting or, at $B = 5$ T, of 0.5% of the Doppler broadening. In other words, the motional Stark effect may indeed be of some importance in accurate calculations of the ion-impact broadening of the Lyman-$\alpha$ line for tokamak edge plasma conditions, but probably not for the separation of the Lorentzian component from the high-velocity wing of the Gaussian component of the fluorescence signal.

In the papers cited concerning the proposed method for $z_{\text{eff}}$ measurements, only very approximate expressions are given for the enhancement of the Lorentzian width by power broadening, which in the high laser field limit actually approaches the light-field induced linear Stark effect associated with the transition dipole matrix element, i.e., the Rabi frequency (Karplus and Schwinger 1948, Griem 1989). As a matter of fact, equation (5) of Bychkov et al. (1987) for the saturation parameter G, and therefore also Derevianko and Oks' (1994) expression for the enhanced Lorenzian width, are clearly inappropriate for the Lyman-$\alpha$ transition unless the radiative widths are replaced by sums of level depopulation rates, including collisional rates. This requirement is obvious for the Lyman-$\alpha$ line, for which the quoted expression gives an infinite Lorentzian width because the radiative width of the $n = 1$ level is zero. Perhaps Doppler-free two-photon polarization spectroscopy should also be considered, which has been successful (Danzmann, Grützmacher and Wende 1986) in measuring Lyman-$\alpha$ Stark profiles under much higher density and lower temperature conditions, for which Stark widths are much larger than in the tokamak edge plasma.

As with many other applications of plasma spectroscopy, more research, experimental, theoretical, and computer modeling, remains to be done before this conceptually very attractive method for the effective charge measurement in high temperature edge-region plasmas can help in solving an important problem in a challenging field of research.

# References

Abdallah, J. Jr., Clark, R.E.H., Keane, C.J., Shepard, T.D. and Suter, L.J. (1993). *J. Quant. Spectrosc. Radiat. Transfer* **50**, 91.

Abramov, V.A. and Lisitsa, V.S. (1977). *Sov. J. Plasma Phys.* **3**, 451.

Adams, M.S., Fedorov, M.V., Krainov, V.P. and Meyerhofer, D.D. (1995). *Phys. Rev. A* **52**, 125.

Adcock, J.C. and Griem, H.R. (1983). *Phys. Rev. Lett.* **50**, 1369.

Adler, H.A., Dengra, A. and Kelleher, D.E. (1993). In *Spectral Line Shapes*, Vol. 7, eds. R. Stamm and B. Talin (Nova Science Publ., Commack, NY), p. 325.

Adler, H.G. and Piel, A. (1991). *J. Quant. Spectrosc. Radiat. Transfer* **45**, 11.

Akhmedov, E.Kh., Godunov, A.L., Zemtsov, Yu.K., Makhrov, V.A., Starostin, A.N. and Taran, M.D. (1985). *Sov. Phys. JETP* **62**, 266.

Al-Saqabi, B.N.I. and Peach, G. (1987). *J. Phys. B* **20**, 1175.

Alastuey, A., Iglesias, C., Lebowitz, J. and Levesque, D. (1984). *Phys. Rev. A* **30**, 2537.

Alexiou, S. (1994a). *J. Quant. Spectrosc. Radiat. Transfer* **51**, 849.

Alexiou, S. (1994b). *Phys. Rev. A* **49**, 106.

Alexiou, S. (1995). *Phys. Rev. Lett.* **75**, 3406.

Alexiou, S. and Maron, Y. (1995). *J. Quant. Spectrosc. Radiat. Transfer* **53**, 109.

Alexiou, S. and Ralchenko, Yu. (1994a). *Phys. Rev. A* **49**, 3086.

Alexiou, S. and Ralchenko, Yu. (1994b). *Phys. Rev. A* **50**, 3553.

Alexiou, S., Weingarten, A., Maron, Y., Sarfaty, M. and Krasik, Ya. E. (1995). *Phys. Rev. Lett.* **75**, 3126.

Ali, A.W. and Griem, H.R. (1965). *Phys. Rev.* **140**, A1046.

Ali, A.W. and Griem, H.R. (1966). *Phys. Rev.* **144**, 366.

Allen, A.W., Blaha, M., Jones, W.W., Sanchez, A. and Griem, H.R. (1975). *Phys. Rev. A* **11**, 477.

Allen, C.W. (1973). *Astrophysical Quantities*, 3rd ed. (Athlone Press, London).

Aller, L.H. (1953). *The Atmospheres of the Sun and Stars* (Ronald Press, New York).

Amus'ya, M. Ya., Buimistrov, V.M., Zon, B.A. and Tsytovich, V.N. (1992). In *Polarization Bremsstrahlung*, eds. V.N. Tsytovich and I.M. Oiringel (Plenum, New York), Chap. 1.

Anderson, P.W. (1949). *Phys. Rev.* **76**, 647.

Anderson, P.W. (1952). *Phys. Rev.* **86**, 809.

Anderson, P.W. and McMillan, W. (1967). In *Theory of Magnetism in Transition Metals*, ed. W. Marshall (Academic Press, New York), p. 50.

Anderson, P.W. and Talman, J.D. (1955). *Bell Tel. Syst. Tech. Public. No.* 3117.

Apruzese, J.P. (1985). *J. Quant. Spectrosc. Radiat. Transfer* **34**, 447.

Apruzese, J.P (1993). *Phys. Rev. E* **47**, 2798.

Apruzese, J.P., Davis, J., Duston, D. and Whitney, K.G. (1980). *J. Quant. Spectrosc. Radiat. Transfer* **23**, 479.

Arnaud, M. and Raymond, J. (1992). *Astrophys. J.* **398**, 394.

Arnaud, M. and Rothenflug, R. (1985). *Astrophys. J. Suppl.* **60**, 425.

Ashbourn, J.P.A. and Ljepojević, N.N. (1995). *Phys. Rev. A* **52**, 4966.

Astapenko, V.A., Kukuskin, A.B. and Lisitsa, V.S. (1992). *J. Phys. B* **25**, 1985.

Athay, G. (1972). *Radiation Transport in Spectral Lines* (Reidel, Dordrecht).

Audebert, P., Geindre, J.P., Gauthier, J.C., Alaterre, Ph., Popovics, C., Cornille, M. and Dubau, J. (1984). *Phys. Rev. A* **30**, R 1582.

Auer, L.H. and Heasly, J.N. (1976). *Astrophys. J.* **205**, 165.

Auer, L.H. and Mihalas, D. (1969). *Astrophys. J.* **158**, 641.

Auger, P. (1925). *J. Phys. Radium* **6**, 205.

Aumayr, F., Hung, J. and Suckewer, S. (1989). *Phys. Rev. Lett.* **63**, 1215.

Aumayr, F., Lee, W., Skinner, C.H. and Suckewer, S. (1991). *J. Phys. B* **24**, 4489.

Avdonina, N.B. and Pratt, R.H. (1993). *J. Quant. Spectrosc. Radiat. Transfer* **50**, 349.

Avrett, E.H. and Hummer, D.G. (1965). *Mon. Not. R. Astron. Soc.* **130**, 295.

Babcock, H.W. (1953). *Astrophys. J.* **118**, 387.

Bacon, M.E., Barnard, A.J. and Curzon, F.L. (1977). *J. Quant. Spectrosc. Radiat. Transfer* **18**, 399.

Bahcall, J.N. and Wolf, R.A. (1968). *Astrophys. J.* **152**, 701.

Bailey, R.E. and Hooper, Jr., C.F. (1972). Communicated by Hooper for figure 2 in section II.2 of Griem (1974).

Baker, E.A.M. and Burgess, D.D. (1977). *J. Phys. B* **10**, L177.

Bar-Shalom, A., Klapisch, M. and Oreg, J. (1988). *Phys. Rev. A* **38**, 1773.

Bar-Shalom, A., Oreg, J. and Goldstein, W.H. (1995). *Phys. Rev. E* **51**, 4882.

Bar-Shalom, A., Oreg, J. Goldstein, W.H., Swartz, D. and Zigler, A. (1989). *Phys. Rev. A* **40**, 3183.

Baranger, M. (1958a). *Phys. Rev.* **111**, 494.

Baranger, M. (1958b). *Phys. Rev.* **112**, 855.

Baranger, M. (1962). In *Atomic and Molecular Processes*, ed. D.R. Bates (Academic Press, New York), Chap. 13.

Baranger, M. and Mozer, B. (1959). *Phys. Rev.* **115**, 521.

Baranger, M. and Mozer, B. (1961). *Phys. Rev.* **123**, 23.

Barnard, A.J. and Cooper, J. (1970). *J. Quant. Spectrosc. Radiat. Transfer* **10**, 695.

Barnard, A.J., Cooper, J. and Shamey, L.J. (1969). *Astron. Astrophys.* **1**, 22.

Barnard, A.J., Cooper, J. and Smith, E.W. (1975). *J. Quant. Spectrosc. Radiat. Transfer* **15**, 429.

Barnes, K.S. (1971). *J. Phys. B* **4**, 1377.

Barnes, K.S. and Peach, G. (1970). *J. Phys. B* **3**, 350.

Barnett, S.M., Huttner, B. and Loudon R. (1992). *Phys. Rev. Lett.* **68**, 3698.

Bartels, H. (1950a). *Z. Phys.* **127**, 243.

Bartels, H. (1950b). *Z. Phys.* **128**, 546.

Bashkin, S. and Stoner, J.O., Jr. (1975). Vol. I; (1978). Vol. II; and (1981). Vol. III. *Atomic Energy Levels and Grotrian Diagrams* (North-Holland, Amsterdam).

Basov, N.G. and Krokhin, O.N. (1964). *Sov. Phys. JETP* **19**, 123.

Bassalo, J.M. and Cattani, M. (1993). *J. Quant. Spectrosc. Radiat. Transfer* **50**, 359.

Bates, D.R. and Damgaard, A. (1949). *Philos. Trans. R. Soc. London, Ser. A* **242**, 101.

Bates, D.R. and McCarroll, R. (1962). *Adv. Phys.* **11**, 39.

Bates, D.R., Kingston, A.E. and McWhirter, R.W.P. (1962a). *Proc. R. Soc. London, Ser. A* **267**, 297.

Bates, D.R., Kingston, A.E. and McWhirter, R.W.P. (1962b). *Proc. R. Soc. London, Ser. A* **270**, 155.

Bauche-Arnault, C., Bauche, J. and Klapisch, M. (1979). *Phys. Rev. A* **20**, 2424.

Bauche-Arnault, C., Bauche, J. and Klapisch, M. (1982). *Phys. Rev. A* **25**, 2641.

Bauche-Arnault, C., Bauche, J. and Klapisch, M. (1985). *Phys. Rev. A* **31**, 2248.

Behar, E., Mandelbaum, P., Schwob, J.L., Bar-Shalom, A., Oreg, J. and Goldstein, W.H. (1995). *Phys. Rev. A* **52**, 3770.

Behringer, K. (1992). *IPF Stuttgart Rep. 92-2.*

Behringer, K. and Fantz, U. (1994). *J. Phys. D* **27**, 2128.

Beiersdorfer, P., Osterheld, A.L., Phillips, T.W., Bitter, M., Hill, K.W. and von Goeler, S. (1995). *Phys. Rev. E* **52**, 1980.

Bekefi, G. (1966). *Radiation Processes in Plasmas* (John Wiley, New York).

Bell, E.W., Guo, X.Q., Forand, J.L., Rinn, K., Swenson, D.R., Thompson, J.S., Dunn, G.H., Bannister, M.E., Gregory, D.C., Phaneuff, R.A., Smith, A.C.H., Müller, A., Timmer, C.A., Wåhlin, E.K., De Paola, B.D. and Belić, D.S. (1994). *Phys. Rev. A* **49**, 4585.

Bell, K.L., Gilbody, H.B., Hughes, J.G., Kingston, A.E., and Smith, F.J. (1983). *J. Phys. Chem. Ref. Data* **12**, 891.

Bely, O. (1966). *Proc. Phys. Soc. London* **88**, 587.

Bely, O. and Faucher, P. (1970). *Astron. Astrophys.* **6**, 88.

Bely, O. and Griem, H.R. (1970). *Phys. Rev. A* **1**, 97.

Bengtson, R.D., Tannich, J.D. and Kepple, P.C. (1970). *Phys. Rev. A* **1**, 532.

Berg, H.F. (1966). *Z. Phys.* **191**, 503.

Berg, H.F., Ali, A.W., Lincke, R. and Griem, H.R. (1962). *Phys. Rev.* **125**, 199.

Berkovsky, M., Dufty, J.W., Calisti, A., Stamm, R. and Talin, B. (1995). *Phys. Rev. E* **51**, 4917.

Berkovsky, M.A., Kelleher, D., Kurilenkov, Yu. K. and Skowronek, M. (1993). *J. Phys. B* **26**, 2475.

Berlin, T.W. and Montroll, E.W. (1952). *J. Chem. Phys.* **20**, 75.

Berman, P.R. and Lamb, W.E. (1971). *Phys. Rev. A* **4**, 319.

Bernard, J.E., Curzon, D.L. and Barnard, A.J. (1981). In *Spectral Line Shapes*, Vol. 1, ed. B. Wende (deGruyter, Berlin), p. 153.

Berrington, K.A., Burke, P.G., Butler, K., Seaton, M.J., Storey, P.J., Taylor, K.T. and Yan, Yu (1987). *J. Phys. B* **20**, 6379.

Bespalov, V.E., Gryaznov, V.K. and Fortov, V.E. (1979). *JETP Lett.* **76**, 140.

Bethe, H.A. (1930). *Ann. Physik (Leipzig)* **5**, 325.

Bethe, H.A. and Salpeter, E.E. (1957). *Quantum Mechanics of One- and Two-Electron Systems* (Springer-Verlag, Berlin).

Beutler, H. (1935). *Z. Phys.* **93**, 177.

Biberman, L.M. (1947). *Zh. Eksp. Teor. Fiz.* **17**, 416. (English transl., ORNL-Tr.-681).

Biberman, L.M. (1948). *Dok. Akad. Nauk. SSSR* **59**, 659 (English transl., Joint Publ. Res. Serv. 5647).

Biberman, L.M. and Norman, G.E. (1960). *Opt. Spectrosc. (USSR)*, **8**, 230.

Biberman, L.M. and Norman, G.E. (1963). *J. Quant. Spectrosc. Radiat. Transfer* **3**, 221.

Biberman, L.M., Norman, G.E. and Ulyanov, K.N. (1962). *Sov. Astron.* **6**, 77.

Biberman, L.M., Vorobev, V.S. and Yakubov, I.T. (1979). *Sov. Phys. Usp.* **22**, 411.

Biberman, L.M., Vorobev, V.S. and Yakubov, I.T. (1987). *Kinetics of Nonequilibrium Low-Temperature Plasmas* (Consultant Bureau, New York).

Bitter, M., Hill, K.W., Sauthoff, N.R., Efthimion, P.C., Meservey, E., Roney, W., von Goeler, S., Horton, R., Goldman, M. and Stodiek, W. (1979). *Phys. Rev. Lett.* **43**, 129.

Bitter, M., Hill, K.W., Zarnstorff, M., von Goeler, S., Hulse, R., Johnson, L.C., Sauthoff, N.R., Sesnic, S., Young, K.M., Tavernier, M., Bely-Dubau, F., Faucher, P., Cornille, M. and Dubau, J. (1985). *Phys. Rev. A* **32**, 3011.

Blagojević, B., Popović, M.V., Konjević, N. and Dimitrijević, M.S. (1994). *Phys. Rev. E* **50**, 2986.

Blagojević, B., Popović, M.V., Konjević, N. and Dimitrijević, M.S. (1996). *Phys. Rev. E* **54**, 743.

Blaha, M. (1969a). *Astron. Astrophys.* **1**, 42.

Blaha, M. (1969b). *Astrophys. J.* **157**, 473.

Blaha, M. (1972). *Astrophys. J. Lett.* **10**, 179.

Blenski, T. and Ishikawa, K. (1995). *Phys. Rev. E* **51**, 4869.

Bliman, S., Cornille, M. and Katsonis, K. (1994). *Phys. Rev. A* **50**, 3134.

Bloembergen, N. (1965). *Nonlinear Optics* (Benjamin, New York).

Böddeker, St. (1995). Dissertation, Ruhr University (Bochum, unpublished).

Böddeker, St., Günter, S., Könies, A., Hitzschke, L. and Kunze, H.-J. (1993). *Phys. Rev. E* **47**, 2785.

Böddeker, St., Kunze, H.-J. and Oks, E. (1995). *Phys. Rev. Lett.* **75**, 4740.

Boercker, D.B. (1987). *Astrophys. J.* **316**, L95.

Boercker, D.B. (1989). In *Spectral Line Shapes*, Vol. 5, ed. J. Szudy (Ossolineum Publ. House, Warsaw), p. 73.

Boercker, D.B. (1993). In *Spectral Line Shapes*, Vol. 7, ed. R. Stamm and B. Talin (Nova Science Publishers, Commack, New York), p. 17.

Boercker, D.B. and Iglesias, C.A. (1984). *Phys. Rev. A* **30**, 2771.

Boercker, D.B., Iglesias, C.A. and Dufty, J.W. (1987). *Phys. Rev. A* **36**, 2254.

Bogen, P. (1970). *Z. Naturforsch.* **25a**, 1151.

Bogen, P., Hey, J.D., Hintz, E., Lie, Y.T., Rusbüldt, D. and Samm, V. (1995). *J. Nucl. Materials* **220-222**, 472.

Bogen, P., Mertens, P., Pasch, E. and Döbele, H.F. (1992). *J. Opt. Soc. Am. B* **9**, 2137.

Boggess, A. (1959). *Astrophys. J.* **129**, 432.

Boley, C.D., Janev, R.K. and Post, D.E. (1984). *Phys. Rev. Lett.* **52**, 534.

Bollé, D. (1981). *Nucl. Phys. A* **353**, 377c.

Bornath, Th. and Schlanges, M. (1993). *Physica A* **196**, 427.

Bornath, Th., Ohde, Th. and Schlanges, M. (1994). *Physica A* **211**, 344.

Bornatici, M., Cano, R., DeBarbieri, O. and Engelmann, F. (1983). *Nucl. Fusion* **23**, 1153.

Böttcher, F., Breger, P., Hey, J.D. and Kunze, H.-J. (1988). *Phys. Rev. A* **38**, 2690.

Böttcher, F., Musielok, J. and Kunze, H.-J. (1987). *Phys. Rev. A* **36**, 2265.

Bracewell, R.N. (1956). *Austr. J. Phys.* **9**, 198.

Bréchot, S. and Van Regemorter, H. (1964). *Ann. Astrophys.* **27**, 432.

Breton, C., DeMichelis, C. and Mattioli, M. (1976). *Nucl. Fusion* **16**, 891.

Brissaud, A. and Frisch, H. (1971). *J. Quant. Spectrosc. Radiat. Transfer* **11**, 1767.

Brissaud, A. and Frisch, H. (1974). *J. Math. Phys.* **15**, 524.

Brown, S.C. (1959). *Basic Data of Plasma Physics* (Wiley, New York); reprinted by the American Vacuum Society in 1994 (AIP, New York).

Burgess, A. (1964). *Astrophys. J.* **139**, 776.

Burgess, A. (1965). *Astrophys. J.* **141**, 1588.

Burgess, A. and Chidichimo, M.C. (1983). *Mon. Not. R. Astron. Soc.* **203**, 1269.

Burgess, A. and Seaton, M.J. (1960). *Mon. Not. R. Astron. Soc.* **120**, 121.

Burgess, A. and Summers, H.P. (1969). *Astrophys. J.* **157**, 1007.

Burgess, A. and Summers, H.P. (1976). *Mon. Not. R. Astron. Soc.* **174**, 345.

Burgess, D.D. (1968). *Phys. Rev.* **176**, 150.

Burgess, D.D. (1970). *J. Phys. B* **3**, L70.

Burgess, D.D. and Lee, R.W. (1982). *J. Physique* **43**, Coll. C2, 413.

Burke, V.M. (1992). *J. Phys. B* **25**, 4917.

Burnett K. and Cooper, J. (1980). *Phys. Rev. A* **22**, 2027.

Burnett, K., Cooper, J., Ballagh, R.J. and Smith, E.W. (1980). *Phys. Rev. A* **22**, 2005.

Burrell, C.F. and Kunze, H.-J. (1972). *Phys. Rev. Lett.* **29**, 1445.

Büscher, S., Glenzer, S., Wrubel, T. and Kunze, H.-J. (1995). *J. Quant. Spectrosc. Radiat. Transfer* **54**, 73.

Büscher, S., Glenzer, S., Wrubel, Th. and Kunze, H.-J. (1996). *J. Phys. B* **29**, 4107.

Bychkov, S.S., Ivanov, R.S. and Stotskii, G.I. (1987). *Sov. J. Plasma Phys.* **13**, 769.

Byron, S., Stabler, R.C. and Bortz, P.I. (1962). *Phys. Rev. Lett.* **8**, 376.

Cacciatore, M., Capitelli, M. and Drawin, H.W. (1976). *Physica* **84C**, 267.

Calisti, A., Khelfaoui, F., Stamm, R., Talin, B. and Lee, R.W. (1990). *Phys. Rev. A* **42**, 5433.

Callaway, J. and Unnikrishnan, K. (1991). *Phys. Rev. A* **44**, 3001.

Cao, Chang-qi and Cao, H. (1993). *J. Phys. B* **26**, 3959.

Cao, H. and Cao, Chang-qi. (1995). *J. Phys. B* **28**, 979.

Cao, H., DiCicco, D. and Suckewer, S. (1993). *J. Phys. B* **26**, 4057.

Cardeñoso, V. and Gigosos, M.A. (1989). *Phys. Rev. A* **39**, 5258.

Carlson, T.A. and Krause, M.O. (1965). *Phys. Rev.* **140**, A1057.

Carlsten, J.L., Szöke, A. and Raymer, M.G. (1977). *Phys. Rev. A* **15**, 1029.

Carson, T., Mayers, D. and Stibbs, D. (1968). *Mon. Not. R. Astron. Soc.* **140**, 483.

Casperson, L.W. and Yariv, A. (1972). *IEEE J. Quantum Electron.* **8**, 80.

Cauble R. and Griem, H.R. (1983). *Phys. Rev.* **27**, 3187.

Cauble, R., Blaha, M. and Davis, J. (1984). *Phys. Rev. A* **29**, 3280.

Censor, Y. (1983). *Proc. IEEE* **71**, 409.

Chambaud, G., Levy, B. and Pernot, P. (1985). *Chem. Phys.* **95**, 299.

Chandrasekhar, S. (1930). *Philos. Mag.* **9** (seventh series), 292.

Chandrasekhar, S. (1931). *Mon. Not. R. Astron. Soc.* **91**, 446.

Chandrasekhar, S. (1950). *Radiative Transfer* (Oxford University Press).

Chandrasekhar, S. (1958). *Astrophys. J.* **128**, 114.

Chappell, W.R., Cooper, J. and Smith, E.W. (1969). *J. Quant. Spectrosc. Radiat. Transfer* **9**, 149.

Chappell, W.R., Cooper, J. and Smith, E.W. (1970). *J. Quant. Spectrosc. Radiat. Transfer* **10**, 1195.

Chiang, W.T. and Griem, H.R. (1978). *J. Phys. B* **11**, L761.

Chung, S., Lin, C.C. and Lee, E.W.P. (1994). *J. Quant. Spectrosc. Radiat. Transfer* **51**, 629.

Chung, Y., Hirose, H. and Suckewer, S. (1989). *Phys. Rev. A* **40**, 7142.

Chung, Y., Lemaire, P. and Suckewer, S. (1988). *Phys. Rev. Lett.* **60**, 1122.

Clark, R.W., Davis, J., Apruzese, J.P. and Giuliani, J.L., Jr. (1995). *J. Quant. Spectrosc. Radiat. Transfer* **53**, 307.

Clausset, F., Stehlé, C. and Artru, M.-C. (1994). *Astron. Astrophys.* **287**, 666.

Cohen-Tannoudji, C., Dupont-Roc, J. and Grynberg, G. (1989). *Photons and Atoms* (John Wiley, New York).

Cohn, A., Bakshi, P. and Kalman, G. (1972). *Phys. Rev. Lett.* **29**, 324.

Collins, L., Kwon, I., Kress, J., Troullier, N. and Lynch, D. (1995). *Phys. Rev. E* **52**, 6202.

Condon, E.V. and Shortley, G.H. (1951). *The Theory of Atomic Spectra*, reprinted in 1970 (Cambridge University Press, Cambridge).

Cooper, J. (1966). *Rep. Prog. Phys.* **29**, 35.

Cooper, J. (1967). *Rev. Mod. Phys.* **39**, 167.

Cooper, J. and Smith, E.W. (1982). *J. Quant. Spectrosc. Radiat. Transfer* **27**, 665.

Cooper, J. and Zoller, P. (1984). *Astrophys. J.* **277**, 813.

Cooper, J., Ballagh, R.J. and Hubeny, I. (1989). In *Spectral Line Shapes*, Vol. 5, ed. J. Szudy (Ossolineum Publ., Warsaw), p. 275.

Cooper, M.S. and DeWitt, H.E. (1973). *Phys. Rev. A* **8**, 1910.

Cooper, W.S. and Hess, R.A. (1970). *Phys. Rev. Lett.* **25**, 433.

Cormack, A.M. (1963). *J. Appl. Phys.* **34**, 2722.

Courant, R. and Hilbert, D. (1953). *Methods of Mathematical Physics* (Interscience Publ., New York), Vol. 1.

Cowan, R.D. (1968). *J. Opt. Soc. Am.* **58**, 808.

Cowan, R.D. (1981). *The Theory of Atomic Structure and Spectra* (University of California Press, Berkeley).

Cox, P.J. and Giuli, R.T. (1968). *Principles of Stellar Structure* (Gordon and Breach, New York), Vol. 1.

Crandall, D.H. (1983). In *The Physics of Ion-Ion and Electron-Ion Collisions*, ed. F. Brouillard and J. Wm. McGowan (Plenum, New York), p. 239.

Crivellari, L., Hubeny, I. and Hummer, D.J., eds. (1991). *Stellar Atmospheres: Beyond Classical Models*, NATO ASI Series C, Math. Phys. Sciences (Kluwer, Dordrecht), Vol. 341.

Cunto, W., Mendoza, C., Ochsenbein, F. and Zeippen, C.J. (1993). *Astron. Astrophys.* **275**, L5.

Dalgarno, A. (1963). *Rev. Mod. Phys.* **35**, 522.

Dalgarno, A. (1966). *Proc. Phys. Soc.* **76**, 422.

Dalgarno, A. and Lynn, N. (1957). *Proc. Phys. Soc. A* **70**, 802.

D'Angelo, N. (1961). *Phys. Rev.* **121**, 505.

Danielsson, M., von Hellermann, M.G., Källne, E., Mandl, W., Morsi, H.W., Summers, H.P. and Zastrow, K.-D. (1992). *Rev. Sci. Instrum.* **63**, 2241.

Danzmann, K. Grützmacher, K. and Wende, B. (1986). *Phys. Rev. Lett.* **57**, 2151.

Däppen, W., Anderson, L.S. and Mihalas, D. (1987). *Astrophys. J.* **319**, 195.

Dasgupta, A. and Whitney, K.G. (1994). *At. Data Nucl. Data Tables* **58**, 77.

DaSilva, L.B., Barbee, T.W. Jr., Cauble, R., Celliers, P., Ciarlo, D., Libby, S., London, R.A., Matthews, D., Mrowka, S., Moreno, J.C., Ress, D., Trebes, J.E., Wan, A.S. and Weber, F. (1995). *Phys. Rev. Lett.* **74**, 3991.

DaSilva, L.B., McGowan, B.J., Kania, D.R., Hammel, B.A., Back, C.A., Hsieh, E., Doyas, R., Iglesias, C.A., Rogers, F.J. and Lee, R.W. (1992). *Phys. Rev. Lett.* **69**, 438.

Davara, G., Gregorian, L., Kroupp, G. and Maron, Y. (1995). Private communication, to be published.

Davidson, S.J., Foster, J.M., Smith, C.C., Warburton, K.A. and Rose, S.J. (1988). *Appl. Phys. Lett.* **52**, 847.

Davis, J. and Blaha, M. (1982). In *Physics of Electronic and Atomic Collisions*, ed. S. Datz (North-Holland, Amsterdam), p. 811.

Davis, J. and Jacobs, V.L. (1980). *J. Quant. Spectrosc. Radiat. Transfer* **24**, 283.

Davis, J., Blaha, M. and Kepple, P.C. (1975). *J. Quant. Spectrosc. Radiat. Transfer* **15**, 1145.

Davis, J., Jacobs, V.L., Kepple, P.C. and Blaha, M. (1977). *J. Quant. Spectrosc. Radiat. Transfer* **17**, 139.

Davis, J., Kepple, P.C. and Blaha, M. (1976). *J. Quant. Spectrosc. Radiat. Transfer* **16**, 1043.

Dawson, J. and Oberman, C. (1962). *Phys. Fluids* **5**, 517.

Debye, P. and Hückel, E. (1923). *Z. Phys.* **24**, 185.

Décoste, R. (1985). *Rev. Sci. Instrum.* **56**, 807.

DeMichelis, C. and Mattioli, M. (1981). *Nucl. Fusion* **20**, 191.

DeMichelis, C. and Mattioli, M. (1984). *Rep. Prog. Phys.* **47**, 1233.

Demtröder, W. (1996). *Laser Spectroscopy*, 2nd ed. (Springer, Berlin).

Demura, A.V. and Sholin, G.V. (1975). *J. Quant. Spectrosc. Radiat. Transfer* **15**, 881.

Demura, A.V. and Stehlé, C. (1995). In *Spectral Line Shapes*, Vol. 8, eds. A.D. May, J.R. Drummond and E. Oks, AIP Conf. Proc. 328 (AIP, New York), p. 177.

Derevianko, A. and Oks, E. (1994). *Phys. Rev. Lett.* **73**, 2059.

Derevianko, A. and Oks, E. (1995). *J. Quant. Spectrosc. Radiat. Transfer* **54**, 137.

Derouard, J. and Sadeghi, N. (1986). *Opt. Commun.* **57**, 239.

DeSilva, A.W. and Goldenbaum, G.C. (1970). In *Methods of Experimental Physics*, Vol. 9a, eds. H.R. Griem and R.H. Lovberg (Academic Press, New York), Chap. 3.

DeSilva, A.W. and Kunze, H.-J. (1994). *Phys. Rev. E* **49**, 4448.

DeSilva, A.W., Baig, T.J., Olivares, I. and Kunze, H.-J. (1992). *Phys. Fluids B* **4**, 458.

Dharma-wardana, M.W.C. and Perrot, F. (1982). *Phys. Rev. A* **26**, 2096.

Dharma-wardana, M.W.C., Perrot, F. and Aers, G.C. (1983). *Phys. Rev. A* **28**, 344.

Dicke, R.H. (1953). *Phys. Rev.* **89**, 472.

Dimitrijević, M.S. and Konjević, N. (1981). In *Spectral Line Shapes*, Vol. 1, ed. B. Wende (de Gruyter, Berlin), p. 211.

Dimitrijević, M.S. and Sahal-Bréchot S. (1990a). *Ann. Phys. (Paris), colloque no. 3, suppl.* **15**, 77.

Dimitrijević, M.S. and Sahal-Bréchot, S. (1990b). *Astron. Astrophys. Suppl. Ser.* **82**, 519.

Dimitrijević, M.S. and Sahal-Bréchot, S. (1992a). *Astron. Astrophys. Suppl. Ser.* **93**, 359.

Dimitrijević, M.S. and Sahal-Bréchot, S. (1992b). *Astron. Astrophys. Suppl. Ser.* **95**, 109.

Dimitrijević, M.S. and Sahal-Bréchot, S. (1992c). *Astron. Astrophys. Suppl. Ser.* **95**, 121.

Dimitrijević, M.S. and Sahal-Bréchot, S. (1992d). *J. Quant. Spectrosc. Radiat. Transfer* **48**, 397.

Dimitrijević, M.S., Feautrier, N. and Sahal-Bréchot, S. (1981). *J. Phys. B* **14**, 2559.

Dirac, P.A.M. (1958). *The Principles of Quantum Mechanics*, 4th Ed. (Oxford University Press, New York); (1967). 4th Ed. (revised).

Dixon, R.H. and Elton, R.C. (1977). *Phys. Rev. Lett.* **38**, 1072.

Dixon, R.H., Seely, J.F. and Elton, R.C. (1978). *Phys. Rev. Lett.* **40**, 122.

Djeniže, S., Skulan, Lj. and Konjević, N. (1995). *J. Quant. Spectrosc. Radiat. Transfer* **54**, 581.

Djeniže, S., Sreckovic, A., Milosavljević, M., Labat, O., Platiska, M. and Purić, J. (1988). *Z. Phys. D* **9**, 129.

Doughty, N.A., Fraser, P.A. and McEachran, R.P. (1966). *Mon. Not. R. Astron. Soc.* **132**, 255.

Drake, G.W.F. (1973). *Astrophys. J.* **184**, 145.

Drake, G.W.F. and Hedgecock, N.E., eds. (1996). *Atomic, Molecular and Optical Physics Reference Book* (AIP, New York).

Drawin, H.W. (1961). *Z. Phys.* **164**, 513.

Drawin, H.W. (1963). *Z. Phys.* **172**, 429.

Drawin, H.W. (1968). In *Plasma Diagnostics*, ed. W. Lochte-Holtgreven (North-Holland, Amsterdam); reprinted by the American Vacuum Society in 1995 (AIP, New York), chapter 14.

Drawin, H.W. (1969). *Z. Phys.* **228**, 99.

Drawin, H.W. (1970). *J. Quant. Spectrosc. Radiat. Transfer* **10**, 33.

Drawin, H.W. (1974). *Z. Naturforsch.* **19**a, 1451.

Drawin, H.W. and Emard, F. (1977). *Physica* **85**C, 333.

Dubau, J. and Kato, T. (1994). *NIFS-Data Rep.* **21**.

Duclos, P. and Cambel, A.B. (1961). *Z. Naturforsch.* **16**a, 711.

Dufty, J.W. (1969). *Phys. Rev.* **187**, 305.

Dufty, J.W., Boercker, D.B. and Iglesias, C.A. (1985). *Phys. Rev. A* **31**, 1681.

Dufty, J.W., Boercker, D.B. and Iglesias, C.A. (1990). *J. Quant. Spectrosc. Radiat. Transfer* **44**, 115.

Dunn, G.H. (1968). *Phys. Rev.* **172**, 1.

Dunn, G.H. (1992). In *Recombination of Atomic Ions*, ed. W.G. Graham (Plenum Press, New York), p. 115.

Durst, R.D., Fonck, R.J., Cosby, G., Evensen, H. and Paul, S.F. (1992). *Rev. Sci. Instrum.* **63**, 4907.

Duston, D. and Davis, J. (1980). *Phys. Rev. A* **21**, 1664.

Duston, D. and Davis, J. (1981). *Phys. Rev. A* **23**, 2602.

Dux, R., Grützmacher, K., de la Rosa, M.I. and Wende, B. (1995). *Phys. Rev. E* **51**, 1416.

D'yachkov, L.G. and Pankratov, P.M. (1994). *J. Phys. B* **27**, 461.

D'yachkov, L.G., Kobzev, G.A. and Pankratov, P.M. (1990). *J. Quant. Spectrosc. Radiat. Transfer* **44**, 123.

D'yachkov, L.G., Kobzev, G.A. and Pankratov, P.M. (1991). *High Temperature* **29**, 311.

Ebeling, W., Foerster, A., Fortov, V.E., Gryaznov, V.K. and Polishuk, A.Ya. (1991). *Thermophysical Properties of Hot Dense Plasmas* (Teubner, Leipzig).

Ebeling, W., Kraeft, W.D. and Kremp, D. (1976). *Theory of Bound States and Ionization Equilibrium in Plasmas and Solids* (Akademie Verlag, Berlin).

Eberhagen, A. and Wunderlich, R. (1970). *Z. Phys.* **232**, 1.

Eberhagen, A., Bernstein, J.M. and Hermannsdorfer, M. (1965). *Z. Naturforsch.* **20a**, 1375.

Ecker, G. (1972). *Theory of Fully Ionized Plasmas* (Academic Press, New York).

Ecker, G. and Kröll, W. (1963). *Phys. Fluids* **6**, 62.

Ecker, G. and Weizel, W. (1958). *Z. Naturforsch.* **13a**, 1093.

Eddington, A.S. (1926). *The Internal Constitution of the Stars* (Cambridge University Press, Cambridge).

Eder, D.C. and Scott, H.A. (1991). *J. Quant. Spectrosc. Radiat. Transfer* **45**, 189.

Edlén, B. (1964). *Handbuch der Physik*, Vol. 27, ed. S. Flügge (Springer, Berlin), p. 80 .

Eidmann, K., Schwanda, W., Földes, I.B., Sigel, R. and Tsakiris, G.D. (1994). *J. Quant. Spectrosc. Radiat. Transfer* **51**, 77.

Einfeld, D. and Sauerbrey, G. (1976). *Z. Naturforsch.* **31a**, 310.

Einstein, A. (1917). *Phys. Z.* **18**, 121.

Eliezer, S., Ghatak, A., Hora, H. and Teller, E. (1986). *Equation of State, Theory and Applications* (Cambridge University Press, Cambridge).

Elton, R.C. (1967). *Astrophys. J.* **148**, 573.

Elton, R.C. (1990). *X-Ray Lasers* (Academic Press, San Diego).

Elton, R.C. and Griem, H.R. (1964). *Phys. Rev.* **135**, A1550.

Elton, R.C. and Köppendörfer, W.W. (1967). *Phys. Rev.* **160**, 194.

Elton, R.C. and Palumbo, L.J. (1974). *Phys. Rev. A* **9**, 1873.

Elton, R.C., Datla, R.U., Roberts, J.R. and Bathia, A.K. (1989). *Phys. Rev. A* **40**, 4142.

Elton, R.C., Datla, R.U., Roberts, J.R. and Bathia, A.K. (1990). *Phys. Scripta* **41**, 449.

Elwert, G. (1952). *Z. Naturforsch.* **7a**, 432.

Engelhardt, W. (1973). *Phys. Fluids* **16**, 217.

Evans, D.E. and Katzenstein, J. (1969). *Rep. Prog. Phys.* **32**, 207.

Evensen, H.T., Durst, R., Fonck, R.J. and Paul, S.F. (1995). *Rev. Sci. Instrum.* **66**, 845.

Faizullov, F.S., Sobolev, N.N. and Kudryavtsev, E.M. (1960). *Opt. Spectrosc. (USSR)* **8**, 311.

Fano, U. (1961). *Phys. Rev.* **124**, 1866.

Fano, U. (1963). *Phys. Rev.* **131**, 259.

Fano, U. and Cooper, J.W. (1968). *Rev. Mod. Phys.* **40**, 441.

Feautrier, N., Tran-Minh, N. and Van Regemorter, H. (1976). *J. Phys. B* **11**, 1871.

Federman, S.R., Beideck, D.J., Schectman, R.M. and York, D.J. (1992). *Astrophys. J.* **401**, 367.

Feldman, U., Seely, J.F. and Bhatia, A.K. (1985). *J. Appl. Phys.* **58**, 3954.

Fill, E.E. and Schöning, T. (1994). *J. Appl. Phys.* **76**, 1423.

Finken, K. (1979). *J. Quant. Spectrosc. Radiat. Transfer* **22**, 397.

Finken, K.H., Buchwald, R., Bertschinger, G. and Kunze, H.-J. (1980). *Phys. Rev. A* **21**, 200.

Finn, G.D. and Mugglestone, D. (1965). *Mon. Not. R. Astron. Soc.* **129**, 221.

Flannery, B.P., Deckman, H.W., Roberge, W.G. and D'Amico, K.L. (1987). *Science* **237**, 1439.

Foley, H.M. (1946). *Phys. Rev.* **69**, 616.

Fonck, R.J. (1985). *Rev. Sci. Instrum.* **56**, 885.

Fonck, R.J. (1990). In *Spectral Line Shapes*, Vol. 6, ed. L. Frommhold and J. Kato, AIP Conf. Proc. 216 (AIP, New York), p. 31.

Fonck, R.J., Cosby, G., Durst, R.D., Paul, S.F., Bretz, N., Scott, S., Synakowski, E. and Taylor, G. (1993). *Phys. Rev. Lett.* **70**, 3736.

Fonck, R.J., Darrow, D.S. and Jaehnig, K.P. (1984). *Phys. Rev. A* **29**, 3288.

Fonck, R.J., Dupperex, P.A. and Paul, S.F. (1990). *Rev. Sci. Instrum.* **61**, 3487.

Fonck, R.J., Goldston, R.J., Kaita, R. and Post, D. (1983). *Appl. Phys. Lett.* **42**, 239.

Fortov, V.E. and Iakubov, I.T. (1990). *Physics of Nonideal Plasmas* (Hemisphere Publ., New York).

Fortov, V.E., Bespalov, V.E., Kulish, M.I. and Kuz, S.I. (1990). In *Strongly Coupled Plasmas*, ed. S. Ichimaru (Elsevier Publ. and Yamada Science Foundation, Tokyo), p. 571.

Foster, J.M., Hoarty, D.J., Smith, C.C., Rosen, P.A., Davidson, S.J., Rose, S.J., Perry, T.S. and Serduke, F.J.D. (1991). *Phys. Rev. Lett.* **67**, 3255.

Frey, H.-D. and Meyer-ter-Vehn, J. (1995). *Inst. Phys. Conf. Ser. No.* **140**: Section 9, 152.

Frieden, B.R. (1983). *J. Opt. Soc. Am.* **73**, 927.

Fritsch, W. and Lin, C.D. (1984). *Phys. Rev. A* **29**, 3039.

Fritsch, W., Gilbody, H.B., Olson, R.E., Cederquist, H., Janev, R.K., Katsonis, K. and Yudin, G. (1991). *Physica Scripta T* **37**, 11.

Fuhr, J.R. and Lesage, A. (1993). *NIST Special Publication* **366**, Suppl. 4.

Fuhr, J.R. and Wiese, W.L. (1995). In *CRC Handbook of Chemistry and Physics*, 76 Ed., ed. D.R. Lide (CRC Press, Baco Raton, Florida), p. 128.

Fuhr, J.R., Martin, G.A. and Wiese, W.L. (1988). *J. Phys. Chem. Ref. Data* **17**, Suppl. 4.

Fujimoto, T. (1973). *J. Phys. Soc. Jap.* **34**, 216 and 1429.

Fujimoto, T. (1979a). *J. Quant. Spectrosc. Radiat. Transfer* **21**, 439.

Fujimoto, T. (1979b). *J. Phys. Soc. Jap.* **47**, 273.

Fujimoto, T. (1980). *J. Phys. Soc. Jap.* **49**, 1561.

Fujimoto, T. (1985). *J. Phys. Soc. Jap.* **54**, 2905.

Fujimoto, T. (1987). *Phys. Rev. A* **35**, 3024.

Fujimoto, T. and Kato, T. (1981). *Astrophys. J.* **246**, 994.

Fujimoto, T. and Kato, T. (1982). *Phys. Rev. Lett.* **48**, 1022.

Fujimoto, T. and Kato, T. (1984). *Phys. Rev. A* **30**, 379.

Fujimoto, T. and Kato, T. (1985). *Phys. Rev. A* **32**, 1663.

Fujimoto, T. and Kato, T. (1987). *Phys. Rev. A* **35**, 3024.

Fujimoto, T. and Kato, T. (1990). Private communication.

Fujimoto, T. and Kawachi, T. (1995). In *Atomic Processes in Plasmas*, 9th APS Topical Conf., ed. W.R. Rowan, AIP Conf. Proc. 322 (AIP, New York), p. 141.

Fujimoto, T. and McWhirter, R.W.P. (1990). *Phys. Rev. A* **42**, 6588.

Fujimoto, T., Koike, F., Sakimoko, K. Oksaka, R., Kawasaki, R., Takiyama, K., Oda, T. and Kato, T. (1992). *Atomic processes relevant to polarization spectroscopy* (National Institute for Fusion Science, Nagoya).

Fujimoto, T., Miyachi, S. and Sawada, K. (1988). *Nucl. Fusion* **28**, 1255.

Fujimoto, T., Sawada, K. and Takahata, K. (1989a). *J. Appl. Phys.* **66**, 2315.

Fujimoto, T., Sawada, K. Takahata, K., Eriguchi, K., Suemitsu, H., Ishii, K., Okasaka, R., Tanaka, H., Maekawa, K., Terumichi, Y. and Tanaka, S. (1989b). *Nucl. Fusion* **29**, 1519.

Furukawa, H. (1995). *Phys. Rev. E* **52**, 2988.

Gabriel, A.H. (1972). *Mon. Not. R. Astron. Soc.* **160**, 99.

Gabriel, A.H. and Jordan, C. (1972). In *Case Studies in Atomic Collisions*, ed. E.W. McDaniel and M.R.C. McDowell (North-Holland, Amsterdam), Vol. 2, p. 209.

Gaisinskii, I.M. and Oks, E.A. (1986). *Sov. J. Plasma Phys.* **12**, 65.

Gaisinskii, I.M. and Oks, E.A. (1987). *Sov. J. Plasma Phys.* **13**, 779.

Gallagher, A. (1996). In *Atomic, Molecular and Optical Physics Reference Book*, eds. G.W. Drake and N.E. Hedgecock, (AIP, New York), Sec. III 18.

Garstang, R.H. (1977). *Rep. Prog. Phys.* **40**, 105.

Gaunt, J.A. (1930). *Proc. R. Soc. London, Ser. A* **126**, 654.

Gavrilenko, V.P. (1993). *JETP* **76**, 236.

Gavrilenko, V.P. and Oks, E.A. (1987). *Sov. J. Plasma Phys.* **13**, 22.

Gavrilov, V.E., Gavrilova, T.V. and Fedorova, T.N. (1985). *Opt. Spectrosc. (USSR)* **59**, 518.

Gawron, A., Hey, J.D., Xu, X.J. and Kunze, H.-J. (1989). *Phys. Rev. A* **40**, 7150.

Gawron, A., Maurmann, S., Böttcher, F., Meckler, A. and Kunze, H.-J. (1988). *Phys. Rev. A* **38**, 4737.

Gel'Medova, L.A. and Shapiro, D.A. (1991). *J. Modern Opt.* **38**, 573.

Geltman, S. (1962). *Astrophys. J.* **136**, 935.

Geltman, S. (1994). *J. Phys. B* **27**, 257.

Gervids, V.I. and Kogan, V.I. (1992). In *Polarization Bremsstrahlung*, eds. V.N. Tsytovich and I.M. Oiringel (Plenum, New York), Chap. 13.

Gianakon, T.A., Fonck, R., Callen, J., Durst, R., Kim, J. and Paul, S. (1992). *Rev. Sci. Instrum.* **63**, 4931.

Gieske, H.-A. and Griem, H.R. (1969). *Astrophys. J.* **157**, 963.

Gilles, D. and Peyrusse, O. (1995). *J. Quant. Spectrosc. Radiat. Transfer* **53**, 647.

Glasstone, S. and Lovberg, R.H. (1960). *Controlled Thermonuclear Reactions* (Van Nostrand, New York).

Glenzer, S. (1995). In *Spectral Line Shapes*, Vol. 8, eds. A.D. May, J.R. Drummond and E. Oks, AIP Conf. Proc. 328 (AIP, New York), p. 134.

Glenzer, S. and Kunze, H.-J. (1996). *Phys. Rev. A* **53**, 2225.

Glenzer, S., Hey, J.D. and Kunze, H.-J. (1994a). *J. Phys. B* **27**, 413.

Glenzer, S., Kunze, H.-J., Musielok, J., Kim, Y.-K. and Wiese, W.L. (1994b). *Phys. Rev. A* **49**, 221.

Glenzer, S., Uzelac, N.I. and Kunze, H.-J. (1992). *Phys. Rev. A* **45**, 8795.

Glenzer, S., Wrubel, Th., Büscher, S., Kunze, H.-J., Godbert, L., Calisti, A., Stamm, R., Talin, B., Nash, J., Lee, R.W. and Klein, L. (1994c). *J. Phys. B* **27**, 5507.

Godbert, L., Calisti, A., Stamm, R., Talin, B., Lee, R. and Klein, L. (1994a). *Phys. Rev. E* **49**, 5644.

Godbert, L., Calisti, A., Stamm, R., Talin, B., Glenzer, S., Kunze, H.-J., Nash, J., Lee, R. and Klein, L. (1994b). *Phys. Rev. E* **49**, 5889.

Golant, V.E. and Fedorov, V.I. (1989). *RF Plasma Heating in Toroidal Devices* (Consultants Bureau, New York).

Golden, L.B. and Sampson, D.H. (1971). *Astrophys. J.* **170**, 181.

Golden, L.B., Sampson, D.H. and Omidvar, K. (1978). *J. Phys. B* **11**, 3235.

Goldsmith, S., Griem, H.R. and Cohen, L. (1984). *Phys. Rev. A* **30**, 2775.

Goldstein, W.H., Whitten, B.L., Hazi, A.U. and Chen, M.H. (1987). *Phys. Rev. A* **36**, 3607.

Goldwire, H.C., Jr. (1968). *Astrophys. J. Suppl.* **17**, 445.

Gordon, F.I. (1993). *Plasma Phys. Contr. Fusion* **35**, 1207.

Gordon, M.H. and Kruger, C.H. (1993). *Phys. Fluids B* **5**, 1014.

Graboske, H.C., Harwood, D.H. and Rogers, F.J. (1969). *Phys. Rev.* **186**, 210.

Green, J. (1964). *J. Quant. Spectrosc. Radiat. Transfer* **4**, 639.

Green, L.C., Rush, P.P. and Chandler, C.D. (1957). *Astrophys. J. Suppl.* **3**, 37.

Greene, R.L. (1976). *Phys. Rev. A* **14**, 1447.

Greene, R.L. (1982a). *J. Quant. Spectrosc. Radiat. Transfer* **27**, 639.

Greene, R.L. (1982b). *J. Phys. B* **15**, 1831.

Gregory, D.C. and Bannister, M.E. (1994). *Oak Ridge National Lab. Rep. ORNL/TM-12729.*

Griem, H.R. (1954). *Z. Phys.* **137**, 18.

Griem, H.R. (1962). *Phys. Rev.* **128**, 997.

Griem, H.R. (1963). *Phys. Rev.* **131**, 1170.

Griem, H.R. (1964). *Plasma Spectroscopy* (McGraw-Hill, New York).

Griem, H.R. (1967). *Astrophys. J.* **148**, 547.

Griem, H.R. (1968a). *Phys. Rev.* **165**, 258.

Griem, H.R. (1968b). *Astrophys. J.* **154**, 1111.

Griem, H.R. (1970). *Comments At. Mol. Phys.* **1**, 145.

Griem, H.R. (1974). *Spectral Line Broadening by Plasma* (Academic Press, New York).

Griem, H.R. (1978). *Phys. Rev. A* **17**, 214.

Griem, H.R. (1979). *Phys. Rev. A* **20**, 606.

Griem, H.R. (1983). In *Handbook of Plasmaphysics*, Vol. 1, eds. A.A. Galeev and R.N. Sudan (North-Holland, Amsterdam), Chap. 3.

Griem, H.R. (1986). *Phys. Rev. A* **33**, 3580.

Griem, H.R. (1988a). *Phys. Rev. A* **38**, 2943.

Griem, H.R. (1988b). *J. Quant. Spectrosc. Radiat. Transfer* **40**, 403.

Griem, H.R. (1989). *Phys. Rev. A* **40**, 3706.

Griem, H.R. (1992). *Phys. Fluids B* **4**, 2346.

Griem, H.R. (1993). In *Spectral Line Shapes*, Vol. 7, ed. R. Stamm and B. Talin (Nova Science Publishers, Commack, NY), p. 3.

Griem, H.R. and Hey, J.D. (1976). *Phys. Rev. A* **14**, 1906.

Griem, H.R. and Kunze, H.-J. (1969). *Phys. Rev. Lett.* **23**, 1279.

Griem, H.R. and Shen, C.S. (1962). *Phys. Rev.* **125**, 196.

Griem, H.R. and Shen, K.Y. (1961). *Phys. Rev.* **122**, 1490.

Griem, H.R. and Tsakiris, G.D. (1982). *Phys. Rev. A* **25**, 1199.

Griem, H.R., Baranger, M., Kolb, A.C. and Oertel, G. (1962). *Phys. Rev.* **125**, 177.

Griem, H.R., Blaha, M. and Kepple, P.C. (1979). *Phys. Rev. A* **19**, 2421.

Griem, H.R., Blaha, M. and Kepple, P.C. (1990). *Phys. Rev. A* **41**, 5600.

Griem, H.R., Huang, Y.W., Wang, J.S. and Moreno, J.C. (1991a). *Phys. Fluids B* **3**, 2430.

Griem, H.R., Iglesias, C.A. and Boercker, D.B. (1991b). *Phys. Rev. A* **44**, 5318.

Griem, H.R., Kolb, A.C. and Shen, K.Y. (1959). *Phys. Rev.* **116**, 4.

Griffin, D.C., Pindzola, M.S. and Bottcher, C. (1987). *Phys. Rev. A* **36**, 3642.

Groebner, R.J., Burrell, K.H. and Seraydarian, R.P. (1990). *Phys. Rev. Lett.* **64**, 3015.

Grützmacher, K. and Wende, B. (1977). *Phys. Rev. A* **16**, 243.

Gryzinski, M. (1959). *Phys. Rev.* **115**, 374.

Gryzinski, M. (1965). *Phys. Rev.* **138**A, 305, 322 and 336.

Guenther, K., Hayess, E., Krebs, K.H., Weixelbaum, L., Wenzel, U., Badalec, J., Jakupka, J., Stoeckel, J., Valovic, M. and Zacek, F. (1989). *J. Nucl. Mater.* **162-164**, 562.

Gündel, H. (1970). *Beiträge aus der Plasmaphysik* **10**, 455.

Gündel, H. (1971). *Beiträge aus der Plasmaphysik* **11**, 1.

Günter, S. (1993). *Phys. Rev. E* **48**, 500.

Günter, S. and Könies, A. (1994a). *Phys. Rev. E* **49**, 4732.

Günter, S. and Könies, A. (1994b). *J. Quant. Spectrosc. Radiat. Transfer* **52**, 819.

Günter, S., Hitzschke, L. and Röpke, G. (1991). *Phys. Rev. A* **44**, 6834.

Günter, S., Stobbe, M., Könies, A. and Halenka, J. (1995). In *Spectral Line Shapes*, Vol. 8, eds. A.D. May, J.R. Drummond and E. Oks, AIP Conf. Proc. 328 (AIP, New York), p. 217.

Gurovich, V.Ts. and Engel'sht, V.S. (1977). *Sov. Phys. JETP* **45**, 232.

Gutierrez, F.A., Jouin, H. and Cormier, E. (1994). *J. Quant. Spectrosc. Radiat. Transfer* **51**, 665.

Hagelstein, P.L. (1986). *Phys. Rev. A* **34**, 924.

Hagelstein, P.L. and Dalhed, S. (1988). *Phys. Rev. A* **37**, 1357.

Hagelstein, P.L. and Jung, R.K. (1987). *At. Data Nucl. Data Tables* **37**, 121.

Hahn, T.D. and Woltz, L.A. (1990). *Phys. Rev. A* **42**, 1450.

Hahn, Y. (1993). *J. Quant. Spectrosc. Radiat. Transfer* **49**, 81.

Hahn, Y. (1994). *J. Quant. Spectrosc. Radiat. Transfer* **51**, 663.

Halenka, J. (1990). *Z. Phys. D* **16**, 1.

Hall, T.A., Djaoui, E., Eason, R.W., Jackson, C.L., Shiwai, B., Rose, S.L., Cole, A. and Apte, P. (1988). *Phys. Rev. Lett.* **60**, 2034.

Hammel, B.A., Keane, C.A., Cable, M.D., Kania, D.R., Kilkenny, J.D., Lee, R.W. and Pasha, R. (1993). *Phys. Rev. Lett.* **70**, 1263.

Hammel, B.A., Keane, C.J., Dittrich, T.R., Kania, D.R., Kilkenny, J.D., Lee, R.W. and Levedahl, W.K. (1994). *J. Quant. Spectrosc. Radiat. Transfer* **51**, 113.

Hammond, P., Read, F.H., Cvejanović, S. and King, C.C. (1985). *J. Phys. B* **18**, L141.

Hanada, K., Ogura, K., Tanaka, H., Maekawa, T., Terumichi, Y. and Tanaka, S. (1990). *Plasma Phys. Contr. Fusion* **32**, 1289.

Hansen, J.P. and McDonald, I. (1976). *Theory of Simple Liquids* (Academic Press, New York).

Hardie, D.J.W. and Olsen, R.E. (1983). *J. Phys. B* **16**, 1983.

Hauer, A. and Baldis, H.A. (1988). In *Applications of Laser Plasmas*, ed. L.J. Radziemski and D.A. Cremers (Marcel Dekker, New York), Chap. 3.

Haynes, D.A. and Hooper, C.F., Jr. (1995). Private communication.

Haynes, D.A., Garber, D.T., Hooper, C.F., Jr., Mancini, R.C., Lee, Y.T., Bradley, D.K., Delettrez, J., Epstein, R. and Jaanimagi, P.A. (1996). *Phys. Rev. E* **53**, 1042.

Heading, D.J., Marangos, J.P. and Burgess, D.D. (1992). *J. Phys. B* **25**, 4745.

Hearn, A.G. (1963). *Proc. Phys. Soc.* **81**, 648.

Hebb, M.H. and Menzel, D.H. (1940). *Astrophys. J.* **92**, 408.

Heckmann, P.H. and Träbert, E. (1989). *Introduction to the Spectroscopy of Atoms* (North-Holland, Amsterdam).

Hegerfeld, G.C. and Kesting, V. (1988). *Phys. Rev. A* **37**, 1488.

Heisenberg, W. (1925). *Z. Phys.* **33**, 879.

Heitler, W. (1954). *The Quantum Theory of Radiation*, 3rd Ed. (Clarendon Press, Oxford); (1984). (Reprinted Dover Publications, New York).

Helbig, V. (1991). *Contrib. Plasma Phys.* **31**, 183.

Helbig, V. and Nick, K.-P. (1981). *J. Phys. B* **14**, 3573.

Helbig, V. and Thomsen, C. (1991). *Spectrochim. Acta* **28**b, 1215.

Henry, R.J.W. (1981). *Phys. Rep.* **68**, 1.

Heroux, L. (1963). *Nature* **198**, 1291.

Herzfeld, K. (1916). *Ann. Phys. (Leipzig)* **51**, 261.

Hess, R. and Burrell, F. (1979). *J. Quant. Spectrosc. Radiat. Transfer* **21**, 23.

Hey, J.D. (1976). *J. Quant. Spectrosc. Radiat. Transfer* **16**, 947.

Hey, J.D. (1977). *J. Quant. Spectrosc. Radiat. Transfer* **17**, 721.

Hey, J.D. (1985). *S. Afr. J. Phys.* **8**, 27.

Hey, J.D. (1989). *J. Quant. Spectrosc. Radiat. Transfer* **41**, 167.

Hey, J.D. (1993). *Am. J. Phys.* **61**, 741.

Hey, J.D. (1994). *Trans. Fusion Technol.* **25**, 315.

Hey, J.D. (1995). Private communication.

Hey, J.D. and Blaha, M. (1978). *J. Quant. Spectrosc. Radiat. Transfer* **20**, 557.

Hey, J.D. and Breger, P. (1980a). *J. Quant. Spectrosc. Radiat. Transfer* **24**, 349.

Hey, J.D. and Breger, P. (1980b). *J. Quant. Spectrosc. Radiat. Transfer* **24**, 427.

Hey, J.D. and Breger, P. (1982). *S. Afr. J. Phys.* **5**, 111.

Hey, J.D. and Breger, P. (1989). *J. Phys. B* **22**, L79.

Hey, J.D. and Griem, H.R. (1975). *Phys. Rev. A* **12**, 169.

Hey, J.D., Lie, Y.T., Rusbüldt, D. and Hintz, E. (1993). *20th EPS Conf. on Controlled Fusion and Plasma Physics (Lisbon)*, **17**C, 1111.

Hey, J.D., Lie, Y.T., Rusbüldt, D. and Hintz, E. (1994). *Contrib. Plasma Phys.* **34**, 725.

Hill, T.L. (1960). *An Introduction to Statistical Thermodynamics* (Addison-Wesley, Reading, MA).

Hirabayashi, A., Nambu, Y., Hasuo, M. and Fujimoto, T. (1988a). *Phys. Rev. A* **37**, 77.

Hirabayashi, A., Nambu, Y., Hasuo, M. and Fujimoto, T. (1988b). *Phys. Rev. A* **37**, 83.

Hitzschke, L. and Röpke, G. (1988). *Phys. Rev. A* **37**, 4991.

Hohenberg, P. and Kohn, W. (1964). *Phys. Rev. B* **136**, 864.

Hoekstra, R. (1995). In *Atomic Processes in Plasmas*, ed. W.L. Rowan, AIP Conf. Proc. 322 (AIP, New York), p. 105.

Höhne, F.E. and Zimmermann, R. (1982). *J. Phys. B* **15**, 2551.

Holland, A., Powell, E.T. and Fonck, R.J. (1991). *Appl. Opt.* **30**, 3740.

Holtsmark, J. (1919). *Ann. Phys. (Leipzig)* **58**, 577.

Holstein, T. (1947). *Phys. Rev.* **72**, 1212.

Holstein, T. (1951). *Phys. Rev.* **83**, 1159.

Hooper, C.F., Jr. (1966). *Phys. Rev.* **149**, 77.

Hooper, C.F., Jr. (1968). *Phys. Rev.* **165**, 215.

Hooper, C.F., Jr., Kilcrease, J.C., Mancini, R.C., Woltz, L.A., Bradley, D.K., Jaanimagi, P.A. and Richardson, M.C. (1989). *Phys. Rev. Lett.* **63**, 267.

Hooper, C.F., Jr., Mancini, R.C., Haynes, D.A. and Garber, D.T. (1995). In *Elementary Processes in Dense Plasmas*, eds. S. Ichimaru and S. Ogata (Addison-Wesley, Reading, Mass.), p. 403.

Hooper, C.F., Jr., Mancini, R.C., Kilcrease, D.P. and Woltz, L.A. (1990). In *Atomic Processes in Plasmas*, eds. Y.K. Kim and R.C. Elton, AIP Conf. Proc. 206 (AIP, New York), p. 204.

Hörmann, H. (1935). *Z. Phys.* **97**, 539.

Huang, Y.W., Böttcher, F., Wang, J.S. and Griem, H.R. (1990a). *Phys. Rev. A* **42**, 2322.

Huang, Y.W., Wang, J.-S., Moreno, J.C. and Griem, H.R. (1990b). *Phys. Rev. Lett.* **65**, 1757.

Huber, D.L. (1969). *Phys. Rev.* **178**, 93.

Hübner, K. (1964). *Z. Naturforsch.* **19**a, 1111.

Huebner, W.F. (1986). In *Physics of the Sun*, ed. P.A. Sturrock, T.E. Holzer, D.M. Mihalas and R.K. Ulrich (Reidel, Dordrecht), Vol. 1, p. 33.

Hummer, D.G. (1962). *Mon. Not. R. Astron. Soc.* **125**, 21.

Hummer, D.G. (1965). *Mem. R. Astron. Soc.* **70**, 1.

Hutcheon, R.J. and McWhirter, R.W.P. (1973). *J. Phys. B* **6**, 2668.

Hutcherson, R.K. and DeSilva, A.W. (1997). *Phys. Rev. E* (in press).

Hutchinson, I.H. (1987). *Principles of Plasma Diagnostics* (Cambridge University Press, Cambridge).

Ida, K. and Hidekuma, S. (1989). *Rev. Sci. Instrum.* **60**, 867.

Ida, K., Hidekuma, S., Miura, Y., Fujita, T., Mori, M., Hoshino, K., Suziki, N. and Yamauchi, T. (1990). *Phys. Rev. Lett.* **65**, 1364.

Iglesias, C.A. and Rogers, F.J. (1995). *Astrophys. J.* **443**, 460.

Iglesias, C.A., Boercker, D.B. and Lee, R.W. (1985). *Phys. Rev. A* **32**, 1906.

Iglesias, C.A., Lebowitz, J. and McGowan, D. (1983). *Phys. Rev. A* **28**, 1667.

Iglesias. E.J. and Griem, H.R. (1988a). *Phys. Rev. A* **38**, 301.

Iglesias, E.J. and Griem, H.R. (1988b). *Phys. Rev. A* **38**, 308.

Iglesias, E.J. and Griem, H.R. (1996). *J. Quant. Spectrosc. Radiat. Transfer* **55**, 383.

Inglis, D.R. and Teller, E. (1939). *Astrophys. J.* **90**, 439.

Irons, F.E. (1973). *J. Phys. B* **6**, 1562.

Irons, F.E. (1979a). *J. Quant. Spectrosc. Radiat. Transfer* **22**, 1.

Irons, F.E. (1979b). *J. Quant. Spectrosc. Radiat. Transfer* **22**, 21.

Irons, F.E. (1979c). *J. Quant. Spectrosc. Radiat. Transfer* **22**, 37.

Ishii, K. and Morita, S. (1984). *Phys. Rev. A* **30**, 2278.

Isler, R.C. (1977). *Phys. Rev. Lett.* **38**, 1359.

Isler, R.C. (1984). *Nucl. Fusion* **24**, 1599.

Isler, R.C. and Olson, R.E. (1988). *Phys. Rev. A* **37**, 3399.

Isler, R.C., Murray, L.E., Kasai, S., Dunlap, J.L., Bates, S.C., Edmonds, P.H., Lazarus, E.A., Ma, C.H. and Murakami, M. (1981). *Phys. Rev. A* **24**, 2701.

Isler, R.C., Neidigh, R.V. and Cowan, R.D. (1977). *Phys. Lett.* **63A**, 295.

Ispolatov, Ya. and Oks, E. (1994). *J. Quant. Spectrosc. Radiat. Transfer* **51**, 129.

Itikawa, Y. (1991). *At. Data Nucl. Data Tables* **49**, 209.

Itikawa, Y., Kato, T. and Sakimoto, K. (1995). *Inst. of Space and Aeronautical Sciences Rep.* **657** (Kanagawa, Japan).

Itikawa, Y., Takayanagi, K. and Iwai, T. (1984). *At. Data Nucl. Data Tables* **31**, 215.

Ivanov, V.V. (1973). *Transfer of Radiation in Spectral Lines.* Natl. Bur. Stand. (U.S.), Spec. Publ. 385 (U.S. Government Printing Office, Washington).

Jackson, J.D. (1962). *Classical Electrodynamics* (John Wiley, New York).

Jackson, J.L. and Klein, L.S. (1969). *Phys. Rev.* **177**, 352.

Jacobs, V.L. and Blaha, M. (1980). *Phys. Rev. A* **21**, 525.

Jacobs, V.L. and Davis, J. (1976). *Phys. Rev. Lett.* **37**, 1390.

Jacobs, V.L., Cooper, J. and Haan, S.L. (1994). *Phys. Rev. A* **50**, 3005.

Jacobs, V.L., Davis, J. and Kepple, P.C. (1976). *Phys. Rev. Lett.* **37**, 1390.

Jacobs, V.L., Davis, J., Kepple, P.C. and Blaha, M. (1977a). *Astrophys. J.* **211**, 605.

Jacobs, V.L., Davis, J., Kepple, P.C. and Blaha, M. (1977b). *Astrophys. J.* **215**, 690.

Jahoda, F. and Sawyer, G.A. (1971). In *Methods of Experimental Physics*, Vol. 9b, eds. H.R. Griem and R.H. Lovberg (Academic Press, New York), Chap. 11.

Jahoda, F.C., Ribe, F.L. and Sawyer, G.A. (1963). *Phys. Rev.* **131**, 24.

Jamelot, G., Jaegle, P., Lemaire, P. and Carillon, A. (1990) *J. Quant. Spectrosc. Radiat. Transfer* **44**, 71.

Janev, R.K. (1995). *Atomic and Molecular Processes in Fusion Edge Plasmas* (Plenum, London).

Janev, R.K., Bransden, B.H. and Gallagher, J.W. (1983). *J. Phys. Chem. Ref. Data* **12**, 829.

Janev, R.K., Ivanovski, G. and Solov'ev, E.A. (1994). *Phys. Rev. A* **49**, R 645.

Janev, R.K., Langer, W.D., Evans, K., Jr. and Post, D.E. (1987). *Elementary Processes in Hydrogen-Helium Plasmas* (Springer, Berlin).

Janev, R.K., Presnyakov, L.P. and Shevelko, V.P. (1985). *Physics of Highly Charged Ions* (Springer, Berlin).

Janicki, C., Décoste R. and Noël, P. (1992). *Rev. Sci. Instrum.* **63**, 4410.

Janicki, C., Décoste, R. and Simm, C. (1989). *Phys. Rev. Lett.* **62**, 3038.

Jefferies, J.T. (1968). *Spectral Line Formation* (Blaisdell, Waltham, MA).

Jefferies, J.T. and Thomas, R.N. (1958). *Astrophys. J.* **127**, 667.

Johnson, L.C. (1972). *Astrophys. J.* **174**, 227.

Johnson, L.C. and Hinnov, E. (1973). *J. Quant. Spectrosc. Radiat. Transfer* **13**, 333.

Johnston, T.W. and Dawson, J.M. (1973). *Phys. Fluids* **16**, 722.

Jones, D.W. and Wiese, W.L. (1984). *Phys. Rev. A* **30**, 2602.

Jones, D.W., Pichler, G. and Wiese, W.L. (1987). *Phys. Rev. A* **35**, 2585.

Jones, D.W., Wiese, W.L. and Woltz, L.A. (1986). *Phys. Rev. A* **34**, 450.

Jones, L.A., Greig, J.R., Oda, T. and Griem, H.R. (1971). *Phys. Rev. A* **4**, 833.

Jones, L.A., Kållne, E. and Thomson, D.B. (1977). *J. Quant. Spectrosc. Radiat. Transfer* **17**, 175.

Joseph, J. and Rohrlich, F. (1958). *Rev. Mod. Phys.* **30**, 354.

Joyce, R.F., Woltz, L.A. and Hooper, C.F., Jr. (1987). *Phys. Rev. A* **35**, 2228.

Jung, Y.-D. (1994). *Phys. Plasmas* **1**, 785.

Kadota, K., Otsuka, M. and Fujita, J. (1980). *Nucl. Fusion* **20**, 209.

Kalkofen, W. (1984). *Methods in Radiative Transfer* (Cambridge University Press, Cambridge).

Karabourniotis, D. (1986). In *Radiative Processes in Discharge Plasmas*, eds. J.M. Proud and L.H. Luessen, NATO ASI Series B: Physics **149**, p. 171.

Karplus, R. and Schwinger, J. (1948). *Phys. Rev.* **73**, 1020.

Karzas, W.J. and Latter, R. (1961). *Astrophys. J. Suppl.* **6**, no. 55, 167.

Kastner, S.O. (1977). *Astron. Astrophys.* **54**, 255.

Kato, S., Nishiguchi, A. and Mima, K. (1994). *Phys. Rev. E* **50**, 2193.

Kato, T. (1994). *At. Data Nucl. Data Tables* **57**, 181.

Kato, T. and Masai, K. (1988). *J. de Physique C1, Suppl. 3*, **49**, 349.

Kato, T., Lang, J. and Berrington, K.E. (1990). *At. Data Nucl. Data Tables* **44**, 133.

Kato, T., Masai, K. and Mizuno, J. (1983). *J. Phys. Soc. Japan* **52**, 3019.

Kato, T., Morita, S., Masai, D. and Hayakawa, S. (1987). *Phys. Rev. A* **36**, 795.

Kato, T., Safranova, U., Shlyaptseva, A., Cornille, M. and Dubau, J. (1995). *NIFS-Data-24*.

Kauffman, R.L. (1991). In "Physics of Laser Plasmas," *Handbook of Plasma Physics*, Vol. 3, ed. A. Rubenchik and S. Witkowski (North-Holland, Amsterdam), chapter 3.

Kawachi, T. and Fujimoto, T. (1995). *Phys. Rev. E* **51**, 1440.

Kawachi, T., Fujimoto, T. and Csanak, G. (1995). *Phys. Rev. E* **51**, 1428.

Kazantsev, S.A. and Henoux, J.-C. (1995). *Polarization Spectroscopy of Ionized Gases* (Kluver Academic Publishers, Dordrecht).

Keane, C.J. and Suckewer, S. (1991). *J. Opt. Soc. Am. B* **8**, 201.

Keane, C.J., Hammel, B.A., Kania, D.R., Kilkenny, J.D., Lee, R.W., Osterheld, A.L., Suter, L.J., Mancini, R.C., Hooper, C.F., Jr. and Delamater, N.D. (1993). *Phys. Fluids B* **5**, 3328.

Keane, C.J., Hammel, B.A., Langer, S.H., Lee, R.W., Calisti, A., Godbert, L., Stamm, R. and Talin, B. (1995). In *Spectral Line Shapes*, Vol. 8, eds. A.D. May, J.R. Drummond and E. Oks, AIP Conf. Proc. 328 (AIP, New York), p. 105.

Keane, C.J., Hammel, B.A., Osterheld, A.L. and Kania, D.R. (1994). *Phys. Rev. Lett.* **72**, 3029.

Keane, C.J., Lee, R.W., Hammel, B.A., Osterheld, A.L., Suter, L.J., Calisti, A., Khelfaoui, F., Stamm, R. and Talin, B. (1990). *Rev. Sci. Instrum.* **61**, 2780.

Kelleher, D.E. (1981). *J. Quant. Spectrosc. Radiat. Transfer* **25**, 191.

Kelleher, D.E. and Wiese, W.L. (1973). *Phys. Rev. Lett.* **31**, 1431.

Kelleher, D.E., Wiese, W.L., Helbig, V., Greene, R.L. and Oza, D.H. (1993). *Physica Scripta T* **47**, 75.

Kelly, R.L. (1987a). *J. Phys. Chem. Ref. Data* **16**, Suppl. 1, 1.

Kelly, R.L. (1987b). *J. Phys. Chem. Ref. Data* **16**, Suppl. 1, 651.

Kepple, P.C. (1972). *Phys. Rev. A* **6**, 1.

Kepple, P.C. (1995). Private communication.

Kepple, P. and Griem, H.R. (1968). *Phys. Rev.* **173**, 317.

Kepple, P.C. and Griem, H.R. (1978). *NRL Memorandum Report* 3634.

Kepple, P.C. and Griem, H.R. (1982). *Phys. Rev. A* **26**, 484.

Kieffer, J.C., Matte, J.P., Pépin, H., Chaker, M., Beaudin, Y., Johnston, T.W., Chien, C.Y., Coe, S., Mourou, G. and Dubau, J. (1992). *Phys. Rev. Lett.* **68**, 480.

Kielkopf, J. (1993). In *Spectral Line Shapes*, Vol. 7, eds. R. Stamm and B. Talin (Nova Science Publ., Commack, New York), p. 271.

Kielkopf, J. (1995). *Phys. Rev. E* **52**, 2013.

Kielkopf, J. and Allard, N.F. (1995). *Astrophys. J.* **450**, L 75.

Kiess, C.C. and Shortley, G.H. (1949). *J. Res. NBS* **42**, 183.

Kilcrease, D.P. (1994). *J. Quant. Spectrosc. Radiat. Transfer* **51**, 161.

Kilcrease, D.P., Mancini, R.C. and Hooper, Jr., C.F. (1993). *Phys. Rev. E* **48**, 3901.

Kim, Y.-K. and Rudd, M.E. (1994). *Phys. Rev. A* **50**, 3954.

Kirkwood, J.G. (1932). *Phys. Z.* **33**, 521.

Kissel, L., Pratt, R.H. and Roy, S.C. (1980). *Phys. Rev. A* **22**, 1970.

Kittel, C. (1963). *Quantum Theory of Solids* (Wiley, New York).

Klarsfeld, S. (1989). *Phys. Rev. A* **39**, 2324.

Klarsfeld, S. and Maquet, A. (1973). *Phys. Lett. B* **43**, 201.

Knyazev, B.A., Chikunov, V.V. and Mel'nikov, P.I. (1992). In *Beams 92 - Proc. of the 9th Int. Conf. on High-Power Particle Beams*, ed. D. Mosher and G. Cooperstein (NTIS, Springfield, Virginia), p. 1043.

Knyazev, B.A., Lebedev, S.V. and Mel'nikov, P.I. (1991a). *Sov. Phys. Tech. Phys.* **36**, 250.

Knyazev, B.A., Mel'nikov, P.I. and Chikunov, V.V. (1991b). *Sov. Tech. Phys. Lett.* **17**, 357.

Kobilarov, R. and Konjević, N. (1990). *Phys. Rev. A* **41**, 6023.

Kobilarov, R., Konjević, N. and Popovich, M.V. (1989). *Phys. Rev. A* **40**, 3871.

Kobzev, G.A., Iakubov, I.T. and Popovich, M.M., eds. (1995). *Transport and Optical Properties of Nonideal Plasma* (Plenum Press, New York and London).

Koch, J.A., McGowan, B.J., DaSilva, L.B., Matthews, D.L., Underwood, J.H., Batson, P.J. and Mrowka, S. (1992). *Phys. Rev. Lett.* **68**, 3291.

Koch, J.A., McGowan, B.J., DaSilva, L.B., Matthews, D.L., Underwood, J.H., Batson, P.J., Lee, R.W., London, R.A. and Mrowka, S. (1994). *Phys. Rev. A* **50**, 1877.

Koenig, M., Malnoult, P. and Nguyen, H. (1988). *Phys. Rev. A* **38**, 2089.

Kogan, V.I. (1960). In *Plasma Physics and the Problems of Controlled Thermonuclear Reactions*, ed. M.A. Leontovich (Pergamon, Oxford), Vol. IV, p. 305.

Kogan, V.I., Kukushkin, A.B. and Lisitsa, V.S. (1992). *Phys. Rep.* **213**, 2.

Kohn, H. (1932). *Phys. Z.* **33**, 957.

Kohn, W. and Sham, L.J. (1965). *Phys. Rev. A* **140**, 1133.

Kohsiek, W. (1977). *J. Quant. Spectrosc. Radiat. Transfer* **17**, 651.

Kolb, A.C. and Griem, H.R. (1958). *Phys. Rev.* **111**, 514.

Könies, A. and Günter, S. (1994). *J. Quant. Spectrosc. Radiat. Transfer* **52**, 825.

Könies, A. and Günter, S. (1995). *Phys. Rev. E* **52**, 6658.

König, R., Kolk, K.-H., Koshelev, K.N. and Kunze, H.-J. (1989). *Phys. Rev. Lett.* **62**, 1750.

Koningstein, J.A. (1972). *Introduction to the Theory of the Raman Effect* (Reidel, Dordrecht).

Konjević, N. and Roberts, J.R. (1976). *J. Phys. Chem. Ref. Data* **5**, 209.

Konjević, N. and Uzelac, N.I. (1990). *J. Quant. Spectrosc. Radiat. Transfer* **44**, 61.

Konjević, N. and Wiese, W.L. (1976). *J. Phys. Chem. Ref. Data* **5**, 259.

Konjević, N. and Wiese, W.L. (1990). *J. Phys. Chem. Ref. Data* **19**, 1307.

Konjević, N., Dimitrijević, M.S. and Wiese, W.L. (1984a). *J. Phys. Chem. Ref. Data* **13**, 619.

Konjević, N., Dimitrijević, M.S. and Wiese, W.L. (1984b). *J. Phys. Chem. Ref. Data* **13**, 649.

Kononov, B. Ya., Koshelev, K.N., Safronova, U.L., Sidel'nikov, Yu. V. and Churilov, S.S. (1980). *Sov. Phys. JETP* **31**, 720.

Kosarev, I.N. and Lisitsa, V.S. (1994). *Sov. Phys. JETP* **79**, 64.

Koshelev, K.N., Sidel'nikov, Yu. V., Decker, G., Kies, V., Malzig, M., Rowekamp, P., Rozmej, F., Schulz, A. and Kunze, H.-J. (1994). *Opt. Spektrosc. (USSR)* **76**, 198.

Kraeft, W.D., Kremp, D., Ebeling, W. and Röpke, G. (1986). *Quantum Statistics of Charged Particle Systems* (Plenum, New York).

Krall, N.A. and Trivelpiece, A.W. (1986). *Principles of Plasma Physics*, 2nd. ed. (San Francisco Press, San Francisco).

Kramers, H.A. (1923). *Philos. Mag.* **46**, 836.

Kramers, H.A. and Heisenberg, W. (1925). *Z. Phys.* **31**, 681.

Krause, M.O., Carlson, T.A. and Dismukes, R.D. (1968). *Phys. Rev.* **170**, 37.

Kress, J.D., Kwon, I. and Collins, L.A. (1995). *J. Quant. Spectrosc. Radiat. Transfer* **54**, 237.

Kruer, W.L. (1988). *The Physics of Laser Plasma Interactions* (Addison-Wesley, Reading, MA).

Kukushkin, A.B. and Lisitsa, V.S. (1992). In *Polarization bremsstrahlung*, eds. V.N. Tsytovich and I.M. Oiringel (Plenum, New York), Chap. 11.

Kunc, J.A. (1987). *Phys. Fluids* **30**, 2255.

Kunc, J.A. and Soon, W.H. (1992). *Astrophys. J.* **396**, 364.

Kunze, H.-J. (1968). In *Plasma Diagnostics*, ed. W. Lochte-Holtgreven, (North-Holland, Amsterdam); reprinted by the American Vacuum Society in 1995 (AIP, New York), chapter 9.

Kunze, H.-J. (1972). *Space Sci. Rev.* **13**, 565.

Kunze, H.-J. (1987). In *Spectral Line Shapes*, Vol. 4, ed. R.J. Exton (Deepak, Hampton, Virginia), p. 23.

Kunze, H.-J., Gabriel, A.H. and Griem, H.R. (1968a). *Phys. Rev.* **165**, 267.

Kunze, H.-J., Gabriel, A.H. and Griem, H.R. (1968b), *Phys. Fluids B* **11**, 662.

Kurucz, R.L. (1970). *Smithsonian Astrophys. Obs. Spec. Rep.* **308**.

Kwok, T., Guberman, S., Dalgarno, A. and Posen, A. (1986). *Phys. Rev. A* **34**, 1962.

Kwon, I., Collins, L.A., Kress, J.D., Troullier, N. and Lynch, D.L. (1994). *Phys. Rev. E* **49**, R4771.

Lablanquie, P. and Morin, P. (1991). *J. Phys. B* **24**, 4349.

LaFontaine, B., Baldis, H.A., Villeneuve, D.M., Dunn, J., Enright, D., Kieffer, J.C., Pépin, H., Rosen, M.D., Matthews, D.L. and Maxon, S. (1994). *Phys. Plasmas* **1**, 2329.

Lamoureux, M., Jacquet, L. and Pratt, R.H. (1989). *Phys. Rev. A* **39**, 6323.

Lamoureux, M., Cauble, R., Kim, L., Perrot, F. and Pratt, R.H. (1987). *J. Quant. Spectrosc. Radiat. Transfer* **37**, 283.

Landau, L.K. and Lifshiftz, E. (1951). *The Classical Theory of Fields* (Addison-Wesley, Reading, Massachusetts).

Landé, A. (1921). *Z. Phys.* **5**, 231.

Landé, A. (1923a). *Z. Phys.* **15**, 189.

Landé, A. (1923b). *Z. Phys.* **19**, 112.

Landman, D.A. (1973a). *Solar Phys.* **30**, 371.

Landman, D.A. (1973b). *Solar Phys.* **31**, 81.

Lang, J. (1994). *Spec. Ed., At. Data Nucl. Data Tables* **57**, Nos. 1/2.

Lang, K.R. (1980). *Astrophysical Formulae* (Springer, New York).

Lange, R. and Schlüter, D. (1992). *J. Quant. Spectrosc. Radiat. Transfer* **48**, 153.

Larkin, A.I. (1960). *Sov. Phys. JETP* **11**, 1363.

Leboucher-Dalimier, E., Poquérusse, A. and Angelo, P. (1993). *Phys. Rev. E* **47**, R1467.

Lee, C.M., Kissel, L. and Pratt, R.H. (1976). *Phys. Rev. A* **13**, 1714.

Lee, R.W. (1979). *J. Phys. B* **12**, 1165.

Lee, R.W. (1985). *J. Appl. Phys.* **58**, 612.

Lee, R.W. (1988). *J. Quant. Spectrosc. Radiat. Transfer* **40**, 561.

Lee, R.W., Whitten, B.L. and Strout, R.E. (1984). *J. Quant. Spectrosc. Radiat. Transfer* **32**, 91.

Lee, Y.T. and More, R.M. (1984). *Phys. Fluids* **27**, 1273.

Leng, Y., Goldhar, J., Griem, H.R. and Lee, R.W. (1995). *Phys. Rev. E* **52**, 4328.

Leo, P.J., Mullamphy, D.F.T., Peach, G. and Whittingham, I.B. (1995). *J. Phys. B* **28**, 4449.

Letokhov, V.S. and Chebotayev, V.P. (1977). *Nonlinear Laser Spectroscopy* (Springer, Berlin).

Levinson, N. (1949). *Danske Videnskab. Selskab. Matematisk Fysiske Medel.* **25** (9).

Levinton, F.M. (1992). *Rev. Sci. Instrum.* **63**, 5157.

Levinton, F.M., Batha, S.A., Yamada, M. and Zarnstorff, M.C. (1993). *Phys. Fluids B* **5**, 2554.

Levinton, F.M., Fonck, R.J., Gammel, G.M., Kaita, R., Kugel, H.W., Powell, E.T. and Roberts, D.W. (1989). *Phys. Rev. Lett.* **63**, 2060.

Lewis, E.L. (1980). *Phys. Rep.* **58**, 1.

Lewis, M. (1961). *Phys. Rev.* **121**, 501.

Lewitt, R.M. (1983). *Proc. IEEE* **71**, 390.

Li, J. and Hahn, Y. (1995). *Phys. Rev. E* **52**, 4281.

Liberman, D.A. (1971). *Phys. Rev. B* **3**, 2081.

Liberman, D.A. (1979). *Phys. Rev. B* **20**, 4981.

Liberman, D.A. (1982). *J. Quant. Spectrosc. Radiat. Transfer* **27**, 335.

Liberman, D.A. and Albritton, J.R. (1995). *J. Quant. Spectrosc. Radiat. Transfer* **53**, 729.

Lifshitz, E.V. (1968). *Sov. Phys. JETP* **26**, 570.

Lindholm, E. (1945). *Ark. Math. Astron. Fys.* **32A**, 17.

Lishiwa, C., Feldstein, J., Stewart, T. and Mushlitz, E. (1985). *J. Chem. Phys.* **83**, 133.

Lisitsa, V.S. (1977). *Sov. Phys. Usp.* **20**, 603.

Lisitsa, V.S. (1994). *Atoms in Plasmas* (Springer, Berlin).

Lisitsa, V.S. and Yakovlenko, S.I. (1975a). *Sov. Phys. JETP* **39**, 759.

Lisitsa, V.S. and Yakovlenko, S.I. (1975b). *Sov. Phys. JETP* **41**, 233.

Ljepojević, N.N., Hutcheon, R.J. and McWhirter, R.W.P. (1984). *J. Phys. B* **17**, 3057.

Ljepojević, N.N., McWhirter, R.W.P. and Volonte, S. (1985). *J. Phys. B* **15**, 3285.

Loboda, P.A., Lykov, V.A. and Popova, V.V. (1994). In *X-Ray Lasers 1994*, eds. D.C. Eder and D.L. Matthews, AIP Conf. Proc. 332 (AIP, New York), p. 505.

Lochte-Holtgreven, W. (1968). *Plasma Diagnostics* (North-Holland, Amsterdam); reprinted by the American Vacuum Society in 1995 (AIP, New York), Chap. 3..

Lorentz, H.A. (1906). *Proc. R. Acad. Sci. (Amsterdam)* **8**, 591.

Lotz, W. (1967a), *Z. Phys.* **206**, 205.

Lotz, W. (1967b). *Astrophys. J. Suppl.* **14**, 207.

Lotz, W. (1968). *Z. Phys.* **216**, 241.

Lotz, W. (1969). *Z. Phys.* **220**, 466.

Lotz, W. (1970). *Z. Phys.* **232**, 101.

Lotz, W. (1973). *J. Opt. Soc. Am.* **57**, 873.

Loudon, R. (1983). *The Quantum Theory of Light*, 2nd Ed. (Clarendon Press, Oxford).

Lunney, J.C. (1983). *J. Phys. B* **16**, 183.

Lyaptsev, A.V. and Kazantsev, S.A. (1993). *Opt. Spectrosc.* **75**, 17.

Macek, J. (1967). *Proc. Phys. Soc. London* **92**, 365.

Magee, N.H., Abdallah, J., Clark, R.E.H., Cohen, J.S., Collins, L.A., Csanak, G., Fontes, C.J., Gauger, A., Keady, J.J., Kilcrease, D.P. and Merts, A.L. (1995). *Astrophys. Applications of Powerful New Databases*, Astron. Soc. Pacific Conf. Ser. Vol. **78**, ed. S.J. Adelman and W.L. Wiese (Book Crafters, Chelsia, Michigan), p. 1.

Malnoult, P., d'Etat, B. and Nguyen, H. (1989). *Phys. Rev. A* **40**, 1983.

Mancini, R.C. and Fontán, C.F. (1985). *J. Quant. Spectrosc. Radiat. Transfer* **34**, 115.

Mancini, R.C., Audebert, P., Geindre, J.P., Rousse, A., Fallies, F., Gauthier, J.C., Mysyrowicz, A., Chambaret, J.P. and Antonetti, A. (1994a). *J. Phys. B* **27**, 1671.

Mancini, R.C., Audebert, P., Geindre, J.P., Rousse, A., Fallies, F., Gauthier, J.C., Mysyrowicz, A., Chambaret, J.P. and Antonetti, A. (1995). In *Atomic Processes in Plasmas*, 9th APS Topical Conference, AIP Conf. Proc. 322, ed. W.L. Rowan (AIP, New York), p. 161.

Mancini, R.C., Hooper, C.F., Jr. and Coldwell, R.L. (1994b). *J. Quant. Spectrosc. Radiat. Transfer* **51**, 201.

Mancini, R.C., Hooper, C.F., Jr., Delamater, N.D., Hauer, A., Keane, C.J., Hammel, B.A. and Nash, J.K. (1992). *Rev. Sci. Instrum.* **63**, 5119.

Mancini, R.C., Joyce, R.F. and Hooper, C.F. (1987). *J. Phys. B* **20**, 2975.

Mancini, R.C., Kilcrease, D.P., Woltz, L.A. and Hooper, Jr., C.F. (1991). *Comp. Phys. Comm.* **63**, 314.

Mandl, W., Wolf, R.C., von Hellermann, M.G. and Summers, H.P. (1993). *Plasma Phys. Contr. Fusion* **35**, 1373.

Mapleton, R.A. (1972). *Theory of Charge Exchange* (Wiley, New York).

Marangos, J.P., Burgess, D.D. and Baldwin, K.G.H. (1988). *J. Phys. B* **21**, 3357.

Margenau, H. (1958). *Phys. Rev.* **109**, 6.

Marjoribanks, R.S., Budnik, F., Kulcsár, G. and Zhao, L. (1995). *Rev. Sci. Instrum.* **66**, 683.

Marjoribanks, R.S., Richardson, M.C., Jaanimagi, P.A. and Epstein, R. (1992). *Phys. Rev. A* **46**, R1747.

Marr, G.V. (1968). *Plasma Spectroscopy* (Elsevier, Amsterdam).

Marss, R.E., Elliot, S.R. and Knapp, D.A. (1994). *Phys. Rev. Lett.* **72**, 4082.

Martin, G.A., Fuhr, J.R. and Wiese, W.L. (1988). *J. Phys. Chem. Ref. Data* **17**, Suppl. 3.

Martin, W.C., Zalubas, R. and Hagan, L. (1978). *Atomic Energy Levels — The Rare Earth Elements*, NSRDS-NBS 60 (U.S. Government Printing Office, Washington, D.C.).

Masai, K. (1994). *J. Quant. Spectrosc. Radiat. Transfer* **51**, 211.

Masnov-Seeuws, F. and McCaroll, R. (1972). *Astron. Astrophys.* **17**, 441.

Mattioli, M., Peacock, N.J. and Summers, H.P. (1989). *Phys. Rev. A* **40**, 3886.

Maurmann, S. and Kunze, H.-J. (1993). *Proc. 6th Int. Conf. on Laser-Aided Plasma Diagnostics* (MIT, Cambridge, MA), p. 49.

Maurmann, S., Kunze, H.-J., Gavrilenko, V. and Oks, E. (1996). *J. Phys. B* **29**, 1525.

Maxon, M.S. and Corman, E.G. (1967). *Phys. Rev.* **163**, 156.

McCormick, K., Söldner, F.X., Eckartt, D., Leuterer, F., Murmann, H., Derfler, H., Eberhagen, A., Gehre, O., Gernhardt, J., v. Gierke, G., Gruber, O., Keilhacker, M., Klüber, O., Lackner, K., Meisel, D., Mertens, V., Röhr, H., Schmitter, K.-H., Steuer, K.H. and Wagner, F. (1987). *Phys. Rev. Lett.* **58**, 491.

McDaniel, E.W. (1964). *Collision Phenomena in Ionized Gases* (Wiley, New York).

McDougall, J. and Stoner, E.C. (1939). *Philos. Trans. R. Soc. London, Ser. A* **237**, 67.

McDowell, M.R.C. and Coleman, J.P. (1970). *Introduction to the Theory of Atomic Collisions* (North-Holland, Amsterdam).

McGowan, J.W. and Clarke, E.M. (1968). *Phys. Rev.* **167**, 43.

McKee, G.R., Fonck, R., Stratton, B., Bell, R., Budny, R., Bush, C., Grek, B., Johnson, D., Park, H., Ramsey, A., Synakowski, E. and Taylor, G. (1995b), *Phys. Rev. Lett.* **75**, 649.

McKee, G.R., Fonck, R.J., Thorson, T.A. and Stratton, B.C. (1995a). *Rev. Sci. Instrum.* **66**, 643.

McLean, E.A., Stamper, J.A., Manka, C.K., Griem, H.R., Droemer, D.W. and Ripin, B.H. (1984). *Phys. Fluids* **27**, 1327.

McQuarrie, D.A. (1976). *Statistical Mechanics* (Harper and Row, New York).

McWhirter, R.W.P. (1965). In *Plasma Diagnostic Techniques*, ed. R.H. Huddlestone and S.L. Leonard (Academic Press, NY), p. 201.

McWhirter, R.W.P. and Hearn, A.G. (1963). *Proc. Phys. Soc.* **82**, 641.

Meinel, E.S. (1988). *J. Opt. Soc. Am. A* **5**, 25.

Melchert, F. (1993). In *The Physics of Electronic and Atomic Collisions*, eds. T. Andersen, B. Fastrup, F. Folkmann, H. Knudsen and N. Andersen, AIP Conf. Proc. No. 295 (AIP, New York), p. 574.

Mendoza, C., Hibbert, A. and Berrington, K.A. (1992). *Rev. Mexicana Astron. Astrofis.* **23**, 19.

Menzel, D.H. (1933). *Proc. Natl. Acad. Sci.* **19**, 40.

Menzel, D.H. and Pekeris, C.L. (1935). *Mon. Not. R. Astron. Soc.* **96**, 77.

Mermin, N.D. (1965). *Phys. Rev. A* **137**, 1141.

Mertens, P. and Bogen, P. (1987). *Appl. Phys. A* **43**, 197.

Merts, A.L., Cowan, R.D. and Magee, N.H. (1976). *The Calculated Power Output from a Thin Iron-Seeded Plasma*, LASL Report, LA-6220-MS (Los Alamos, NM).

Meulenbroeks, R.F.G., van Beek, A.J., van Helvoort, A.J.G., van der Sanden, M.C.M. and Schram, D.C. (1994). *Phys. Rev. E* **49**, 4397.

Mewe, R. (1972). *Astron. Astrophys.* **20**, 215.

Midha, J.M. and Gupta, S.C. (1994). *J. Quant. Spectrosc. Radiat. Transfer* **52**, 897.

Mihajlov, A.A., Ermolaev, A.M. and Dimitrijević, M.S. (1993). *J. Quant. Spectrosc. Radiat. Transfer* **50**, 227.

Mihalas, D. (1978). *Stellar Atmospheres*, 2nd. ed. (Freeman, San Francisco).

Mihalas, D. and Mihalas, B.W. (1984). *Foundations of Radiation Hydrodynamics* (Oxford University Press, Oxford).

Mijatović, Z., Konjević, N., Ivković, M. and Kobilarov, R. (1995). *Phys. Rev. E* **51**, 4891.

Milne, E.A. (1928). *Mon. Not. R. Astron. Soc.* **88**, 493.

Milne, E.A. (1930). *Mon. Not. R. Astron. Soc.* **90**, 769.

Mitchell, A.C.G. and Zemanski, M.W. (1934). *Resonance Radiation and Excited Atoms*, reprinted in 1961 and 1971. (Cambridge University Press, Cambridge).

Mitchell, K.B., van Hulsteyn, D.B., McCall, G.H., Lee, P. and Griem, H.R. (1979). *Phys. Rev. Lett.* **42**, 232.

Miyachi, S., Sawada, K., Takahata, K., Suemitsu, H., Fujimoto, T., Okasaka, R., Ishii, K., Hirata, Y., Fukao, M., Tanaka, H., Maekawa, T., Terumichi, Y. and Takaka, S. (1989). *J. Quant. Spectrosc. Radiat. Transfer* **42**, 355.

Moore, C.A., Davis, G.P. and Gottscho, R.A. (1984). *Phys. Rev. Lett.* **52**, 538.

Moore, C.E. (1949–58). *Atomic Energy Levels*, Vols. I-III, Natl. Bur. Stand. (U.S.), Circ. **467** (U.S. Government Printing Office Washington, D.C.); reissued 1971 as NSRDS-NBS **35**, Vols. I-III.

Moores, D.L., Golden, L.B. and Sampson, D.H. (1980). *J. Phys. B* **13**, 385.

Moran, T.G. (1995). *Phys. Rev. E* **51**, 3464.

More, R.M. (1982). *J. Quant. Spectrosc. Radiat. Transfer* **27**, 345.

More, R.M. (1983). In *Atomic and Molecular Physics of Controlled Thermonuclear Fusion*, eds. C. Joachain and D. Post (Plenum, New York), p. 399.

More, R.M. (1985). In *Advances in Atomic and Molecular Physics*, eds. D.R. Bates and I. Esterman (Academic Press, New York), Vol. 21, p. 305.

More, R.M. (1986). In *Atoms in Unusual Situations*, ed. J.P. Briand (Plenum, New York), p. 166.

More, R.M. (1989). In *Theory of Atoms in Dense Plasmas*, ed. R. Marrus (Plenum, New York), p. 419.

More, R.M. (1991). In *Handbook of Plasma Physics*, Vol. 3, eds. A. Rubenchik and S. Witkowski (Elsevier-North Holland, Amsterdam), Chap. 2.

More, R.M. (1994). In *Laser Interactions with Atoms, Solids, and Plasmas*, ed. R.M. More (Plenum, New York).

More, R.M. and Warren, K.H. (1991). *Ann. Phys. (NY)* **207**, 282.

Moreno, J.C., Griem, H.R., Goldsmith, S. and Knauer, J. (1989). *Phys. Rev. A* **39**, 6033.

Moreno, J.C., Griem, H.R., Lee, R.W. and Seely, J.F. (1993). *Phys. Rev. A* **47**, 374.

Mostovych, A.N., Chan, L.Y., Kearny, K.J., Garren, D., Iglesias, C.A., Klapisch, M. and Rogers, F.J. (1995). *Phys. Rev. Lett.* **75**, 1530.

Mostovych, A.N., Kearny, K.J. and Stamper, J.A. (1990). In *Strongly Coupled Plasma Physics*, ed. S. Ichimaru (Elsevier Publ. and Yamada Science Foundation, Tokyo), p. 589.

Mott, N.F. (1961). *Philos. Mag.* **6**, 287.

Mott, N.F. and Massey, H.S. (1965). *The Theory of Atomic Collisions* (Clarendon, Oxford).

Mozer, B. and Baranger, M. (1960). *Phys. Rev.* **118**, 626.

Mukamel, S. (1995). *Principles of Nonlinear Optical Spectroscopy* (Oxford University Press, Oxford).

Munster, A. (1969). *Statistical Thermodynamics* (Springer, Berlin), Vol. 1.

Murillo, M.S. and Weisheit, J.C. (1995). *J. Quant. Spectrosc. Radiat. Transfer* **54**, 271.

Musielok, J., Böttcher, F., Griem, H.R. and Kunze, H.-J. (1987). *Phys. Rev. A* **36**, 5683.

Musielok, J., Wiese, W.L. and Veres, J. (1995). *Phys. Rev. A* **51**, 3588.

Myers, B.R. and Levine, M.A. (1978). *Rev. Sci. Instrum.* **49**, 610.

Nagayama, Y. (1987). *J. Appl. Phys.* **62**, 2702.

Nee, T.-J. A. (1987). *J. Quant. Spectrosc. Radiat. Transfer* **38**, 213.

Nee, T.-J. A. and Griem, H.R. (1976). *Phys. Rev. A* **14**, 1853.

Nguyen, H., Koenig, M., Benredjem, D., Caby, M. and Coulaud, G. (1986). *Phys. Rev. A* **33**, 1279.

Nick, K.P., Richter, J. and Helbig, V. (1984). *J. Quant. Spectrosc. Radiat. Transfer* **32**, 1.

Nienhus, G. (1973). *Physica* **66**, 245.

Nilsen, J., Koch, J.A., Scofield, J.H., McGowan, B.J., Moreno, J.C. and DaSilva, L.B. (1993). *Phys. Rev. Lett.* **70**, 3713.

Nilsen, J., Vainshtein, L.A., Ivanov, I.A., Safranova, U.A., Dubau, J. and Cornille, M. (1994). *J. Quant. Spectrosc. Radiat. Transfer* **52**, 729.

Nishimura, H., Kiso, T., Shiraga, H., Endo, T., Fujita, K., Sunahara, A., Takabe, H., Kato, Y. and Nakai, S. (1995). *Phys. Plasmas* **2**, 2063.

Norman, G.E. (1963). *Opt. Spectrosc. (USSR)* **14**, 277.

Numano, M. (1995). *J. Quant. Spectrosc. Radiat. Transfer* **53**, 527.

O'Brien, J.T. and Hooper, Jr., C.F. (1972). *Phys. Rev. A* **5**, 867.

Oda, T. and Kiriyama, S. (1980). *J. Phys. Soc. Japan* **49**, 385.

Oertel, G.K. and Shomo, L.P. (1968). *Astrophys. J. Suppl.* **16**, 175.

Ohmura, T. and Ohmura, H. (1961). *Phys. Rev.* **121**, 513.

Oks, E. (1995). *Plasma Spectroscopy — The Influence of Microwave and Laser Fields* (Springer, Berlin).

Oks, E. and Sholin, G.V. (1977). *Sov. Phys. Tech. Phys.* **21**, 144.

Oks, E.A., Böddeker, St. and Kunze, H.-J. (1991). *Phys. Rev. A* **44**, 8338.

Oks, E., Derevianko, A. and Ispolatov, Ya. (1995). *J. Quant. Spectrosc. Radiat. Transfer* **54**, 307.

Olivares, I. and Kunze, H.-J. (1993). *Phys. Rev. E* **47**, 2006.

Olson, R.E. (1987). *Phys. Rev. A* **36**, 1519.

Omont, A. (1977). *Prog. Quantum Electron.* **5**, 69.

Omont, A. and Meunier, J. (1968). *Phys. Rev.* **169**, 92.

Oppenheimer, J.R. (1927). *Z. Phys.* **43**, 27.

Osterbrock, D.E (1962). *Astrophys. J.* **135**, 195.

Osterhold, M., Himmel, G. and Schlüter, H. (1989). *J. Quant. Spectrosc. Radiat. Transfer* **41**, 425.

Oza, D.H., Greene, R.L. and Kelleher, D.E. (1986). *Phys. Rev. A* **34**, 4519.

Oza, D.H., Greene, L.R. and Kelleher, D.E. (1988a). *Phys. Rev. A* **37**, 531.

Oza, D.H., Greene, L.R. and Kelleher, D.E. (1988b). *Phys. Rev. A* **38**, 2544.

Pal, R. and Griem, H.R. (1979). *Phys. Fluids* **22**, 1790.

Pal'chikov, V.G. and Shevelko, V.P. (1995). *Reference Data on Multicharged Ions* (Springer, Berlin).

Paschen, F. and Back, E. (1912). *Ann. Phys. (Leipzig)* **39**, 897.

Paschen, F. and Back, E. (1913). *Ann. Phys. (Leipzig)* **40**, 960.

Payne, H.E., Anantharamaiah, K.R. and Erickson, W.C. (1994). *Astrophys. J.* **430**, 690.

Peach, G. (1970). *Mem. R. Astron. Soc.* **73**, 1.

Peach, G. (1981). *Adv. Phys.* **30**, 367.

Peach, G. (1996). In *Atomic, Molecular and Optical Physics Reference Book*, eds. G.W.F. Drake and N.E. Hedgecock (AIP, New York), Sec. V.37.

Peacock, N.J. and Norton, B.A. (1975). *Phys. Rev. A* **11**, 2142.

Penetrante, B.M. and Bardsley, J.N. (1991). *Phys. Rev. A* **43**, 3100.

Pengelly, R.M. (1964). *Mon. Not. R. Astron. Soc.* **127**, 145.

Pengelly, R.M. and Seaton, M.J. (1964). *Mon. Not. R. Astron. Soc.* **127**, 165.

Penney, C.M. (1969). *J. Opt. Soc. Am.* **59**, 34.

Percival, I.C. and Seaton, M.J. (1958). *Philos. Trans. R. Soc. London, Ser. A* **251**, 113.

Pérez, C., Aparicio, J.A., de la Rosa, I., Mar, S. and Gigosos, M.A. (1995). *Phys. Rev. E* **51**, 3764.

Perrot, F. (1982). *Phys. Rev. A* **26**, 1035.

Perrot, F. and Dharma-wardana, M.W.C. (1995). *Phys. Rev. E* **52**, 5352.

Perry, T.S., Davidson, S.J., Serduke, F.J.D., Bach, D.R., Smith, C.C., Foster, J.M., Doyas, R.D., Ward, R.A., Iglesias, C.A., Rogers, F.J., Abdallah, J., Stewart, R.E., Kilkenny, J.D. and Lee, R.W. (1991). *Phys. Rev. Lett.* **67**, 3784.

Pert, G.J. (1994). *J. Opt. Soc. Am. B* **11**, 1425.

Pert, G.J. (1995). *Phys. Rev. E* **51**, 4778.

Petrini, D. (1969). *Astron. Astrophys.* **1**, 139.

Peyrusse, O., Kieffer, J.C., Côte, C.Y. and Chaker, M. (1993). *J. Phys. B* **26**, L511.

Pfennig, H. and Trefftz, E. (1966). *Z. Naturforsch.* **21**a, 317.

Phelps, A.V. (1958). *Phys. Rev.* **110**, 1362.

Phelps, A.V. (1990). *J. Phys. Chem. Ref. Data* **19**, 653.

Phelps, A.V. (1991). *J. Phys. Chem. Ref. Data* **20**, 557.

Phelps, A.V. (1992). *J. Phys. Chem. Ref. Data* **21**, 883.

Piel, A. and Slupek, J. (1984). *Z. Naturforsch.* **39**a, 1041.

Pindzola, M.S., Griffin, D.C. and Bottcher, C. (1986). *Phys. Rev. A* **33**, 3787.

Pittman, T.L. and Fleurier, C. (1986). *Phys. Rev. A* **33**, 1291.

Pittman, T.L., Voigt, P. and Kelleher, D.E. (1980). *Phys. Rev. Lett.* **45**, 723.

Planck, M. (1924). *Ann. Phys. (Leipzig)* **75**, 673.

Podgoretskii, M.I. and Stepanov, A.W. (1961). *Sov. Phys. JETP* **13**, 393.

Podivilov, E.V. and Shapiro, D.A. (1992). *Sov. Phys. JETP Lett.* **56**, 449.

Pohlmeyer, B.A., Dinklage, A. and Kunze, H.-J. (1996). *J. Phys. B* **29**, 221.

Polishchuk, A. Ya. and Meyer-ter-Vehn, J. (1994a). *Phys. Rev. E* **49**, 1563.

Polishchuk, A. Ya. and Meyer-ter-Vehn, J. (1994b). *Phys. Rev. E* **49**, 663.

Pollock, E.L. and London, R.A. (1993). *Phys. Fluids B* **5**, 4495.

Popović, M.M. and Dordević, D.S. (1993). In *Strongly Coupled Plasma Physics*, ed. H.M. Van Horn and S. Ichimaru (Univ. of Rochester Press, Rochester, N.Y.), p. 273.

Posener, D.W. (1959). *Aust. J. Phys.* **12**, 184.

Post, D.E., Jensen, R.V., Tarter, C.B., Grassberger, W.H. and Lokke, W.A. (1977). *At. Data Nucl. Data Tables* **20**, 397.

Post, D.E., Mikkelsen, D.R., Hulse, R.A., Stewart, L.D. and Weisheit, J.C. (1981). *J. Fusion Energy* **1**, 129.

Pradhan, A.K. (1978). *Mon. Not. R. Astron. Soc.* **184**, 89P.

Preissing, N.D., Campos, O., Kunze, H.-J., Osterheld, A.L. and Walling, R.S. (1993). *Phys. Rev. E* **48**, 3867.

Preobrazhenskii, N.G. (1971). *Spectroscopy of Optically Thick Plasmas*, in Russian (Nauka, Novosibirsk).

Preobrazhenskii, N.G. and Pikalov, V.V. (1982). *Unstable Solutions in Problems of Plasma Diagnostics*, in Russian (Nauka, Novosibirsk).

Purić, J., ts Djeniže, S., Srecković, A., Platisa, M. and Labat, J. (1988). *Phys. Rev. A* **37**, 498.

Racah, G. (1942). *Phys. Rev.* **62**, 438.

Racah, G. (1943). *Phys. Rev.* **63**, 367.

Radon, J. (1917). *Ber. Verh. Saechs. Akad. Wiss., Leipzig, Math. Phys. Kl.* **69**, 262.

Radtke, R. and Günther, K. (1986). *Contrib. Plasma Phys.* **26**, 143.

Radtke, R., Günther, K. and Spanke, R. (1986). *Contrib. Plasma Phys.* **26**, 151.

Ragozin, E.N. (1985). *Sov. J. Plasma Phys.* **11**, 577.

Ragozin, E.N., König, R. and Kuz'micheva, M. Yu. (1993). *J. Quant. Spectrosc. Radiat. Transfer* **49**, 39.

Rautian, S.G. and Sobel'man, I.I. (1967). *Sov. Phys. Usp.* **9**, 701.

Read, F.H. (1984). *J. Phys. B* **17**, 3965.

Reif, F. (1965). *Fundamentals of Statistical and Thermal Physics* (McGraw-Hill, New York).

Rice, J.E., Marmar, E.S., Terry, J.L., Källne, E. and Källne, J. (1986). *Phys. Rev. Lett.* **56**, 50.

Rickert, A., Eidmann, K., Meyer-ter-Vehn, J., Serduke, F.J.D. and Iglesias, C.A. (1995). *Third international opacity workshop and code comparison study*, Max-Planck Institute for Quantum Optics Report, MPQ 204.

Riley, D. and Willi, O. (1995). *Phys. Rev. Lett.* **75**, 4039.

Riley, D., Willi, O., Rose, S.J. and Afshar-Rad, T. (1989). *Europhys. Lett.* **10**, 125.

Roberts, J.R. and Voigt, P.A. (1971). *J. Res. Natl. Bur. Stand., Sec. A* **75**, 291.

Rogers, F.J. (1977). *Phys. Lett. A* **61**, 358.

Rogers, F.J. (1979). *Phys. Rev. A* **19**, 375.

Rogers, F.J. (1981). *Phys. Rev. A* **24**, 1531.

Rogers, F.J. (1984). *Phys. Rev. A* **29**, 868.

Rogers, F.J. (1986). *Astrophys. J.* **310**, 723.

Rogers, F.J. (1990). *Astrophys. J.* **352**, 689.

Rogers, F.J. (1991). In *High Pressure Equations of State: Theory and Applications*, ed. S. Eliezer and R. A. Ricci (North-Holland, Amsterdam), p. 77.

Rogers, F.J. (1994). In *The equation of state in astrophysics*, IAU Colloquium 147, ed. G. Chabrier and E. Schatzman (Cambridge University Press), p. 16.

Rogers, F.J. and Iglesias, C.A. (1992). *Astrophys. J. Suppl.* **79**, 507.

Röhr, H., Steuer, K.H. and ASDEX Team (1988). *Rev. Sci. Instrum.* **59**, 1875.

Rohrlich, F. (1959). *Astrophys. J.* **129**, 441 and 449.

Rosenbluth, M.N. and Rostoker, N. (1962). *Phys. Fluids* **5**, 776.

Rosseland, S. (1924). *Mon. Not. R. Astron. Soc.* **84**, 525.

Rotenberg, M., Bivins, R., Metropolis, N. and Wooten, K. (1959). *The 3-j and 6-j Symbols* (The Technology Press of MIT, Cambridge, MA).

Royer, A. (1980). *Phys. Rev. A* **22**, 1625.

Rozsnyai, B.F. (1979). *J. Quant. Spectrosc. Radiat. Transfer* **22**, 337.

Rozsnyai, B.F. and Lamoureux, M. (1990). *J. Quant. Spectrosc. Radiat. Transfer* **43**, 381.

Rubenchik, A. and Witkowski, S. (1991). *Handbook of Plasmaphysics*, Vol. 3, Physics of Laser Plasma, eds. M.N. Rosenbluth and R.Z. Sagdeev (North Holland, Amsterdam).

Rudge, M.R.H. and Schwartz, S.B. (1965). *Proc. Phys. Soc. London* **86**, 773.

Rudge, M.R.H. and Schwartz, S.B. (1966). *Proc. Phys. Soc. London* **88**, 563.

Safronova, U.I., Shlyaptseva, A.S., Kato, T., Masai, K. and Vainshtein, L.A. (1995). *At. Data Nucl. Data Tables* **60**, 1.

Saha, M.N. (1920). *Philos. Mag. (Ser. 6)* **40**, 472.

Saha, M.N. (1921). *Proc. R. Soc. London, Ser. A* **99**, 135.

Sahal-Brechót, S. (1969a). *Astron. Astrophys.* **1**, 91.

Sahal-Brechót, S. (1969b). *Astron. Astrophys.* **2**, 322.

Sahal-Brechót, S. (1991). *Astron. Astrophys.* **245**, 322.

Sahal-Brechót, S. and Segre, E.R. (1971). *Astron. Astrophys.* **13**, 161.

Sakurai, J.J. (1967). *Advanced Quantum Mechanics* (Addison-Wesley, Reading, MA.).

Salzborn, E. (1990). In *The Physics of Electronic and Atomic Collisions*, ed. A. Dalgarno, R.S. Freund, P.M. Koch, M.S. Lubell, and T.B. Lucatoro, AIP Conf. Proc. No. 205 (AIP, New York), p. 290.

Salzmann, D. and Szichman, H. (1987). *Phys. Rev. A* **35**, 807.

Salzmann, D., Stein, J., Goldberg, I.B. and Pratt, R.H. (1991). *Phys. Rev. A* **44**, 1270.

Salzmann, D., Yin, R.Y. and Pratt, R.H. (1985). *Phys. Rev. A* **32**, 3627.

Sampson, D.H. (1959). *Astrophys. J.* **129**, 734.

Sampson, D.H. and Zhang, H.L. (1987). *Phys. Rev. A* **36**, 3590.

Sampson, D.H. and Zhang, H.L. (1988). *Phys. Rev. A* **37**, 3765.

Sampson, D.H., Zhang, H.L. and Mohanti, A.K. (1988). *Phys. Rev. A* **38**, 4569.

Sanchez, A., Blaha, M. and Jones, W.W. (1973). *Phys. Rev. A* **8**, 774.

Saraph, H., Seaton, M.J. and Shemming, J. (1969). *Philos. Trans. R. Soc. London, Ser. A* **264**, 77.

Sarfaty, M., Maron, Y., Krasik, Ya.E., Weingarten, A., Arad, R., Shpitalnik, R., Fruchtman, A. and Alexiou, S. (1995). *Phys. Plasmas* **2**, 2122.

Sarid, E., Maron, Y. and Troyansky, L. (1993). *Phys. Rev. E* **48**, 1364.

Sasaki, S., Ohkouchi, Y., Takamura, S. and Kato, T. (1994). *J. Phys. Soc. Japan* **63**, 2942.

Sasaki, S., Takamura, S., Masuzaki, S., Watanabe, S., Kato, T. and Kadota, K. (1995). National Institute for Fusion Science of Japan, Report 346.

Sawada, K. and Fujimoto, T. (1994). *Phys. Rev. E* **49**, 5565.

Sawada, K., Eriguchi, K. and Fujimoto, T. (1993). *J. Appl. Phys.* **73**, 8122.

Scheuer, P.A.G. (1960). *Mon. Not. R. Astron. Soc.* **120**, 231.

Schiff, L.I. (1955). *Quantum Mechanics*, 2nd. Ed. (McGraw-Hill, New York).

Schlanges, M. and Bornath, Th. (1993). *Physica A* **192**, 262.

Schlüter, D. (1968). *Z. Phys.* **210**, 80.

Schnapauff, R. (1968). *Z. Astrophys.* **68**, 431.

Schöning, T. (1993a). *Astron. Astrophys.* **267**, 300.

Schöning, T. (1993b). *J. Phys. B* **26**, 899.

Schöning, T. (1994a). *Astron. Astrophys.* **282**, 994.

Schöning, T. (1994b). *J. Phys. B* **27**, 4501.

Schöning, T. (1995). Private communication.

Schorn, R.P., Hintz, E., Rusbüldt, D., Aumayr, F., Schneider, M., Unterreiter, E. and Winter, H. (1991). *Appl. Phys. B* **52**, 71.

Schorn, R.P., Wolfrum, E., Aumayr, F., Hintz, E., Rusbüldt, D. and Winter, H. (1992). *Nucl. Fusion* **32**, 351.

Schott, G.A. (1912). *Electromagnetic Radiation* (Cambridge University Press, Cambridge).

Schulz, A. and Koshelev, K.N. (1995). *J. Quant. Spectrosc. Radiat. Transfer* **54**, 361.

Schwanda, W. and Eidmann, K. (1992). *Phys. Rev. Lett.* **69**, 3507.

Schweer, B., Mank, G., Pospieszczyk, D., Brosda, B. and Pohlmeyer, B. (1992). *J. Nucl. Mater.* **196-198**, 174.

Scott, F.R., Neidigh, R.V., McNally, J.R. and Cooper, W.S. (1970). *J. Appl. Phys.* **41**, 5327.

Seaton, M.J. (1955a). *Mon. Not. R. Astron. Soc.* **115**, 279.

Seaton, M.J. (1955b). *Proc. R. Soc. London, Ser. A* **231**, 37.

Seaton, M.J. (1955c). *Proc. Phys. Soc., London* **68**, 457.

Seaton, M.J. (1962). In *Atomic and Molecular Processes*, ed. D.R. Bates (Academic Press, New York), chapter 11.

Seaton, M.J. (1964a). *Planet. Space Sci.* **12**, 55.

Seaton, M.J. (1964b). *Mon. Not. R. Astron. Soc.* **127**, 191.

Seaton, M.J. (1987a). *J. Phys. B* **20**, 6363.

Seaton, M.J. (1987b). *J. Phys. B* **20**, 6431.

Seaton, M.J. (1988). *J. Phys. B* **21**, 3033.

Seaton, M.J. (1989). *J. Phys. B* **22**, 3603.

Seaton, M.J. (1990). *J. Phys. B* **23**, 3255.

Seaton, M.J. (1995). *J. Phys. B* **28**, 565.

Seaton, M.J. and Storey, P.J. (1976). In *Atomic Processes and Applications*, ed. P.G. Burke and B. Moiseiwitsch (North-Holland, Amsterdam), chapter 6.

Seaton, M.J., Yan, Y., Mihalas, D. and Pradhan, A.K. (1994). *Mon. Not. R. Astron. Soc.* **266**, 805.

Seaton, M.J., Zeippen, C.J., Tully, J.A., Pradhan, A.K., Mendoza, C., Hibbert, A. and Berrington, K.A. (1992). *Rev. Mexicana Astron. Astrof.* **23**, 19.

Seidel, J. (1979). *Z. Naturforsch.* **34a**, 1385.

Seidel, J. (1981). In *Spectral Line Shapes*, Vol. 1, ed. B. Wende (de Gruyter, Berlin), p. 3.

Seidel, J. (1985). In *Spectral Line Shapes*, Vol. 3, ed. F. Rostas (de Gruyter, Berlin), p. 69.

Seidel, J. (1987). In *Spectral Line Shapes*, Vol. 4, ed. R.J. Exton (Deepak, Hampton, VA), p. 57.

Seidel, J., Arndt, S. and Kraeft, W.D. (1995). *Phys. Rev. E* **52**, 5387.

Sheffield, J. (1975). *Plasma Scattering of Electromagnetic Radiation* (Academic Press, New York).

Shepard, T.D., Back, C.A., Kalantar, D.H., Kauffman, R.L., Keane, C.J., Klem, D.E., Lasinski, B.F., MacGowan, B.J., Powers, L.V., Suters, L.J., Turner, R.E., Failor, B.H. and Hsing, W.W. (1995). *Rev. Sci. Instrum.* **66**, 749.

Shepard, T.D., Back, C.A., Kalantar, D.H., Kauffman, R.L., Keane, C.J., Klem, D.E., Lasinski, B.F., MacGowan, B.J., Powers, L.V., Suters, L.J., Turner, R.E., Failor, B.H. and Hsing, W.W. (1996). *Phys. Rev. E* **53**, 358.

Shestakov, A.I. and Eder, D.C. (1989). *J. Quant. Spectrosc. Radiat. Transfer* **42**, 483.

Shevelko, V.P. and Vainshtein, L.A. (1993). *Atomic Physics for Hot Plasmas* (Inst. of Phys. Publ., Bristol and Philadelphia).

Shimamura, I. and Fujimoto, T. (1990). *Phys. Rev. A* **42**, 2346.

Sholin, G.V. (1961). *Sov. Phys. Dokl.* **15**, 1040.

Sholin, G.V., Demura, A.V. and Lisitsa, V.S. (1974). *Sov. Phys. JETP* **37**, 1057.

Sholin, G.V., Lisitsa, V.S. and Kogan, V.I. (1971). *Sov. Phys. JETP* **32**, 758.

Shore, B.W. (1990). *The Theory of Coherent Atomic Excitation*, Vol. 1, Simple atoms and fields; Vol. 2, Multilevel atoms and incoherence, (Wiley Interscience, New York).

Shore, B.W. and Menzel, D.H. (1968). *Principles of Atomic Spectra* (John Wiley, New York).

Skobelev, I. Yu., Vinogradov, A.V. and Yukov, E.A. (1978). *Physica Scripta* **18**, 78.

Slattery, W.L., Doolen, G.D. and DeWitt, H.E. (1982). *Phys. Rev. A* **26**, 2255.

Smith, E.W. (1968). *Phys. Rev.* **166**, 102.

Smith, E.W. and Hooper, Jr., C.F. (1967). *Phys. Rev.* **157**, 126.

Smith, E.W., Cooper, J. and Roszman, L.J. (1973). *J. Quant. Spectrosc. Radiat. Transfer* **13**, 1523.

Smith, E.W., Cooper, J. and Vidal, C.R. (1969). *Phys. Rev.* **185**, 140.

Smith, E.W., Cooper, J., Chappell, W.R. and Dillon, T. (1971). *J. Quant. Spectrosc. Radiat. Transfer* **11**, 1547.

Snyder, S.C., Lassahn, G.D. and Reynolds, L.D. (1993). *Phys. Rev. E* **48**, 4124.

Snyder, S.C., Murphy, A.B., Hofeldt, D.L. and Reynolds, L.D. (1995). *Phys. Rev. E* **52**, 2999.

Snyder, S.C., Reynolds, L.D., Fincke, J.R., Lassahn, G.D., Grandy, J.D. and Repetti, T.E. (1994). *Phys. Rev. E* **50**, 519.

Sobel'man, I.I. (1992). *Atomic Spectra and Radiative Transitions*, 2nd Ed. (Springer Verlag, Berlin).

Sobel'man, I.I., Vainshtein, L.A. and Yukov, E.A. (1981). *Excitation of Atoms and Broadening of Spectral Lines* (Springer, Berlin); (1995), second (revised) edition.

Sobolev, V.V. (1957). *Sov. Astron.* **1**, 678.

Sobolev, V.V. (1963). *A Treatise on Radiative Transfer* (Van Nostrand, Princeton, NJ).

Sobolev, V.V. (1972). *Scattering of Light by Planetary Atmospheres* (Nauka, Moscow).

Soltwisch, H. and Kusch, H.J. (1979). *Z. Naturforsch.* **34a**, 300.

Sommerfeld, A. (1928). *Z. Phys.* **47**, 1.

Sommerfeld, A. (1931). *Ann. Phys. (Leipzig)* **11**, 257.

Sommerville, W.B. (1967). *Astrophys. J.* **149**, 811.

Spitzer, L., Jr. (1962). *Physics of Fully Ionized Plasmas*, 2nd ed. (Wiley-Interscience, New York).

Spitzer, L., Jr. and Greenstein, J.L. (1951). *Astrophys. J.* **114**, 407.

Springer, P.T., Fields, D.F., Wilson, B.G., Nash, J.K., Goldstein, W.H., Iglesias, C.A., Rogers, F.J., Swenson, J.K., Chen, M.H., Bar-Shalom, A. and Stewart, R.E. (1992). *Phys. Rev. Lett.* **69**, 3735.

Springer, P.T., Fields, D.F., Wilson, B.G., Nash, J.K., Goldstein, W.H., Iglesias, C.A., Rogers, F.J., Swenson, J.K., Chen, M.H., Bar-Shalom, A. and Stewart, R.E. (1994). *J. Quant. Spectrosc. Radiat. Transfer* **51**, 371.

Stacey, D.N. and Cooper, J. (1969). *Phys. Lett.* **30A**, 49.

Stamm, R., Smith, E.W. and Talin, B. (1984). *Phys. Rev. A* **30**, 2039.

Stamm, R., Talin, B., Pollock, E.L. and Iglesias, C.A. (1986). *Phys. Rev. A* **34**, 4144.

Stefanović, I., Ivković, M. and Konjević, N. (1995). *Phys. Scripta* **52**, 178.

Stehlé, C. (1986). *Phys. Rev. A* **34**, 4153.

Stehlé, C. (1990). *J. Quant. Spectrosc. Radiat. Transfer* **44**, 135.

Stehlé, C. (1994a). *Astron. Astrophys.* **292**, 699.

Stehlé, C. (1994b). *Astron. Astrophys. Suppl.* **104**, 509.

Stehlé, C. (1995). In *Spectral Line Shapes*, Vol. 8, ed. A. David May, J.R. Drummond and E. Oks, AIP Conf. Proc. 328 (AIP, New York), p. 36.

Stehlé, C., Voslamber, D. and Feautrier, N. (1989). *J. Phys. B* **22**, 3657.

Stein, J. and Salzmann, D. (1992). *Phys. Rev. A* **45**, 3943.

Stern, R.A. and Cheo, P.K. (1969). *Phys. Rev. Lett.* **23**, 1426.

Stewart, J.C. and Pyatt, K.D. (1966). *Astrophys. J.* **144**, 1203.

Stewart, J.C., Peek, J.M. and Cooper, J. (1973). *Astrophys. J.* **179**, 983.

Stilley, J.L. and Callaway, J. (1970). *Astrophys. J.* **160**, 245.

Stix, T. (1992). *Waves in Plasmas*, 2nd ed. (AIP, New York).

Stratton, B.C., Fonck, R.J., McKee, G., Thorson, T., Hammett, G., Phillips, C.K. and Synakowski, E.J. (1994). *Nucl. Fusion* **34**, 734.

Strömgren, B. (1932). *Z. Astrophys.* **4**, 118.

Suckewer, S. (1991). *Phys. Fluids B* **3**, 2437.

Suemitsu, H., Iwaki, K., Takemoto, Y. and Yoshida, E. (1990). *J. Phys. B* **23**, 1129.

Summers, H.P. (1974). *Mon. Not. R. Astron. Soc.* **169**, 663.

Summers, H.P. (1977). *Mon. Not. R. Astron. Soc.* **178**, 101.

Summers, H.P. (1992). *Rev. Sci. Instrum.* **63**, 10.

Summers, H.P., Thomas, P., Giannella, R., von Hellermann, M., Dickson, W., Lawson, K., Mandl, W., Briden, P. and members of Exp. Div. II (1991). *Suppl. Z. Phys. D* **21**, 17.

Swanson, D. (1989). *Plasma Waves* (Academic Press, New York).

Synakowski, E.J., Bell, R.E. and Bush, C.E. (1995). *Rev. Sci. Instrum.* **66**, 649.

Szöke, A. and Javan, A. (1963). *Phys. Rev. Lett.* **10**, 521.

Talin, B., Calisti, A., Godbert, L., Stamm, R., Lee, R.W. and Klein, L. (1995). *Phys. Rev. A* **51**, 1918.

Tallents, G.J. (1985). *J. Phys. B* **18**, 3299.

Tarter, C.B. (1977). *J. Quant. Spectrosc. Radiat. Transfer* **17**, 531.

Theimer, O. (1957). *Z. Naturforsch.* **12a**, 518.

Theimer, O. and Kepple, P. (1970). *Phys. Rev. A* **1**, 957.

Thomas, L.H. (1927). *Proc. Cambridge Philos. Soc.* **23**, 713 and 829.

Thomas, R.N. (1957). *Astrophys. J.* **125**, 260.

Thomas, R.N. and Athay, R.G. (1961). *Physics of the Solar Chromosphere* (Interscience, New York).

Thomsen, C. and Helbig, V. (1991). *Spectrochim. Acta* **46B**, 1215.

Thomson, J.J. (1912). *Philos. Mag.* **23**, 449.

Tighe, R.J. and Hooper, Jr., C.F. (1976). *Phys. Rev. A* **14**, 1514.

Tighe, R.J. and Hooper, Jr., C.F. (1977). *Phys. Rev. A* **15**, 1773.

Tinschert, K., Müller, A., Hoffmann, G., Salzborn, E. and Younger, S.M. (1991). *Phys. Rev. A* **43**, 3522.

Tolman, R.C. (1938, 1979). *Principles of Statistical Mechanics* (Dover, New York); unabridged and unaltered republication of first edition, published by Oxford University Press.

Tonks, L. and Langmuir, I. (1929). *Phys. Rev.* **34**, 876.

Townes, C.H. and Schawlow, A.L. (1955). *Microwave Spectroscopy* (McGraw-Hill, New York).

Tran-Minh, N. and Van Regemorter, H. (1972). *J. Phys. B* **5**, 903.

Tran-Minh, N., Feautrier, N. and Van Regemorter, H. (1976). *J. Quant. Spectrosc. Radiat. Transfer* **16**, 849.

Traving, G. (1960). *Über die Theorie der Druckverbreiterung von Spektrallinien* (Braun, Karlsruhe).

Truong-Bach and Drawin, H.W. (1982). *J. Quant. Spectrosc. Radiat. Transfer* **27**, 627.

Tsytovich, V.N. and Oiringel, I.M., eds. (1992). *Polarization Bremsstrahlung* (Plenum Press, New York).

Uchikawa, S., Griem, H.R. and Düchs, D. (1980). *Calculation of Impurity Radiation from Near-Corona-Equilibrium Plasmas*, Max-Planck Inst. of Plasmaphysik Rep. IPP-6/99.

Unnikrishnan, K., Callaway, J. and Oza, D.H. (1990). *Phys. Rev. A* **42**, 6602.

Unno, W. (1952a). *Publ. Astron. Soc. Japan* **3**, 158.

Unno, W. (1952b). *Publ. Astron. Soc. Japan* **4**, 100.

Unsöld, A. (1938). *Ann. Phys. (Leipzig)* **33**, 607.

Unsöld, A. (1948). *Z. Astrophys.* **24**, 355.

Unsöld, A. (1955). *Physik der Sternatmosphären*, 2nd ed. (Springer, Berlin).

Uschmann, I., Förster, E., Nishimura, H., Fujita, K., Kato, Y. and Nakai, S. (1995). *Rev. Sci. Instrum.* **66**, 734.

Uzelac, N.I. and Konjević, N. (1989). *J. Phys. B* **22**, 2517.

Uzelac, N.I., Stefanović, I. and Konjević, N. (1991). *J. Quant. Spectrosc. Radiat. Transfer* **46**, 447.

Uzelac, N.I., Glenzer, S., Konjević, N., Hey, J.D. and Kunze, H.-J. (1993). *Phys. Rev. E* **47**, 3623.

Vainshtein, L.A. and Safronova, V.I. (1978). *At. Data Nucl. Data Tables* **21**, 49.

Vainshtein, L.A. and Safronova, V.I. (1980). *At. Data Nucl. Data Tables* **25**, 311.

Vainshtein, L.A., Sobel'man, I.I. and Presnyakov, L.P. (1962). *Sov. Phys. JETP* **16**, 370.

van der Mullen, J.A.M. (1990). *Phys. Report* **191**, 109.

van der Mullen, J.A.M., Benoy, D.A., Fey, F.H.A.G., van der Sijde, B. and Vlček, J. (1994). *Phys. Rev. E* **50**, 3925.

van der Mullen, J.A.M., van der Sijde, B. and Schram, D.C. (1983). *Phys. Lett. A* **96**, 239.

Van Regemorter, H. (1962). *Astrophys. J.* **132**, 906.

Van Regemorter, H. and Hoang-Binh, D. (1993). *Astron. Astrophys.* **277**, 623.

Vance, R.L. and Gallup, G.A. (1980). *J. Chem. Phys.* **73**, 894.

Vidal, C.R., Cooper, J. and Smith, E.W. (1973). *Astrophys. J. Suppl.* **25**, 37.

Vinogradov, A.V. and Sobel'man, I.I. (1973). *Sov. Phys. JETP* **36**, 1115.

Vinogradov, A.V., Skobelev, I. Yu. and Yukov, E.A. (1975). *Sov. J. Quantum Electron.* **5**, 630.

Vitel, Y. and Skowronek, M. (1987a). *J. Phys. B* **20**, 6477.

Vitel, Y. and Skowronek, M. (1987b). *J. Phys. B* **20**, 6493.

Voigt, W. (1912). *Münch. Ber.*, 603.

von Goeler, S., Fishman, H., Ignat, D., Jones, S., Roney, P., Stevens, J., Bernabei, S., Davis, W., Kaita, R., Paoletti, F., Petravich, G. and Rimini, F. (1995). *Phys. Plasmas* **2**, 207.

von Goeler, S., Stodiek, W., Eubank, H., Fishman, H., Grebenshchikov, S. and Hinnov, E. (1975). *Nucl. Fusion* **15**, 301.

von Hellermann, M.G., Mandl, W. and Summers, H.P. (1990). *Rev. Sci. Instrum.* **61**, 3479.

von Hellermann, M., Mandl, W., Summers, H.P., Boileau, A., Hoekstra, R., de Heer, F.J. and Frieling, J. (1991). *Plasma Phys. Contr. Fusion* **33**, 1805.

Voslamber, D. (1972). *Phys. Lett.* **40A**, 266.

Voslamber, D. (1993). In *Spectral Line Shapes*, Vol. 7, ed. R. Stamm and B. Talin (Nova Science Publ., Commack, NY), p. 321.

Voslamber, D. and Yelnik, J.-B. (1978). *Phys. Rev. Lett.* **41**, 1233.

Vriens, L. and Smeets, A.H.M. (1980). *Phys. Rev. A* **22**, 940.

Walling, R.S. and Weisheit, J.C. (1988). *Phys. Rep.* **162**, 1.

Wang, J.S. and Griem, H.R. (1983). *Phys. Rev. A* **27**, 2249.

Wang, J.S., Griem, H.R. and Iglesias, E.J. (1989). *Phys. Rev. A* **40**, 4115.

Wang, J.S., Griem, H.R., Hess, R. and Rowan, W.L. (1988). *Phys. Rev. A* **38**, 4761.

Wang, J.S., Griem, H.R., Huang, Y.W. and Böttcher, F. (1992). *Phys. Rev. A* **45**, 4010.

Wark, J.S., Djaoui, A., Rose, S.J., He, H., Renner, O., Missalla, T. and Foerster, E. (1994). *Phys. Rev. Lett.* **72**, 1826.

Watanabe, T. (1965a). *Phys. Rev.* **138**, A1573.

Watanabe, T. (1965b). *Phys. Rev.* **140**, AB5.

Weber, E.W. and Humpert, H.J. (1981). *Phys. Lett.* **83**A, 386.

Webster, D.L., Hansen, W.W. and Duveneck, F.B. (1933). *Phys. Rev.* **43**, 839.

Weisheit, J.C. (1975). *J. Phys. B* **8**, 2556.

Weisheit, J.C. (1979). *J. Quant. Spectrosc. Radiat. Transfer* **22**, 585.

Weisheit, J.C. and Shore, B.W. (1974). *Astrophys. J.* **194**, 519.

Weiss, A., Keady, J.J. and Magee, N.H. (1990). *At. Data Nucl. Data Tables* **45**, 209.

Weisskopf, V. (1933). *Z. Phys.* **85**, 451.

Welch, B.L., Griem, H.R., Terry, J., Kurtz, C., LaLombard, B., Lipshultz, B., Marmar, E. and McCracken, G. (1995). *Phys. Plasmas* **2**, 4246.

West, W.P., Thomas, D.M., de Grassie, J.S. and Zheng, S.B. (1987). *Phys. Rev. Lett.* **58**, 2758.

West, W.P., Thomas, D.M., Ensberg, E.S., deGrassie, J.S. and Baur, F.F. (1986). *Rev. Sci. Instrum.* **57**, 1552.

Whiting, E.E. (1968). *J. Quant. Spectrosc. Radiat. Transfer* **8**, 1379.

Wiese, W.L. (1968). In *Methods of Experimental Physics* Vol. 7A, ed. B. Bederson and W.L. Fite (Academic Press, New York), p. 117.

Wiese, W.L. and Konjevic, N. (1992). *J. Quant. Spectrosc. Radiat. Transfer* **47**, 185.

Wiese, W.L., Fuhr, J.R. and Deters, T.M. (1995). Atomic Transition Probabilities of Carbon, Nitrogen and Oxygen, Monograph 7, *J. Phys. Chem. Ref. Data*.

Wiese, W.L., Kelleher, D.E. and Helbig, V. (1975). *Phys. Rev. A* **11**, 1854.

Wiese, W.L., Kelleher, D.E. and Paquette, D.R. (1972). *Phys. Rev. A* **6**, 1132.

Wiese, W.L., Smith, M.W. and Glennon, B.M. (1966). *Atomic Transition Probabilities*, Vol. I, (*H* through *Ne*), NSRDS-NBS No. 4 (U.S. Government Printing Office, Washington, D.C.).

Wiese, W.L., Smith, M.W. and Miles, B.M. (1969). *Atomic Transition Probabilities*, Vol. II, (*Na* through *Ca*), NSRDS-NBS No. 22 (U.S. Government Printing Office, Washington, D.C.).

Wigner, E. (1931). *Phys. Z.* **32**, 450.

Wilson, R. (1962). *J. Quant. Spectrosc. Radiat. Transfer* **2**, 477.

Wimmel, H.K. (1961). *J. Quant. Spectrosc. Radiat. Transfer* **1**, 1.

Winhart, G., Eidmann, K., Iglesias, C.A., Bar-Shalom, A., Mínquez, E., Rickert, A. and Rose, S.J. (1995). *J. Quant. Spectrosc. Radiat. Transfer* **54**, 437.

Winter, H. (1982). *Comments At. Mol. Phys.* **12**, 165.

Wishart, A.W. (1979). *Mon. Not. R. Astron. Soc.* **187**, 59.

Wolfrum, E., Hoekstra, R., de Heer, F.J., Morgenstern, R. and Winter, H. (1992). *J. Phys. B* **25**, 2597.

Woolley, v.d.R. and Allen, C.W. (1948). *Mon. Not. R. Astron. Soc.* **108**, 292.

Woltz, L.A. and Hooper, Jr., C.F. (1988). *Phys. Rev. A* **38**, 4788.

Woltz, L.A., Jacobs, V.L., Hooper, Jr., C.F. and Mancini, R.C. (1991). *Phys. Rev. A* **44**, 1281.

Wróblewski, D., Burrell, K.H., Lao, L.L., Politzer, P. and West, W.P. (1990). *Rev. Sci. Instrum.* **61**, 3552.

Wróblewski, D., Huang, L.K. and Moos, H.W. (1988b). *Rev. Sci. Instrum.* **59**, 2341.

Wróblewski, D., Huang, L.K., Moos, H.W. and Phillips, P.E. (1988a). *Phys. Rev. Lett.* **61**, 1724.

Wrubel, T., Glenzer, S., Büscher, S. and Kunze, H.-J. (1996a). *J. Atmos. Terr. Phys.* **58**, 1077.

Wrubel, T., Glenzer, S., Büscher, S., Kunze, H.-J. and Alexiou, S. (1996b). *Astron. Astrophys.* **306**, 1023.

Wulff, H. (1958). *Z. Phys.* **150**, 614.

Xu, J., Ida, K. and Fujita, J. (1995). National Institute for Fusion Science (Nagoya) Report 349.

Yakovlev, D.G., Yasevich, V. Yu. and Ljublin, B.V. (1989). *J. Phys. B* **22**, 3851.

You, L. and Cooper, J. (1994). *Phys. Rev. A* **50**, 5264.

Younger, S.M. (1981). *J. Quant. Spectrosc. Radiat. Transfer* **26**, 329.

Younger, S.M. and Wiese, W.L. (1978). *J. Quant. Spectrosc. Radiat. Transfer* **22**, 161.

Yu Yan and Seaton, M.J. (1987). *J. Phys. B* **20**, 6409.

Zaidi, H.R. (1968). *Phys. Rev.* **173**, 132.

Zaidi, H.R. (1972). *Can. J. Phys.* **50**, 2792.

Zangers, J. and Meiners, D. (1989). *J. Quant. Spectrosc. Radiat. Transfer* **42**, 25.

Zastrow, K.-D., Källne, E. and Summers, H.P. (1990). *Phys. Rev. A* **41**, 1427.

Zel'dovich, Ya. B. and Raizer, Yu. P. (1966). *Physics of Shockwaves and High Temperature Hydrodynamic Phenomena* (Academic Press, New York), Vol. 1.

Zemansky, M.W. (1930). *Phys. Rev.* **36**, 919.

Zemansky, M.W. (1932). *Phys. Rev.* **42**, 843.

Zemtsov, Yu. K. and Starostin, A.N. (1993). *JETP* **76**, 186.

Zhang, H.L. and Sampson, D.H. (1993). *Phys. Rev. A* **47**, 208.

Zhidkov, A.G., Tkachev, A.N. and Yakovlenko, S.I. (1986). *Sov. Phys. JETP* **64**, 261.

Zhu, Z.Y., Wang, J.S., Iglesias, E.J., Maffei, K.C. and Griem, H.R. (1983). *Nucl. Fusion* **23**, 1686.

Zigler, A., Jacobs, V.L., Newman, D.A., Burkhalter, P.G., Nagel, D.J., Luk, T.S., McPherson, A., Boyer, K. and Rhodes, C.K. (1992). *Phys. Rev. A* **45**, 1569.

Zimmerman, G.B. and More, R.M. (1980). *J. Quant. Spectrosc. Radiat. Transfer* **23**, 517.

Zwanzig, R.W. (1961). *Lectures in Theoretical Physics* (Wiley-Interscience, New York), Vol. III.

Zwicker, H. (1968). In *Plasma Diagnostics*, ed. W. Lochte-Holtgreven (North Holland, Amsterdam); reprinted by the American Vacuum Society in 1995 (AIP, New York), chapter 4.

# Index